CISM International Centre for Mechanical Sciences

Courses and Lectures

Volume 605

For more than 40 years the book series edited by CISM, "International Centre for Mechanical Sciences: Courses and Lectures", has presented groundbreaking developments in mechanics and computational engineering methods. It covers such fields as solid and fluid mechanics, mechanics of materials, micro- and nanomechanics, biomechanics, and mechatronics. The papers are written by international authorities in the field. The books are at graduate level but may include some introductory material.

More information about this series at https://link.springer.com/bookseries/76

Holm Altenbach · Artur Ganczarski
Editors

Advanced Theories for Deformation, Damage and Failure in Materials

Springer

Editors
Holm Altenbach
Otto-von-Guericke-Universität Magdeburg
Magdeburg, Germany

Artur Ganczarski
Politechnika Krakowska im. Tadeusza Kościuszki
Kraków, Poland

ISSN 0254-1971 ISSN 2309-3706 (electronic)
CISM International Centre for Mechanical Sciences
ISBN 978-3-031-04352-9 ISBN 978-3-031-04354-3 (eBook)
https://doi.org/10.1007/978-3-031-04354-3

This Springer imprint is published by the registered company Springer Nature Switzerland AG
The registered company address is: Gewerbestrasse 11, 6330 Cham, Switzerland

Preface

The course *Advanced Theories for Deformation, Damage and Failure in Materials* was organized during a difficult time. Originally it was planed as a normal course in Udine within the week May 4–8th, 2020 with the lecturers Holm Altenbach (Otto-von-Guericke-Universität Magdeburg, Germany), Artur Ganczarski (Cracow University of Technology, Poland), Frédéric Barlat (Pohang University of Science and Technology), René de Borst (University of Sheffield, UK), Ramesh Talreja (Texas A&M University, College Station, TX, USA), and Ewald Werner (Technische Universität München, Germany). Unfortunately, this concept could not be realized because the COVID 19 pandemic did not allow a direct event. In addition, by personal reasons two lecturers rejected the invitation.

The question immediately arose, how we could realize the planned event a successfully. Special thanks to Thomas Seifert (Hochschule Offenburg, Germany), who immediately agreed to take over part of the lectures. Finally, the course was modified, whereby the overall continuum mechanical framework was marked out with the lectures on "Creep and Damage of Materials at Elevated Temperatures". In the part "Time-dependent and Time-independent Models of Cyclic Plasticity for Low-cycle and Thermomechanical Fatigue Life Assessment" some general statements were specified in a very special manner: instead of three-dimensional statements only one-dimensional statements were discussed.

The contributions "Anisotropic Plasticity During Non-proportional Loading" and "Anisotropy of Yield/Failure Criteria—Comparison of Explicit and Implicit Formulations" showing how the limit states of materials can be estimated. In addition, the contribution "Damage and Failure of Composite Materials" demonstrates the possibility to extend continuum mechanics to continuum damage mechanics of composite materials.

Overall the course was successful and the online format was adopted. Unfortunately, the personal conversation was missing, which was always very important for CISM courses. With these lecture notes it seems that we have reflected the course content well. We hope that the audience, but also those who are interested, will get suggestions for further scientific work.

Finally, we hope that the online format will not become the default. The CISM has always been a place for personal encounters and scientific exchange. This can only happen face-to-face.

Magdeburg, Germany Holm Altenbach
Kraków, Poland Artur Ganczarski

Contents

Chapter 1
Creep and Damage of Materials at Elevated Temperatures

Holm Altenbach

Abstract Creep and damage mechanics are branches of engineering mechanics based partly on continuum mechanics and some engineering approaches to material modeling. The developments in this research field were motivated by some failure cases in the 19th century. The structural failure occurred even if the loading stresses were below the yield point of the material. The reason for that were the elevated temperatures resulting in creep behaviour of the structure. The first theories have been formulated as uniaxial equations with only a few parameters. Later, these equations were extended to three-dimensional equations, substituting the scalar stress and strain by their tensorial counterparts. In addition, for better comparison to one-dimensional results, equivalent statements for the applied stresses were introduced. Up to now, there is no creep mechanics theory which is as strict as continuum mechanics. However, there are many engineering theories through which more and more solutions for practical cases can be obtained. Below, a state of the art report of creep and damage mechanics for metallic materials and structures composed from these materials is given. Some remarks concerning the equivalent stress statements are added. In the further chapters, some parts of this contribution are extended.

1.1 Motivation and Some Historical Remarks

1.1.1 Motivation

Material modeling and simulation is one of the top topics which is in the focus of many researchers and research teams. However, more and more politicians are taking this question into account. The reason for this is the feared climate crisis, which led to the demand to position oneself better in terms of total energy consumption, but also in terms of energy sources. There are various approaches to solving the problems, one

H. Altenbach (✉)
Lehrstuhl für Techniche Mechanik, Institut für Mechanik, Fakultät für Maschinenbau,
Otto-von-Guericke-Universität Magdeburg, Universitätsplatz 2, 39106 Magdeburg, Germany
e-mail: holm.altenbach@ovgu.de

H. Altenbach and A. Ganczarski (eds.), *Advanced Theories for Deformation, Damage and Failure in Materials*, CISM International Centre for Mechanical Sciences 605, https://doi.org/10.1007/978-3-031-04354-3_1

of which is surely the loan building. In addition, the better use of the existing potential of the materials or the new development of materials with the aim of substituting conventional materials is a major challenge.

Considering the challenges, some years ago the following statement of the European Commission was published (de Baas, 2017):

- The future of the European industry is associated with a strong material modelling capacity.
- An efficient modeling approach is needed to shorten the development process of materials-enabled products.

Among the actual areas of activities which the Commission highlighted was research and innòvation w.r.t *Key Enabling Technologies* in material research and development. The modeling of materials should be directed to

- Nanotechnologies,
- Advanced Materials and Manufacturing,
- Biotechnology, and
- Coal and Steel.

The last item is surprising: coal and steel are old materials and the first one belongs to the fossil energy sources which should be excluded. On the other hand, this is a motivation for all of us to study the material more and more and to finally find better mathematical descriptions of the material behavior which are the base of any computer simulation. This is not as costly as complex experiments.

The description of the material behavior can be performed using the *Materials Science Approach* or the *Continuum Mechanics Approach*. Both approaches have advantages and disadvantages. The Material Science Approach results in a precise description of the deformation and damage mechanisms, however the constitutive equations are mostly one-dimensional and an extension to three-dimensional equations is difficult. In Continuum Mechanics the constitutive equations are phenomenologically based as usual resulting in partly rough models, but an extension to three-dimensional equations is rational based or can be realized by engineering assumptions.

The material behavior is very complex, especially if we are looking at the whole lifetime of the materials. It is obvious that any constitutive description can perform only in an approximate sense and the range of applicability is limited. Such ranges are shown in Fig. 1.1.

1.1.2 Creep

Let us discuss an example of material behavior which is named *Creep*. Creep can be time-dependent material behavior or as an inelastic behavior resulting, for example, in permanent strains. There are different definitions of creep in the literature. One definition is coming from Material Science:

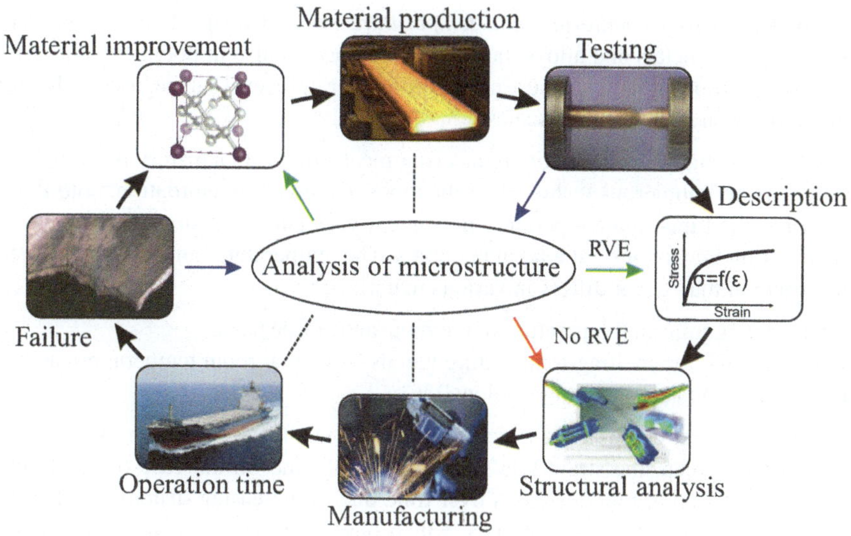

Fig. 1.1 Interaction scheme (Prygorniev, 2015)

Definition 1 (*Creep*) Creep is the tendency of a solid material to move slowly and deform permanently under the influence of mechanical stresses. It can occur as a result of long-term exposure to levels of stress that are still below the yield stress. Creep is more severe in materials that are subjected to heat for long periods.

Remark 1 (*Creep verses plasticity*) In contrast to plasticity in the case of creep behavior the stress level is moderate (lower the yield stresses). However, the temperature level is elevated. As usual the temperature level is characterized with the help of the melting temperature T_{melt}. Creeping processes in metals start if the temperature level is higher $0.3T_{\text{melt}}$. The effects of creep deformation become noticeable at 45% of the melting point for ceramics. Note that in both cases the temperature should be in K.

Remark 2 (*Creep deformation mechanisms*) With respect to different temperature and loading levels and taking into account the microstructure of different materials various deformation mechanisms can be obtained (Cotrell, 1967): bulk diffusion (Nabarro-Herring creep), climb (the strain is accomplished by climb), climb-assisted glide (the climb is an enabling mechanism, allowing dislocations to get around obstacles), grain boundary diffusion (Coble creep), thermally activated glide (e.g., via cross-slip), …. In Frost and Ashby (1982) deformation maps (stress normalized using the shear modulus versus homologous temperature with contours of strain rate) are presented. They allow the representation of the dominant deformation mechanisms in a material loaded under a given set of conditions. They are developed for metallurgical as well as geological materials having a crystalline microstructure.

Remark 3 (*Various materials and temperature ranges*) First creep studies were performed for metals and alloys, however, creep occurs also in polymers, concrete, ceramics, composites, etc. In these cases, the temperature level can be lower or higher which result in different time scales.

Creep mechanics is a part of engineering mechanics with a history of more than 100 years and numerous technical applications. Creep (also retardation, cold flow) describes the time- and temperature-dependent viscoelastic or, in general, inelastic deformation under constant load in materials. The temperature range in which creep deformation may occur differs in various materials:

- tungsten requires a temperature in the thousands of degrees,
- plastics and low-melting-temperature metals—creep at room temperature, and
- ice will creep at temperatures below 0 °C.

There are three creep stages which can be observed in many situations. In primary, or transient, creep, the strain rate is a function of time. In the case of most pure materials, the strain rate decreases over time due to increasing dislocation density or evolving grain size. For materials, which have large amounts of solid solution hardening, the strain rate increases over time due to a thinning of solute drag atoms as dislocations move. In the secondary, or steady-state, creep, dislocation structure and grain size have reached equilibrium, and therefore the strain rate is constant. In tertiary creep, the strain rate exponentially increases with stress. This can be due to necking phenomena, internal cracks, or voids, which all decrease the cross-sectional area and increase the true stress on the region, further accelerating deformation and leading to fracture.

In applications three questions arise. The first one is: a suitable material description must be found first of all. This task is not trivial, since the different concepts, based on considerations of material physics, materials science and continuum mechanics have advantages and disadvantages. It is important to ensure that the effort and benefit are in an appropriate relationship and that the identification of the parameters in the equations describing the material behavior can be solved in a satisfactory manner. It should be noted that not every conceivable experiment to determine material parameters can actually be carried out in the laboratory. The second one: another problem is related to the fact that a suitable structural mechanical description must be made. Components are geometrically complex structures. Hence, their geometrical description and the implementation are usually associated with the introduction of models. These simplify reality and thus enable practical problems to be analyzed with less effort. However, the use of certain models is allowed within the limits of the simplifications. For example, it is known that thin-walled components (they are typical for creep mechanics applications) can often be analyzed with two-dimensional equations. Which theory has to be used e.g. in the case of plates (Kirchhoff, Mindlin, Reissner, Ambarcumyan, von Kármán, …) has been the subject of numerous studies. The third one is related to the selection of a suitable numerical analysis method (finite element method, boundary element method, etc.). This is also important, but it is not supposed to be content of this presentation. For further reading concerning numerical aspects see Naumenko and Altenbach (2020).

The time-dependent changes of the state of the structure is an important feature for the safety analysis. In micrographs the following observations can be made and the indicated actions should be realized:

- isolated cavities (only observation),
- oriented cavities (observation with fixed inspection intervals),
- micro cracks (limited operating time until repair),
- macro cracks and fracture (repair).

It is obvious that a precise prediction of the behavior of a structure under creep conditions has a great economical effect.

1.1.3 Brief Historical Overview

There are many publications related to *Creep Mechanics* presenting the state of the art and a historical survey. The history up to 1970 is well described in Odqvist and Hult (1962), Odqvist (1981). First creep observations were already published in the second half of the 19th century. The starting point of research were accidents with load stresses well below the yield point. Systematic investigations were at first summarized in da Costa Andrade (1910), da Costa Andrade and Porter (1914).

The Norton-Bailey law maybe is the most important creep law, which is a power law in the sense of mathematics (Norton, 1929; Bailey, 1935). On first industrial applications from energy machine construction (for example, gas turbines) was reported in Stodola (1933). Because in general mechanical loads are multiaxial, the response should also be multiaxial. Suitable equations were suggested by Odqvist (1935) (see also Odqvist 1974). The material was assumed to be isotropic, which can be represented using stress and strain tensor invariants (as a particular case the von Mises equivalent stress and strain). A consistent tensorial description which also includes anisotropy was suggested by Reiner (1943), Reiner (1945), Prager (1945). Mismatches with experimental results led to modifications of the creep equations (Nadai, 1938; Soderberg, 1938).

In the 50th of the last century stability problems and elements of the geometrically nonlinear theory were considered (Hoff, 1953, 1957). Due to the increasing use of polymers as structural materials, analogies between viscoelasticity and creep became the focus of researchers (see, for example, Rabotnov, 1980). Last but least, creep also was established for concrete which resulted in a new direction of *Creep Mechanics* (Arutyunyan, 1966).[1]

There are a lot of scientific activities in the field of *Creep Mechanics* and *Creep Modeling and Testing*. Among them, the International Union for Theoretical and Applied Mechanics (IUTAM) organizes an IUTAM symposium *Creep in Structures* since 1960 every 10 years:

[1] From this direction we have now a new branch in theory of modeling the constitutive behavior of materials—mechanics of growing solids (Manzhirov et al., 2017).

- 1960—Stanford/U.S.A. (Hoff, 1962),
- 1970—Göteborg/Sweden (Hult, 1972),
- 1980—Leicester/U.K. (Ponter and Hayhurst, 1981),
- 1990—Kraków/Poland (Życzkowski, 1991),
- 2000—Nagoya/Japan (Murakami and Ohno, 2001), and
- 2012—Paris/France[2] (Altenbach and Kruch, 2013).

Probably the next symposium will be

- 2023—Magdeburg/Germany.[3]

1.2 Creep Model

1.2.1 Basic Model

There are different ways to describe the creep behavior. Among them, there purely phenomenologically and mechanism-based approaches. In dependence from research and application topic one can start from

- considerations of the materials science and physics,
- macroscopic observations, and
- *Continuum Mechanics* based methods.

That means the creep modeling is an interdisciplinary task using tools from (*Physics*, *Material Science* and/or *Mechanics*). The final constitutive equations not always coincide. Each approach has advantages and disadvantages:

- *Physics* and *Material Science* based approaches and creep equations based on them are most suitable to characterize processes at the microlevel. In the focus are, for example, the structure-properties relations. At the same time, three-dimensional generalization is often complicated.
- The phenomenological description within engineering mechanics is not always strict enough to meet all aspects of the modeling requirements. On the other hand, the one-dimensional equations can be easily extended to the corresponding three-dimensional creep equations which can be implemented into existing commercial finite element codes.

The last item (extension of classical one-dimensional models) is discussed, for example, in Naumenko and Altenbach (2016). The following items are presented:

- Since the loading state is generally three-dimensional, a suitable three-dimensional description of the material behavior is needed. Starting with one-dimensional relations, the constitutive relations are presented for scalar variables, like stress and

[2] By some special reasons, the title of this symposium was changed.

[3] The announcement is given on `https://iutam.org/event/iutam-symposium-on-creep-in-structures/` or `http://www.iutam-symposium.ovgu.de`.

strain rate. The corresponding three-dimensional equations are similar—instead the scalar variables the tensorial counterparts are introduced.

- In addition, for the use of simple experimental results in the material parameters identification some hypotheses concerning the equivalence of the one-dimensional and three-dimensional states should be introduced. Examples are the equivalent stress of von Mises, Tresca, etc. Further information on equivalence hypotheses are given, for example, in Kolupaev (2018).
- The anisotropy in material behavior is also important for creep processes. We have to distinguish
  - initial isotropic materials which may show anisotropic behavior (for example, damage-induced anisotropy) in the tertiary creep stage,
  - initial isotropic materials which may show anisotropic behavior (for example, orthotropy) after technological treatments (among them, rolling).
  - There are also à priori anisotropic materials (initial anisotropy).

Creep tests play an important role in creep modeling:

- The simplest creep test is an analogy to the tensile test for which only a normal stress is considered in the tensile direction resulting from a constant load (force). The observations allow the conclusion that even in the case of constant stresses the normal strains are increasing, which is named creep.
- Another simple creep test is the torsion test, where a shear stress can be noticed as a result of a constant acting torsional moment. From this test one concludes that even in the case of constant shear stresses the shear strains are increasing (again creep).

Both tests can be superposed, in this case, one gets a complex stress state with the following stress tensor $\boldsymbol{\sigma}$ (Naumenko and Altenbach, 2016)

$$\boldsymbol{\sigma} = \sigma \boldsymbol{k} \otimes \boldsymbol{k} + \tau \left(\boldsymbol{e}_{\varphi} \otimes \boldsymbol{k} + \boldsymbol{k} \otimes \boldsymbol{e}_{\varphi} \right) \tag{1.1}$$

where $\boldsymbol{k}$ is a unit vector in the normal creep direction and $\boldsymbol{e}_{\varphi}$ denotes a unit vector in the circumferential direction. σ and τ are the normal stress in the $\boldsymbol{k}$-direction and the shear stress in the circumferential direction $\boldsymbol{e}_{\varphi}$, $\otimes$ is the dyadic product.

Let us analyze the complex model. The inelastic material behavior is often assumed to be independent from the hydrostatic stress state which means we assume incompressibility. In addition to the stress tensor, the stress deviator $\boldsymbol{s}$ needs to be introduced

$$\boldsymbol{s} = \boldsymbol{\sigma} - \frac{1}{3}\boldsymbol{\sigma} \cdot\cdot \boldsymbol{E}\boldsymbol{E} = \sigma \left(\boldsymbol{k} \otimes \boldsymbol{k} - \frac{1}{3}\boldsymbol{E} \right) + \tau \left(\boldsymbol{e}_{\varphi} \otimes \boldsymbol{k} + \boldsymbol{k} \otimes \boldsymbol{e}_{\varphi} \right) \tag{1.2}$$

with $\boldsymbol{E}$ as the second rank unit tensor. Furthermore, to ensure a better comparison of one-dimensional and three-dimensional states, an equivalence hypothesis should be found. In the simplest case, the von Mises equivalent stress σ_{vM} can be suggested as

$$\sigma_{\mathrm{vM}} = \sqrt{\frac{3}{2}\boldsymbol{s} \cdot\cdot \boldsymbol{s}} = \sqrt{\sigma^2 + 3\tau^2}. \tag{1.3}$$

The double scalar product "$\cdot\cdot$" will be used as in Altenbach (2018).

If the material in tests shows a significant secondary creep stage, the starting point for modeling is the secondary stage. After that we extend the equations to the other two creep stages (primary and tertiary creep stages). The uniaxial description is based on the secondary stage with the following assumption

$$\dot{\varepsilon}^{\mathrm{cr}}_{\mathrm{min}} = f(\sigma, T). \tag{1.4}$$

$\dot{\varepsilon}^{\mathrm{cr}}_{\mathrm{min}}$ is the minimal creep rate, σ is the applied stress responsible for the creep and T is the temperature which is further assumed to be constant in order to simplify the model

$$\dot{\varepsilon}^{\mathrm{cr}}_{\mathrm{min}} = f(\sigma). \tag{1.5}$$

The experimental verification is straightforward. From the literature several approximations for the function of the minimal creep strain rate are known. The power law (Bailey, 1935; Norton, 1929) is the one used the most. However, there are also reasons to use other approximations like the exponential function or a hyperbolic sine (Nadai, 1938) function. Some comments concerning the best choice of the function of the minimal creep strain rate are given in the relevant literature (see, for example, Kowalewski et al., 1994; Betten, 2001).

The extension to the primary and tertiary stages can be realized as follows. The ansatz for the secondary creep (1.5) can be used introducing one or several hardening variables. Let us assume that H denotes a hardening variable so we now introduce

$$\dot{\varepsilon}^{\mathrm{cr}} = f(\sigma, H, T) \qquad \text{or} \qquad \dot{\varepsilon}^{\mathrm{cr}} = f(\sigma, H) \tag{1.6}$$

which should be completed with an evolution equation for H

$$\dot{H} = \dot{H}(\sigma, H, T) \qquad \text{or} \qquad \dot{H} = \dot{H}(\sigma, H). \tag{1.7}$$

For tertiary creep, for example, a damage variable ω can be introduced

$$\dot{\varepsilon}^{\mathrm{cr}} = f(\sigma, H, \omega, T) \qquad \text{or} \qquad \dot{\varepsilon}^{\mathrm{cr}} = f(\sigma, H, \omega) \tag{1.8}$$

and a damage evolution equation should be postulated

$$\dot{\omega} = \dot{\omega}(\sigma, H, \omega, T) \qquad \text{or} \qquad \dot{\omega} = \dot{\omega}(\sigma, H, \omega) \tag{1.9}$$

Remark 4 (*Variables describing mechanisms*) Mechanisms like hardening, damage, softening, etc. can be presented by inner variables and their evolution equations.

During the primary or tertiary creep several mechanisms can be observed. In this case we have H_i and ω_j with the number of mechanisms during the primary creep i and the number of mechanisms during the tertiary creep j.

Remark 5 (*Creep models without significant secondary creep*) In the case when in experiments no significant secondary creep behaviour is established, one can introduce equations for the remaining parts not based on secondary creep equations. Examples are given in Naumenko and Kostenko (2009), Naumenko et al. (2011a), Naumenko and Gariboldi (2014), Eisenträger et al. (2018), Naumenko et al. (2020).

The procedure presented here is not limited to uniaxial behavior. By introducing suitable tensor variables for the stress and the strain rate as well as corresponding equivalent variables, multi-axial constitutive and evolution laws can be established. The first question is what kind of equivalence hypotheses need be introduced. It should be noted that equivalence concepts for the stresses and the strains are always just engineering hypothesises, therefore, any concept and potential modifications must be examined again to determine whether the assumptions made are valid (Kolupaev et al., 2009, 2013; Altenbach et al., 2014; Kolupaev, 2018)

The concept has to be changed, if the anisotropy must be included since anisotropy tensors should be justified and described. For the damage-induced anisotropy, evolution laws have to be added and presented mathematically in a proper manner. Models become very complex and the identification effort is increasing dramatically.

1.2.2 Continuum Damage Mechanics

Let us start with the classical Kachanov-Rabotnov approach. In 1958, Lazar Kachanov published his pioneering paper (Kachanov, 1958). He introduced the continuity as a damage variable, which was a distributed field variable in the range $1 \geq \psi \geq 0$ (note that 1 and 0 are not realistic). One year later, Rabotnov published his first work in damage mechanics (Rabotnov, 1959) introducing damage as a distributed field variable within the range $0 \leq \omega \leq 1$ (0 and 1 are again not realistic). The developments in *Creep* and *Creep-Damage Mechanics* are summarized in the monograph Rabotnov (1969). It is obvious that continuity and damage are conjugate and scalar variables. There are many suggestions concerning the extension of this damage concept introducing, for example,

- damage vector (Kachanov, 1974),
- damage tensor (for example, Betten, 2001) or
- load dependency (strength differential effect, Casey and Jahedmotlagh, 1984)

With respect to the identification of the material parameters in the constitutive or evolution equations one needs experimental observations. There are a lot of papers in the literature presenting one experimental procedure. In Lemaitre and Dufailly (1987) several methods are discussed:

- direct measurements
  - the observation of micrographic pictures (for example, cross section area changes),
  - the measurement of the variation of the density,
- non direct measurements.[4]
  - destructive
    - the measurement of the variations of the elasticity modulus,
    - the measurement of the ultrasonic waves propagation,
    - the measurement of the cyclic plasticity or creep responses,
  - non destructive
    - the measurement of the variation of the micro-hardness,
    - the measurement of the electrical potential.

Since continuum damage mechanics is a branch of the rational continuum mechanics the starting point for the governing equations are five balance equations

- mass,
- momentum,
- moment of momentum,
- energy, and
- entropy (in the form of inequality/equality).

The entropy balance plays a special role because damage is a dissipative process. The balance of mass in solid mechanics is as usual a conservation law. Even in the case of damage mechanics, where the loss of mass can be observed (for example, in the case of corrosion processes), we assume that the density is constant and the loss of mass can be ignored.

The inelastic deformation and damage processes are combined. The damage processes are characterized by

- formation, growth and coalescence of voids on grain boundaries,
- microcracks in particles of the second phase,
- decohesion at particle/matrix interfaces, and
- surface cracks under cyclic loading.

The last part of the "material's history" is related to softening and ageing:

- increase of inelastic strain rate and
- damage (formation of cracks and final fracture of solids)

The basics of damage mechanics are presented in Krajcinovic and Lemaitre (1987), Lemaitre and Chaboche (1990), Krajcinovic (1995), Krajcinovic (1996), Lacy et al. (1997), Skrzypek and Ganczarski (1999), Altenbach and Skrzypek (1999), Lemaitre

[4] The classification of the measurement method is taken from the original paper of Lemaitre and Dufailly (1987).

and Desmorat (2005), Murakami (2012), Menzel and Sprave (2020), among others. Let us summarize the basic ideas. The damage rate and the inelastic strain rate are determined by

- the stress level,
- the accumulated damage, and
- the temperature.

In the case of multi-axial loading the kind of stress state has a significant influence on the damage growth and the different stress states corresponding to the same von Mises equivalent stress leading to different equivalent *tertiary creep* rates while the equivalent strain rate in the *secondary stage* is approximately the same.

Various damage variables were introduced in the literature:

- scalar-valued - the models are limited by isotropic damage,
- vector-valued - now orientation of microcracks can be taken into account,
- tensor-valued
  - 2nd rank tensor - the models are limited by orthotropic material behaviour,
  - 4th rank tensor - the reduced (effective) stiffness can be presented, and
  - 8th rank tensor - necessary for the correct mapping of the 4th rank elasticity tensor into the 4th rank elasticity-damage tensor.

Note that the number of parameters increases dramatically with the rank of the damage tensor.

A detailed discussion of damage variables (tensors) is given in Chaboche (1987), Chaboche (1988a), Chaboche (1988b), Lü and Chen (1989), Betten (1992), Chaboche (1993), Zheng and Betten (1995), Voyiadjis and Kattan (2009), Olsen-Kettle (2019). The damage variables can be introduced with the help of microstructural observations. Among these observations is the directional effect of creep damage

- during a cyclic torsion test on copper, voids nucleate and grow predominantly on those grain boundaries, which are perpendicular to the first principal direction of the stress tensor (Hayhurst, 1972), and
- initially isotropic material subjected to constant or monotonic loading, but the influence of the damage anisotropy on the observed creep behavior is not significant (Naumenko and Altenbach, 2016).

Summarizing the state of the art we have the following models

- phenomenological (with the pionering works of Kachanov, 1958; Rabotnov, 1959),
- micromechanically consistent (for example, Rodin and Parks, 1986),
- mechanism based (for example, Hayhurst, 1994), and
- based on dissipation (see, for example, Sosnin, 1973; Sosnin et al., 1976).

The three-dimensional extension can be performed as follows:

- the creep process is determined by the effective stress tensor taking into account the damage

$$\tilde{\boldsymbol{\sigma}} = \boldsymbol{f}(\boldsymbol{\sigma}, \omega)$$

- and the creep potential for the damaged material has the same form as for the secondary creep.

As an example we can take the Norton-Bailey-Odqvist creep potential

$$W(\tilde{\boldsymbol{\sigma}}) = \frac{\sigma_0}{n+1}\left(\frac{\tilde{\sigma}_{\mathrm{vM}}}{\sigma_0}\right)^{n+1}$$

with

$$\tilde{\boldsymbol{\sigma}} = \frac{\sigma_{\mathrm{I}}}{1-\omega}\boldsymbol{n}_{\mathrm{I}} \otimes \boldsymbol{n}_{\mathrm{I}} + \sigma_{\mathrm{II}}\boldsymbol{n}_{\mathrm{II}} \otimes \boldsymbol{n}_{\mathrm{II}} + \sigma_{\mathrm{III}}\boldsymbol{n}_{\mathrm{III}} \otimes \boldsymbol{n}_{\mathrm{III}}, \qquad \sigma_{\mathrm{I}} > 0.$$

σ_0, n are material parameters, σ_{I} is the first (maximal) principal stress, $\sigma_{\mathrm{II}}, \sigma_{\mathrm{III}}$ are the other two principal stresses and $\boldsymbol{n}_{\mathrm{I}}, \boldsymbol{n}_{\mathrm{II}}, \boldsymbol{n}_{\mathrm{III}}$ are the principal directions and $\tilde{\sigma}_{\mathrm{vM}}$ is the the von Mises equivalent effective stress

$$\tilde{\sigma}_{\mathrm{vM}} = \sqrt{\frac{3}{2}\tilde{\boldsymbol{s}} \cdot\cdot \tilde{\boldsymbol{s}}}$$

with

$$\tilde{\boldsymbol{s}} = \tilde{\boldsymbol{\sigma}} - \frac{1}{3}\tilde{\boldsymbol{\sigma}} \cdot\cdot \boldsymbol{E}\boldsymbol{E}.$$

The constitutive equation can be given for the creep rate (Leckie and Hayhurst, 1977)

$$\dot{\boldsymbol{\varepsilon}}^{\mathrm{cr}} = \dot{\lambda}(\tilde{\sigma}_{\mathrm{vM}})\left[\boldsymbol{s} + \frac{\omega}{1-\omega}\sigma_{\mathrm{I}}\left(\boldsymbol{n}_{\mathrm{I}} \otimes \boldsymbol{n}_{\mathrm{I}} - \frac{1}{3}\boldsymbol{E}\right)\right].$$

The generalized von Mises-type secondary creep equation is

$$\dot{\boldsymbol{\varepsilon}}^{\mathrm{cr}} = \frac{3}{2}a\left(\frac{\sigma_{\mathrm{vM}}}{1-\omega}\right)^{n}\frac{\boldsymbol{s}}{\sigma_{\mathrm{vM}}}.$$

a is a material parameter and n is the creep exponent. The derivation is based on the strain equivalence principle (Lemaitre, 1984) and the effective stress tensor takes the form $\tilde{\boldsymbol{\sigma}} = \boldsymbol{\sigma}/(1-\omega)$. The damage evolution equation is formulated for the damage rate

$$\dot{\omega} = \dot{\omega}(\boldsymbol{\sigma}, \omega)$$

with the damage evolution equation

$$\dot{\omega} = \frac{b(\sigma_{\mathrm{eq}}^{\omega})^k}{(1-\omega)^l}. \tag{1.10}$$

Here b, k, l are material parameters. The equivalent stress $\sigma_{\mathrm{eq}}^{\omega}(\boldsymbol{\sigma})$ is different from σ_{vM} since damage is not the same at tension and compression. A possible suggestion for the damage equivalent stress assuming isotropic materials is

$$\sigma_{\text{eq}}^{\omega} = \sigma_{\text{eq}}^{\omega}[I_1(\boldsymbol{\sigma}), I_2(\boldsymbol{\sigma}), I_3(\boldsymbol{\sigma})]$$

with the stress tensor invariants $I_1(\boldsymbol{\sigma})$, $I_2(\boldsymbol{\sigma})$ and $I_3(\boldsymbol{\sigma})$.

Remark 6 Various suggestions for the invariants can be made. A brief discussion is given, for example, in Życzkowski (1991), Kolupaev et al. (2013).

The time to creep fracture in the case of Eq. (1.10) can be estimated by

$$t_* = \frac{(\sigma_{\text{eq}}^{\omega})^{-k}}{(l+1)b}.$$

However, there are also other suggestions in the literature:

- Sdobyrev (1959) (taking into account the von Mises stress and the first principal stress)

$$\sigma_{\text{I}} + \sigma_{\text{vM}} = 2f(\log t_*),$$

- Trunin (1965) (taking into account the different damage behavior at tension and compression)

$$2\sigma_{\text{eq}}^{*} = (\sigma_{\text{vM}} + \sigma_{\text{max t}})\, a^{1-2\eta}, \quad \eta = 3\sigma_{\text{m}}/(\sigma_{\text{vM}} + \sigma_{\text{max t}}),$$

  with

$$\sigma_{\text{max t}} = \frac{\sigma_{\text{I}} + |\sigma_{\text{I}}|}{2}, \quad \sigma_{\text{m}} = \frac{1}{3}(\sigma_{\text{I}} + \sigma_{\text{II}} + \sigma_{\text{III}}),$$

- Hayhurst (1972)

$$t_* = A(\alpha\sigma_{\max t} + \beta I_1 + \gamma\sigma_{\text{vM}})^{-\chi},$$

  which is based on isochronous rupture loci with $\sigma_{\text{eq}}^{\omega} = \alpha\sigma_{\max t} + \beta I_1 + \gamma\sigma_{\text{vM}}$. $A, \xi, \alpha, \beta, \gamma$ are fitting parameters. However, in this case the maximal tension stress, the first invariant of the stress tensor (that means the hydrostatic behavior) and the von Mises stress have an influence on the damage behavior.

Micromechanically-consistent models were suggested, for example, in Rodin and Parks (1986). The creep potential is now

$$W(\boldsymbol{\sigma}, \rho, n) = \frac{\dot{\varepsilon}_0\sigma_0}{n+1} f\,[\zeta(\boldsymbol{\sigma}), \rho, n] \left(\frac{\sigma_{\text{vM}}}{\sigma_0}\right)^{n+1}$$

with the material parameters $\dot{\varepsilon}_0, \sigma_0$ and the measure of damage $\rho = a^3 N/V$. N is the number of cracks (voids) in a volume V, a is the averaged radius of a crack, $\zeta(\boldsymbol{\sigma})$ is a function representing the influence of the kind of stress state, for example,

$$\zeta(\boldsymbol{\sigma}) = \frac{\sigma_{\text{I}}}{\sigma_{\text{vM}}}.$$

The creep rate in this case is equal to

$$\dot{\boldsymbol{\varepsilon}}^{\mathrm{cr}} = \dot{\varepsilon}_0 \left(\frac{\sigma_{\mathrm{vM}}}{\sigma_0}\right)^n \left[\frac{3}{2}\left(f - \frac{\zeta f_{,\zeta}}{n+1}\right)\frac{\boldsymbol{s}}{\sigma_{\mathrm{vM}}} + \frac{f_{,\zeta}}{n+1}\boldsymbol{n}_I \otimes \boldsymbol{n}_I\right].$$

Mechanism-based models were presented, for example, by Hayhurst (1994). Hayhurst and co-workers suggested several models taking into account different materials with their specific mechanisms. In Perrin and Hayhurst (1996) the creep constitutive equations for a 0.5Cr-0.5Mo-0.25V ferritic steel (temperature range 600-675° includes several effects:

- the cavitated area fraction (Cane, 1981)

$$A_{\mathrm{f}} = D\varepsilon_{\mathrm{vM}} \left(\frac{\sigma_{\mathrm{I}}}{\sigma_{\mathrm{vM}}}\right)^{\mu} \tag{1.11}$$

  with the parameters D, μ, where μ is related to the multiaxial stress state, and the equivalent von Mises strain $\varepsilon_{\mathrm{vM}}$,
- based on Eq. (1.11) the evolution of cavitation damage, which describes the intergranular cavitation, can be presented

$$\dot{\omega} = C\varepsilon_{\mathrm{vM}} \left(\frac{\sigma_{\mathrm{I}}}{\sigma_{\mathrm{vM}}}\right)^{\mu}, \tag{1.12}$$

  with the fitting parameter C,
- the particle coarsening evolution (see, for example, Dyson and Oosgerby 1993; Lifshitz and Slyozov 1961)

$$\dot{\phi} = \left(\frac{K_{\mathrm{c}}}{3}\right)(1-\phi)^4$$

  with the parameter K_{c} as a function of the initial particle spacing and temperature, and
- initial strain hardening (primary creep stage) due to the formation of a dislocation substructure (Dyson and Oosgerby, 1993; Kowalewski et al., 1994)

$$\dot{H} = \frac{h_{\mathrm{c}}\dot{\varepsilon}^{\mathrm{cr}}_{\mathrm{vM}}}{\sigma_{\mathrm{vM}}}\left(1 - \frac{H}{H_*}\right).$$

  h_{c} is a material parameters and H_* is the saturation value.

The creep rate can be presented by a hyperbolic sin function

$$\dot{\varepsilon}^{\mathrm{cr}}_{\mathrm{vM}} = A \sinh \frac{B\sigma_{\mathrm{vM}}(1-H)}{(1-\phi)(1-\omega)}$$

with the parameters A and B. The advantage of the hyperbolic sin function in comparison with the Norton law is that it is valid for a wide range of stress.

The presented model is correct if the temperature is constant. Otherwise some changes are necessary:

$$A = A_0 B \exp\left(-\frac{Q_A}{RT}\right), \quad B = B_0 \exp\left(-\frac{Q_B}{RT}\right),$$
$$K_c = \frac{K_{c0}}{B^3} \exp\left(-\frac{Q_{K_c}}{RT}\right), \quad D = D_0 \exp\left(-\frac{Q_D}{RT}\right).$$

The models based on dissipation (see, for example, Sosnin 1973; Sosnin et al. 1976) can be introduced as follows

$$q = \int_0^t \sigma \dot{\varepsilon}^{\mathrm{cr}} \mathrm{d}\tau, \qquad q = \int_0^t \boldsymbol{\sigma} \cdot\cdot\, \dot{\boldsymbol{\varepsilon}}^{\mathrm{cr}} \mathrm{d}\tau$$

with the evolution equation

$$\dot{q} = f_\sigma(\sigma) f_T(T) f_q(q)$$

The creep rate in this case is

$$\dot{\boldsymbol{\varepsilon}}^{\mathrm{cr}} = \frac{3}{2} \frac{P}{\sigma_{\mathrm{vM}}} \boldsymbol{s}, \quad P = \boldsymbol{\sigma} \cdot\cdot\, \dot{\boldsymbol{\varepsilon}}^{\mathrm{cr}} = \sigma_{\mathrm{vM}} \dot{\varepsilon}^{\mathrm{cr}}_{\mathrm{vM}}$$

with $\dot{q} \equiv P = f_\sigma(\sigma_{\mathrm{vM}}) f_T(T) f_q(q)$

There are more models in the literature. For further reading Naumenko and Altenbach (2016), Altenbach and Eisenträger (2020), Altenbach and Öchsner (2020), Altenbach and Knape (2020), Eisenträger and Altenbach (2020) can be recommended. In addition, various problems based on the introduced models were solved. Some results of the author's team are presented in Altenbach and Naumenko (1997), Altenbach et al. (1997), Altenbach et al. (2000a), Altenbach et al. (2000b), Altenbach et al. (2001a), Altenbach et al. (2001b), Altenbach (2002), Altenbach et al. (2002), Altenbach and Naumenko (2002), Altenbach et al. (2004), Naumenko and Altenbach (2005), Altenbach and Naumenko (2007), Altenbach et al. (2008a), Altenbach et al. (2008b), Naumenko et al. (2009), Naumenko et al. (2011b), Gorash et al. (2012), Naumenko and Altenbach (2020).

1.3 Continuum Mechanics

In the previous section some elements of continuum damage mechanics were introduced. Below one can find some general statements with respect to the governing equations of continuum mechanics in the one-dimensional and in the three-dimensional case.

1.3.1 Preliminary Remarks

Let us start with a definition[5].

Definition 2 (*Continuum mechanics*) Continuum mechanics is a branch of mechanics that deals with the analysis of the kinematics and the mechanical behavior of materials modeled as a continuous mass rather than as discrete particles.

The French mathematician Augustin-Louis Cauchy was (maybe) the first to formulate such models in the 19th century. The research in the area continues today and is related to solids, fluids, and gases. There are some doubts up to now:

- What are the reasons - stresses or strains, etc.?
- What is the starting point - Euler, Bernoulli, or Newton?

The first question is easy to answer. In analogy to materials testing, input (e.g. stresses) and output (e.g. strains) can be swapped. This results in stress-controlled and strain-controlled tests. The answer to the second question is fundamental. Newton based "his mechanics" on four axioms. Strictly speaking, point mechanics was thus established (see, for example, Feynman et al. 1964, 1965).

This part of the lecture notes is based on the German textbook (Altenbach, 2018). The first definition is here the definition of the continuum.

Definition 3 (*Continuum—first definition*) The continuum is a manifold of points filling the space or a part of the space continuously at each moment t. Material properties are prescribed to the points.

It seems that is a purely mathematical definition, but beyond this definition we will have a lot of mechanical applications. With respect to this definition any field (mechanical, thermal, electric, magnetic, chemical, etc.), but also coupled fields (for example, the thermo-mechanical field) can be presented in rational description.

Let us introduce some assumptions:

- Space: Euclidean three-dimensional space or lower/higher dimensional spaces
  - two-dimensional, one-dimensional spaces and
  - four- or higher dimensional spaces (Minkowski spaces)

[5] https://en.wikipedia.org/wiki/Continuum_mechanics, 31.12.2021

are assumed.

- Time: $t \geq 0$ and the time is a monotonic increasing function (note that the initial moment cannot be established, but fixed by our considerations, in this sense we have a pseudo-time).
- Body: In continuum mechanics models we assign a region in three-dimensional Euclidean space to the material body $\mathcal{B}$ with the volume V and a surface A.
- Mass: This is the quantity of matter in a physical body and now, for the first time, we have material points instead of geometrical points, if we prescribe the density to each point (it is also a measure of the body's inertia or the resistance to acceleration).
- Homogenity: This means that in each point we have the same properties (this assumption is not necessary, but results in simplifications).
- Isotropy: Uniformity in all orientations, which is not necessary, but results in simplifications.

Continuum mechanics should be valid

- for any continua (solids, fluids, gases, etc.) and
- scale-independent (macro, meso, micro, nano, etc.).

The basic equations are

- the material independent equations and
- the material dependent equations.

Finally, based on the governing equations and suitable boundary and initial conditions an initial-boundary value problem can be established and solved. For further reading we recommend Truesdell and Toupin (1960), Haupt (2002), Truesdell and Noll (2004), Dill (2006), Lai et al. (2010), Hutter and Jöhnk (2010), Eremeyev et al. (2013).

1.3.2 Brief Historical Outline

Mechanics is one of the oldest sciences. The beginning may be related to Archimedes of Syracuse (c. 287–c. 212 BC), who was a Greek mathematician, physicist, engineer, astronomer, and inventor. He was known for Archimedes' principle, Archimedes' screw, the basics of hydrostatics, the lever, and infinitesimals. He introduced the geometrical model of a rhombicuboctahedron, which is now used as an elementary cell of metal foam with big porosity. His earliest remaining writings regarding levers date from the 3rd century: "*give me a place to stand, and I shall move the earth with a lever*". This is a remark of Archimedes who formally stated the correct mathematical principle of levers (quoted by Pappus of Alexandria). In ancient Egypt, constructors used the lever to move and uplift obelisks weighting more than 100 t. The force applied at end points of the lever is proportional to the ratio of the length of the lever arm measured between the fulcrum (pivoting point) and application point of the force applied at each end of the lever. Mathematically, this is expressed by $M = Fd$ (M - torque, F - force, d distance).

For the first time, material properties were included by Galileo Galilei (*15 February 1564$^{\text{Julian calendar}}$ in Pisa, † 8 January 1642$^{\text{Julian calendar}}$ in Arcetri near Florence) and Robert Hooke (*18 July 1635$^{\text{Julian calendar}}$ in Freshwater, Isle of Wight, † 3 March 1703$^{\text{Julian calendar}}$ in London). The last one introduced the linear elastic behavior:

Hooke' law: *ceiiinosssssttuv* (1676), was a Latin anagram, whose solution was published in 1678 as *Ut tensio, sic vis* which means *As the extension, so the force*.

There are several applications of this law—one is the elastic spring with $F = -k\Delta x$ (k is the spring rate and Δx is the elongation). This is a similar law to $\sigma = E\varepsilon$ with the stress σ, the strain ε and the Young's modulus E. It is obvious, that the spring model is a rheological model for elasticity.

Isaac Newton (* 25 December 1642$^{\text{Julian calendar}}$ in Woolsthorpe-by-Colsterworth in Lincolnshire, † 20 March 1726$^{\text{Julian calendar}}$ in Kensington) summarized the knowledge of that time: the Newtonian Mechanics is based on four axioms:

- $\frac{\mathrm{d}}{\mathrm{d}t}(m\boldsymbol{v}) = \mathbf{0}$ (statics),
- $\frac{\mathrm{d}}{\mathrm{d}t}(m\boldsymbol{v}) = \boldsymbol{F}$ (kinetics),
- actio = reactio, and
- superposition.

However, it seems that this type of mechanics is valid for point-bodies only. No rotations can be observed for them. But from the beam theory of Euler (Leonhard Euler * 15 April 1707 in Basel, † 7 September 1783$^{\text{Julian calendar}}$ in Sankt Petersburg) and Bernoulli (Daniel Bernoulli *29 January 1700$^{\text{Julian calendar}}$ in Groningen, †17 March 1782 in Basel) it was known that the requirements from the boundary conditions can be satisfied with the introduction of independent moments. As the result, the second Eulerian law of motion (balance of moment of momentum) was established as independent from the first law (balance of momentum) (see Truesdell 1964; Naue 1984). As a conclusion, one has the independence of translations and rotations and the force and moment actions in the general case.

The classical continuum mechanics is based on Cauchy's contributions (Augustin-Louis Cauchy *21 August 1789 in Paris, †23 May 1857 in Sceaux). He introduced the stress definition (around 1822).

Definition 4 (*Stress*) Stress σ is a measure of the average amount of force F exerted per unit area A. It is a measure of the intensity of the total internal forces acting within a body across imaginary internal surfaces, as a reaction to external applied forces and body forces.

For the completeness of the governing equations constitutive equations must be added. The constitutive equations in the case of Cauchy's elasticity contain the

Young's modulus, introduced by Thomas Young (*13 June 1773 in Milverton, Somersetshire, †10 May 1829 in London). It should be mentioned that the concept was developed in 1727 by Leonhard Euler and the first experiments that used the concept of Young's modulus in its current form were performed by Giordano Riccati (*25 February 1709 in Castelfranco Veneto, province Trévise/Vénéti, †20 July 1790 in Treviso) in 1782 - predating Young's work by 25 years. The definition of this material parameter can be given as follows.

Definition 5 (*Elastic modulus*) An elastic modulus, or modulus of elasticity, is the mathematical description of a material's tendency to be deformed elastically (i.e., non-permanently) when a force is applied to it. The elastic modulus of an object is defined as the slope of its stress-strain curve in the elastic deformation region:

$$E \equiv \frac{\text{stress}}{\text{strain}}.$$

Remark 7 (*Poisson's ratio*) Cauchy's constitutive equations contain in the case of the symmetric stress tensor and isotropic material behaviour a second material parameter. There was a long discussion concerning this parameter - is it fixed (for example 0.25) or individual for each material. Finally, Poisson (Siméon Denis Poisson *21 June 1781 in Pithiviers/Département Loiret, †25 April 1840 in Paris) introduced the second parameter individual for each material:

$$\nu = -\frac{\varepsilon_{\text{transverse}}}{\varepsilon_{\text{axial}}}.$$

The Poisson ratio must be between -1.0 and $+0.5$ because of the positive definiteness of the strain energy. Classicical structural materials have Poisson's ratio values ranging between 0.0 and 0.5 and a perfectly incompressible material would have a Poisson's ratio of exactly 0.5. Some polymer foams, origami folds, certain cells, etc. can exhibit the negative Poisson ratio. These materials are named auxetic materials.

At the end of the nineteenth century there were several discussions w.r.t. the extension of the classical (Cauchy) continuum mode, which considered translations of the material points, but does not account independent rotations. In this direction a more or less complete model was suggested by the Cosserat brothers (Eugène Maurice Pierre Cosserat, *4 March 1866 in Amiens, †31 May 1931 in Toulouse, and François Nicolas Cosserat, *26 October 1852 in Douai, †22 March 1914 in France) in 1909 (Cosserat and Cosserat, 1909). They extended the classical continuum model with the independent balance of momentum and balance of moment of momentum. The theory was not used within the next 50 years since constitutive equations were not presented.

The great mathematician David Hilbert (*23 January 1862 in Königsberg, †14 February 1943 in Göttingen) delivered a lecture at the International Congress of Mathematicians in Paris (1900) with the title *Mathematical Problems*. Here, he formulated twenty-three problems in mathematics, which were all unsolved at the time (ten were presented during the lecture, the others were added later). The 6th problem is the mathematical treatment of the axioms of physics:

(a) axiomatic treatment of probability with limit theorems for foundation of statistical physics, and
(b) the rigorous theory of limiting processes "which lead from the atomistic view to the laws of motion of continua".

This problem, which is related to continuum mechanics, is unsolved up to now.

1.3.3 One-Dimensional Case

The basic elements of continuum mechanics are:

- kinematics (motion and deformation),
- actions (mechanical, thermal, etc.),
- balances (mass, linear momentum, angular momentum, energy, and entropy),
- constitutive equations (individual response of the "material"), and
- initial and boundary conditions.

Below, we present these elements at first for the one-dimensional case.[6] Three-dimensional equations will be discussed later.

Kinematics. We consider a rod to illustrate the main ideas of continuum mechanics in a simple manner. The rod is characterized by his length and the tension stiffness (product of the Young's modulus and the cross-section area). Let us consider the following assumptions of the theory of rods.

Definition 6 (*Rod*) A rod is a straight structural member with cross-section dimensions much less than the axial length.

Rods can be subjected to different types of loadings including tension (compression), bending and torsion. In the last cases the terms bar or beam are used.

As usual in continuum mechanics, at least two configurations have to be introduced: the reference or initial configuration (at $t = t_0$) and the deformed or actual configuration (at $t > t_0$). The deformed or configuration of a rod can be described by specifying the deformed rod axis, the actual cross-section area and triads of unit vectors to characterize the orientation of cross-sections. In order to define the deformed line only one coordinate is required. The problem to compute a deformed configuration for given loads is therefore one-dimensional.

The theory of roads can be introduced by the direct approach (see, for example, the pioneering work of Ericksen and Truesdell, 1958) or starting from the three-dimensional equations as, for example, in Green et al. (1974). The direct approach considers a rod as a deformable line that means as a one-dimensional continuum.

[6] This part is detailed presented in Naumenko and Altenbach (2016).

Every cross section behaves like a rigid body in the sense that translations and cross-section rotations are basic degrees of freedom for every point of the line. In our case we consider only translations since the rod is assumed to be under tension/compression. The mechanical interactions between two neighboring cross sections can be (normal and/or shear) forces and (bending and/or twisting) moments. Here, we assume the special case with longitudinal forces only. The general theory based on the direct approach was suggested, for example, in Altenbach et al. (2013).

Let us consider a rod subjected to a tensile/compression force only. Let $\boldsymbol{i}$ be the unit vector designating the direction of the straight rod axis. To describe the positions of the cross sections of the rod in the reference configuration the vector $\boldsymbol{R} = X\boldsymbol{i}$ with the coordinate X is introduced. The corresponding position in the actual configuration is defined by the vector $\boldsymbol{r} = x\boldsymbol{i}$ with the coordinate x.

The motion of the rod is given by

$$x = \Phi(X, t). \tag{1.13}$$

The basic problem of continuum mechanics is to find the function Φ for all points of $0 \leq X \leq \ell_0$, for the given time interval $t_0 \leq t \leq t_n$. For $t = t_0$ we have $X = \Phi(X, t_0)$. The displacement u is defined by

$$u = x - X. \tag{1.14}$$

To analyze the motion it is useful to introduce the change of Φ with respect to the reference coordinate X and the time t. The deformation gradient[7,8] F is

$$F = \frac{\partial \Phi}{\partial X}. \tag{1.15}$$

The velocity field v is the time derivative of Φ

$$v = \frac{\partial \Phi}{\partial t} = \dot{u}. \tag{1.16}$$

F in the one-dimensional theory is the local stretch λ

$$F \equiv \lambda = \frac{\mathrm{d}x}{\mathrm{d}X}.$$

Here $\mathrm{d}X$ and $\mathrm{d}x$ are line elements in the infinitesimal neighborhood of a cross section in the reference and actual configurations, respectively. The local strain ε is defined as in strength of materials (Altenbach, 2020; Beer et al., 2009)

[7] The deformation gradient is usually not introduced within the one-dimensional theory of rods.

[8] After Lurie the deformation gradient should be named gradient of the position vector (Lurie, 2005).

$$\varepsilon = \frac{\mathrm{d}x - \mathrm{d}X}{\mathrm{d}X} = \lambda - 1 = \frac{\partial u}{\partial X}. \tag{1.17}$$

A simplification can be introduced for the case of a homogeneous rod.

Definition 7 (*Homogeneous rod*) If the material properties and the cross section area do not depend on the position X then the rod is called homogeneous.

Then the stretch and the strain can be computed as follows

$$\lambda = \frac{\ell}{\ell_0}, \quad \varepsilon = \frac{\ell - \ell_0}{\ell_0}. \tag{1.18}$$

With the help of Eq. (1.18) the strains can be evaluated from experimental data of uni-axial tests since we assume homogeneous behavior in the measuring range. If a rod is non-homogeneous then local strains should be evaluated. A strain gauge should be placed in a position along the rod to estimate the local strain value. The strain gauge must be small. In this case the measured strain can be assumed to be constant in the gauge range, otherwise the measured strain would depend on the length of the strain gauge. A similar assumption can be applied to motivate Eq. (1.17). The value of the line element $\mathrm{d}X$ should be "small enough" such that the strain over $\mathrm{d}X$ is constant. The infinitesimal volume, area and line elements are introduced such that density, stress, strain, etc. can be assumed uniform over the infinitesimal elements. This belongs to the basics in Continuum Mechanics.

The introduced configurations are named Lagrangian (reference) and Eulerian (actual) configurations. If Φ is bi-unique invertible, then

$$F^{-1} = \frac{\mathrm{d}X}{\mathrm{d}x}. \tag{1.19}$$

Let us consider that f is a field like density, energy, displacement, velocity, strain, stress, etc. Then $f = f(X, t)$ is the Lagrangian or material description, alternatively $f = f(x, t)$ is the Eulerian or spatial description. The derivatives of f w.r.t. X and x can be computed as

$$\frac{\partial f}{\partial X} \equiv f'^0, \quad \frac{\partial f}{\partial x} \equiv f' \tag{1.20}$$

and the relation between the derivatives is

$$\frac{\partial f}{\partial X} = F\frac{\partial f}{\partial x} \quad \text{or} \quad f'^0 = Ff'. \tag{1.21}$$

The invertibility of the motion Φ is equivalent to

$$X = \Phi^{-1}(x, t), \tag{1.22}$$

and both descriptions are equivalent in the sense that if f is known as a function of X and t, one may use the transformation (1.22) to find

$$f(X,t) = g(x,t).$$

Let us introduce the derivative of the velocity with respect to the reference coordinate

$$v'^0 = \dot{F}. \tag{1.23}$$

The derivative of the velocity with respect to the actual coordinate is

$$v' = \dot{F}F^{-1}. \tag{1.24}$$

Let us introduce the fundamental theorem of calculus (Apostol, 1967). Assuming $f(x)$ is continuous for $a \leq x \leq b$

$$\int_a^b f'(x)\mathrm{d}x = f(b) - f(a). \tag{1.25}$$

If $f(x)$ has n jumps at points $x_k, k = 1, 2, \ldots, n$ within $a \leq x \leq b$ and $f'(x)$ is continuous between the jump points

$$\int_a^b f'(x)\mathrm{d}x = f(b) - f(a) + \sum_{k=1}^{n} [\![f(x_k)]\!], \quad [\![f(x_k)]\!] \equiv f(x_k^+) - f(x_k^-). \tag{1.26}$$

Now, we assume that the velocity field is given as a function of the spatial coordinate and the time, i.e. $v(x,t)$. The material time derivative of a field $f(x,t)$ is

$$\frac{\mathrm{d}}{\mathrm{d}t} f = \frac{\partial}{\partial t} f + v f'. \tag{1.27}$$

Balance Equations. Basic balance equations of continuum mechanics are applied directly to the deformable line. The first balance equation in the case of solids is as usual the conservation of mass. The mass of an infinitesimal part of the rod is

$$\mathrm{d}m = \rho A \mathrm{d}x = \rho_0 A_0 \mathrm{d}X. \tag{1.28}$$

Here, ρ and ρ_0 are the density in the actual and the initial configuration, respectively. The conservation of mass (1.28) takes the form

$$F \rho A = \rho_0 A_0. \tag{1.29}$$

A and A_0 are the cross section areas in the actual and the reference configurations. The change of the volume is

$$J = \frac{\mathrm{d}V}{\mathrm{d}V_0} = \frac{A\mathrm{d}x}{A_0\mathrm{d}X} = \frac{A}{A_0}F. \tag{1.30}$$

$\mathrm{d}V$ and $\mathrm{d}V_0$ are infinitesimal volume elements in both configurations. The conservation of mass (1.28) yields

$$\frac{\rho_0}{\rho} = J. \tag{1.31}$$

$J > 0$ and if $\rho = \rho_0$ one obtains $J = 1$. In addition, $0 < J < 1$ is related to decreasing density, $J > 1$ - to increase of density.

The second balance equation is the balance of momentum. The momentum of an infinitesimal part of the rod is

$$\mathrm{d}p = v\mathrm{d}m = v\rho A\mathrm{d}x.$$

Consider a part of the rod, for example, between x_1 and x_2. The momentum for this part in the actual configuration is

$$p_{\mathrm{II}} = \int\limits_{x_1}^{x_2} v\rho A\mathrm{d}x. \tag{1.32}$$

The balance of momentum or the first law of dynamics states that the rate of change of momentum of a body is equal to the total force acting on the body. To introduce the forces acting on the part II of the rod let us cut it by two cross sections with the coordinates x_1 and x_2. The parts I and III belong to the environment of the part II on the left and the right and the corresponding mechanical actions can be modeled by two longitudinal forces: $\boldsymbol{N}_{\mathrm{I-II}}$ - the action of the part I on the part II and $\boldsymbol{N}_{\mathrm{III-II}}$ the action of the part III on the part II. Similarly, the actions on the parts I and III can be introduced. For example, $\boldsymbol{N}_{\mathrm{II-I}}$ is the action of the part II on the part I. The following abbreviations can be introduced

$$\begin{aligned} \boldsymbol{N}_{\mathrm{II-I}} &= \boldsymbol{N}_{(\boldsymbol{i})}(x_1) = N(x_1)\boldsymbol{i}, \quad \boldsymbol{N}_{\mathrm{I-II}} = \boldsymbol{N}_{(-\boldsymbol{i})}(x_1) = -N(x_1)\boldsymbol{i}, \\ \boldsymbol{N}_{\mathrm{III-II}} &= \boldsymbol{N}_{(\boldsymbol{i})}(x_2) = N(x_2)\boldsymbol{i}, \; \boldsymbol{N}_{\mathrm{II-III}} = \boldsymbol{N}_{(-\boldsymbol{i})}(x_2) = -N(x_2)\boldsymbol{i}. \end{aligned} \tag{1.33}$$

The balance of momentum for the part II is[9]

$$\frac{\mathrm{d}}{\mathrm{d}t}\int\limits_{x_1}^{x_2} v\rho A\mathrm{d}x = N(x_2) - N(x_1). \tag{1.34}$$

With the fundamental theorem of calculus (1.25)[10]

[9] Body forces like the force of gravity are not included here for the sake of brevity.

[10] Here and in the following derivations we assume that N and other field variables are smooth functions. In the case of finite jumps one should apply Eq. (1.26) and introduce the jump conditions.

$$N(x_2) - N(x_1) = \int_{x_1}^{x_2} N' \mathrm{d}x. \tag{1.35}$$

Applying Eq. (1.28) one may evaluate the rate of change of momentum as follows

$$\frac{\mathrm{d}}{\mathrm{d}t} \int_{x_1}^{x_2} v\rho A \mathrm{d}x = \frac{\mathrm{d}}{\mathrm{d}t} \int_{X_1}^{X_2} v\rho_0 A_0 \mathrm{d}X = \int_{x_1}^{x_2} \dot{v}\rho A \mathrm{d}x. \tag{1.36}$$

With Eqs (1.35) and (1.36) the integral form of the balance of momentum is

$$\int_{x_1}^{x_2} (\dot{v}\rho A - N')\mathrm{d}x = 0. \tag{1.37}$$

The local form of Eq. (1.37) is valid for any part of the rod. Since x_1 and x_2 are arbitrary, the integral (1.37) is zero if

$$\rho A \dot{v} = N'. \tag{1.38}$$

Multiplying both parts of Eq. (1.38) by F yields

$$F \rho A \dot{v} = F N'. \tag{1.39}$$

With the conservation of mass (1.29) and the relation between the derivatives (1.21), Eq. (1.39) takes the following form

$$\rho_0 A_0 \dot{v} = N'^0. \tag{1.40}$$

Equation (1.40) is the local form for the balance of momentum with respect to the reference configuration.

Remark 8 (*Moment of momentum in the one-dimensional theory*) The balance of moment of momentum cannot be discussed within the one-dimensional theory of rods at tension/compression.

The last two balances are the first and second law of thermodynamics. The balance of energy (first law) can be introduced as follows. The total energy E for any part of the rod is defined as the sum of the kinetic energy K and the internal energy U

$$\mathrm{E} = \mathrm{K} + \mathrm{U} \tag{1.41}$$

with

- the kinetic energy

$$\mathrm{K} = \int\limits_{x_1}^{x_2} \rho \mathcal{K} A \mathrm{d}x, \qquad \mathcal{K} = \frac{1}{2} v^2$$

- the internal energy

$$\mathrm{U} = \int\limits_{x_1}^{x_2} \rho U A \mathrm{d}x,$$

where $\mathcal{K}$ and U are densities of the kinetic and the internal energy, respectively. If we have additional, not only mechanical effects, the energy balance equation states that the rate of change of the energy of a body is equal to the mechanical power plus the rate of change of non-mechanical energy, for example heat, supplied into the body. Now we have

$$\frac{\mathrm{d}}{\mathrm{d}t}\mathrm{E} = \mathrm{L} + \mathrm{Q}, \tag{1.42}$$

where L is the mechanical power and Q is the rate of change of non-mechanical energy supply. The mechanical power of internal forces (1.33) is defined as follows:

- the mechanical power

$$\mathrm{L} = N(x_2)v(x_2) - N(x_1)v(x_1) = \int\limits_{x_1}^{x_2} (Nv)' \mathrm{d}x,$$

- rate of change of energy supply through the cross sections of the parts I, II and III of the rod can be defined by analogy to Eq. (1.33)

$$\begin{aligned} \mathrm{Q}_{\mathrm{II-I}} &= \mathcal{Q}_{(i)}(x_1) = -\mathcal{Q}(x_1), \quad \mathrm{Q}_{\mathrm{I-II}} = \mathcal{Q}_{(-i)}(x_1) = \mathcal{Q}(x_1), \\ \mathrm{Q}_{\mathrm{III-II}} &= \mathcal{Q}_{(i)}(x_2) = -\mathcal{Q}(x_2), \; \mathrm{Q}_{\mathrm{II-III}} = \mathcal{Q}_{(-i)}(x_2) = \mathcal{Q}(x_2), \end{aligned} \tag{1.43}$$

- rate of change of the energy supply through the volume of the part II is

$$\mathrm{Q}_{V_{\mathrm{II}}} = \int\limits_{x_1}^{x_2} \rho r A \mathrm{d}x,$$

where r is the density of the energy supply. The total rate of energy supply into the part II is

$$\mathcal{Q}(x_1) - \mathcal{Q}(x_2) + \int\limits_{x_1}^{x_2} \rho r A \mathrm{d}x = \int\limits_{x_1}^{x_2} (-\mathcal{Q}' + \rho r A) \mathrm{d}x. \tag{1.44}$$

Considering the energy balance equation (1.42)

$$\frac{\mathrm{d}}{\mathrm{d}t}\int_{x_1}^{x_2}\left(\rho\frac{1}{2}v^2+\rho U\right)A\mathrm{d}x=\int_{x_1}^{x_2}(N'v+Nv'-Q'+\rho rA)\mathrm{d}x \tag{1.45}$$

and the mass conservation equation (1.28)

$$\begin{aligned}\frac{\mathrm{d}}{\mathrm{d}t}\int_{x_1}^{x_2}\left(\rho\frac{1}{2}v^2+\rho U\right)A\mathrm{d}x&=\frac{\mathrm{d}}{\mathrm{d}t}\int_{X_1}^{X_2}\left(\rho_0\frac{1}{2}v^2+\rho_0 U\right)A_0\mathrm{d}X\\&=\int_{X_1}^{X_2}\left(\dot{v}v+\dot{U}\right)\rho_0A_0\mathrm{d}X=\int_{x_1}^{x_2}\left(\dot{v}v+\dot{U}\right)\rho A\mathrm{d}x,\end{aligned} \tag{1.46}$$

we get the from the energy balance equation (1.45)

$$\int_{x_1}^{x_2}\left[(\rho A\dot{v}-N')v+\rho A\dot{U}-Nv'+Q'-\rho rA\right]\mathrm{d}x=0 \tag{1.47}$$

with the balance of momentum (1.38)

$$\int_{x_1}^{x_2}\left(\rho A\dot{U}-Nv'+Q'-\rho rA\right)\mathrm{d}x=0, \tag{1.48}$$

and finally the local form of the energy balance

$$\rho A\dot{U}=Nv'-Q'+\rho Ar. \tag{1.49}$$

Multiplying both sides of (1.49) by F and using the conservation of mass (1.29) as well as the relation between the derivatives (1.21) provides the local form of the energy balance per unit length of the rod in the reference configuration

$$\rho_0A_0\dot{U}=Nv'^0-Q'^0+\rho_0A_0r. \tag{1.50}$$

Here, the second law of thermodynamics will be presented as the Clausius-Planck inequality

$$\frac{\mathrm{d}}{\mathrm{d}t}\mathrm{S}-\frac{\mathrm{Q}}{T}\geq 0, \tag{1.51}$$

with T - the absolute temperature. The entropy of the part II of the rod is

$$\mathrm{S} = \int\limits_{x_1}^{x_2} \rho S A \mathrm{d}x, \tag{1.52}$$

where S is the entropy density. With Eqs. (1.43) and (1.25)

$$\int\limits_{x_1}^{x_2} \left[\rho \dot{S} A + \left(\frac{Q}{T} \right)' - \frac{\rho A r}{T} \right] \mathrm{d}x \geq 0. \tag{1.53}$$

The local form is

$$\left(\frac{Q}{T} \right)_{II} = -\frac{Q(x_2)}{T(x_2)} + \frac{Q(x_1)}{T(x_1)} + \int\limits_{x_1}^{x_2} \frac{\rho A r}{T} \mathrm{d}x = -\int\limits_{x_1}^{x_2} \left[\left(\frac{Q}{T} \right)' - \frac{\rho A r}{T} \right] \mathrm{d}x. \tag{1.54}$$

Inserting (1.52) and (1.54) into (1.51) provides the integral form of the entropy inequality. Since x_1 and $x_2 > x_1$ are arbitrary the local form of the entropy inequality can be given as follows

$$\rho \dot{S} A \geq -\left(\frac{Q}{T} \right)' + \frac{\rho r A}{T}. \tag{1.55}$$

By multiplying the both sides of (1.55) by T

$$\rho \dot{S} T A \geq -Q' + Q \frac{T'}{T} + \rho r A. \tag{1.56}$$

From the energy balance equation (1.49)

$$\rho A r - Q' = \rho A \dot{U} - N v' \tag{1.57}$$

one gets the dissipation inequality

$$N v' - \rho A \dot{U} + \rho \dot{S} T A - Q \frac{T'}{T} \geq 0 \tag{1.58}$$

or dividing (1.58) by the cross section area provides the following local form of the dissipation inequality

$$\sigma v' - \rho \dot{U} + \rho \dot{S} T - q \frac{T'}{T} \geq 0, \quad \sigma = \frac{N}{A}, \quad q = \frac{Q}{A}, \tag{1.59}$$

where σ is called Cauchy stress or true stress (which is related to the actual cross-section area) and q is the heat supply through the infinitesimal cross section. With the Helmholtz free energy density $\Psi = U - ST$ from Eq. (1.59) follows

$$\sigma v' - \rho\dot{\Psi} - \rho S\dot{T} - q\frac{T'}{T} \geq 0. \tag{1.60}$$

With Eq. (1.24) it follows that

$$\sigma \dot{F} F^{-1} - \rho\dot{\Psi} - \rho S\dot{T} - q\frac{T'}{T} \geq 0. \tag{1.61}$$

Multiplying both sides of (1.58) by F and using the conservation of mass (1.29) as well as the relation between the derivatives (1.21), the local form of the dissipation inequality per unit length of the rod in the reference configuration can be obtained

$$N v'^0 - \rho_0 A_0 \dot{U} + \rho_0 A_0 \dot{S} T - Q\frac{T'^0}{T} \geq 0. \tag{1.62}$$

Dividing by A_0 yields

$$P v'^0 - \rho_0 \dot{U} + \rho_0 \dot{S} T - \tilde{q}\frac{T'^0}{T} \geq 0, \tag{1.63}$$

where $P = N/A_0$ is the engineering stress related to the initial cross-section area,[11] $\tilde{q} = Q/A_0$ is the heat flow. In terms of the free energy the inequality takes the following form

$$P v'^0 - \rho_0 \dot{\Psi} - \rho_0 S\dot{T} - \tilde{q}\frac{T'^0}{T} \geq 0. \tag{1.64}$$

With Eq. (1.23) the velocity derivative can be replaced by the rate of the deformation gradient leading to

$$P \dot{F} - \rho_0 \dot{\Psi} - \rho_0 S\dot{T} - \tilde{q}\frac{T'^0}{T} \geq 0. \tag{1.65}$$

Taking into account that the normal force is $N = PA_0 = \sigma A$ the following relation between the stress measures can be established

$$P = \sigma J F^{-1}. \tag{1.66}$$

Similarly, with the heat flux $Q = qA = \tilde{q} A_0$

$$\tilde{q} = q J F^{-1}. \tag{1.67}$$

Constitutive Equations. Up to now, we have established four balance equations and a kinematical equation, but the number of unknowns (displacement, stress, strain, density, heat flux, energy, entropy) in these equations is greater than the number of equations. Additional equations can be introduced, if we take the material behavior

[11] In the literature this stress is often named first Piola-Kirchhoff stress.

into account. Below, we will discuss one of the simplest cases: thermo-elastic material behavior. Let us assume for elasticity that the stress is a function of the strain and elastic behavior is reversible under adiabatic (thermodynamic process without transferring heat between the thermodynamic system and its environment) or isothermal (the temperature T of a system remains constant) conditions. The constitutive assumption in this case is

$$\sigma = \sigma(F, J, T, T') \quad \Rightarrow \quad \Psi = \Psi(F, J, T, T')$$

and the inequality (1.61) takes the following form

$$\left(\sigma F^{-1} - \rho\frac{\partial\Psi}{\partial F}\right)\dot{F} - \rho\frac{\partial\Psi}{\partial J}\dot{J} - \rho\left(\frac{\partial\Psi}{\partial T} + S\right)\dot{T} - \rho\frac{\partial\Psi}{\partial T'}\dot{T}' - q\frac{T'}{T} \geq 0. \quad (1.68)$$

This inequality is valid if the following conditions are fulfilled

$$\sigma = \rho F\frac{\partial\Psi}{\partial F}, \quad \frac{\partial\Psi}{\partial J} = 0, \quad S = -\frac{\partial\Psi}{\partial T}, \quad \frac{\partial\Psi}{\partial T'} = 0, \quad -q\frac{T'}{T} \geq 0. \quad (1.69)$$

It is obvious that the constitutive equation for the stress yields

$$F\frac{\partial\Psi(F)}{\partial F} = \frac{\partial\Psi(\varepsilon_{\mathrm{H}})}{\partial\varepsilon_{\mathrm{H}}}, \quad \varepsilon_{\mathrm{H}} = \ln F = \ln\lambda.$$

Let us introduce ε_{H} as Hencky strain (sometimes called true strain). With Eq. (1.31) one gets

$$\sigma = \rho\frac{\partial\Psi}{\partial\varepsilon_{\mathrm{H}}} = \frac{\rho}{\rho_0}\frac{\partial\rho_0\Psi}{\partial\varepsilon_{\mathrm{H}}} = \frac{1}{J}\frac{\partial\rho_0\Psi}{\partial\varepsilon_{\mathrm{H}}} \quad (1.70)$$

and from Eqs (1.69) and (1.70) it follows that the free energy density must be formulated as a function of the Hencky strain and the temperature and under isothermal conditions the work done by the stress $J\sigma$ on the infinitesimal change of the Hencky strain is the total differential of the strain energy density function

$$J\sigma\mathrm{d}\varepsilon_{\mathrm{H}} = \mathrm{d}(\rho_0\Psi). \quad (1.71)$$

The stress measure $J\sigma$ is the Kirchhoff stress.[12] For adiabatic processes, i.e. for processes without heat transfer with the environment, one may use the local energy balance equation (1.49) to show that

$$J\sigma\mathrm{d}\varepsilon_{\mathrm{H}} = \mathrm{d}(\rho_0 U). \quad (1.72)$$

[12] Taking into account Eq. (1.66) the relation between the Cauchy, the first Piola and the Kirchhoff stress can be established. This is the reason that sometimes the Kirchhoff stress is named second Piola stress because the first Piola stress (engineering stress) is mapped in another configuration.

The relationship between the stress measures (1.66) is

$$P = \frac{\partial \rho_0 \Psi(F)}{\partial F}. \tag{1.73}$$

For many structural materials small values of strain ε can be assumed such that $\varepsilon^2 \ll \varepsilon < 1$. The stress is proportional to the strain ($\varepsilon_H \approx \varepsilon$) and Eq. $(1.69)_1$ can be linearized

$$\sigma = \frac{\partial \rho_0 \Psi(\varepsilon)}{\partial \varepsilon} \tag{1.74}$$

and the following linear constitutive equations can be established

$$\sigma = E(\varepsilon - \varepsilon^{\text{th}}), \quad \varepsilon^{\text{th}} = \alpha_{\text{th}}\Theta \tag{1.75}$$

with $\Theta = T - T_0$. The free energy follows from Eq. (1.74)

$$\rho_0 \Psi = \frac{1}{2} E\varepsilon^2 - E\alpha_{\text{th}}\Theta\varepsilon + f(T) \tag{1.76}$$

with the unknown function $f(T)$. The determination of the function $f(T)$ is given in Naumenko and Altenbach (2016).

Within the linearized theory the deformation has minor influence on the heat transfer such that the heat equation can be solved independently providing the temperature $T(x, t)$. The balance of momentum (1.38) with the constitutive equation (1.75) yields

$$\rho A\ddot{u} = [EA(u' - \alpha_{\text{th}}\Theta)]',$$

which can be found in any textbook of strength of materials (Altenbach, 2020; Beer et al., 2009).

1.3.4 Three-Dimensional Case

Introduction. In this part we discuss the three-dimensional equations of continuum mechanics. Let us start with a definition.

Definition 8 (*Continuum—second definition*) A continuum is an ensemble of material points with properties which changes continuously (distributed uniformly throughout, and completely fills the space it occupies)

This definition is close to **Definition** 3. However, both definitions are connected with some open questions:

- *Are there any restrictions concerning the problem of the dimension?* It seems as if the answer was no because we can introduce one-, two- and three-dimensional theories without problems. In this case, the theories are related to the assumed space, and one gets, for example, theories for deformable lines or surfaces. Using the ideas of Minkowski, who discussed n-dimensional spaces, one cane introduce a four-dimensional space with three directions and the time which is the fourth coordinate.

- *Are there any restrictions for the degrees of freedom?* It seems again as if the answer was no because the classical continuum (Cauchy continuum) is based on three independent degrees of freedom (three translations), the micropolar continuum (Eremeyev et al., 2013) is based on six degrees of freedom (three translations and three rotations), and the Mindlin theory of plate contains five degrees of freedom (three translations and two rotations). There are other suggestions in the literature.
- *Are there different continuum theories for solids, liquids, and gases?* The continuum theory must be unique with respect to the basics. Differences can be established, taking the material properties into account.
- *Are there restrictions concerning the external actions?* No since we have continuum theories for pure mechanical or thermo-mechanical actions, but during the last 30 years theories were developed which also took electrical and magnetic fields into account. Now there are also continuum theories for chemical or biological problems.
- *Are there limitations for the scale?* No, at the moment we have theories for macro-, micro, and nano-scale, based on the same theoretical foundations.

Let us here introduce the three-dimensional space $\mathbb{R}^3$ with the arbitrary reference point 0. The (pseudo)time is in the range $t_0 \leq t < \infty$. The basic model is the body $\mathcal{B}$ with the volume V and the surface A. Then, the mass of the body is defined with the help of the density

$$\rho = \rho(\boldsymbol{X}, t), \qquad \rho > 0$$

as

$$m = \int_V \rho(\boldsymbol{X}, t) \mathrm{d}V.$$

Homogeneity (equal properties in all material point $\boldsymbol{X}$) and isotropy (no directional dependence of the properties) can be introduced for making the models simpler.

Kinematics. The kinematics are the part of mechanics describing the motion of material points, the change of the distance between points, etc. This is a pure mathematical description and the reasons for motion or change of the distance are disregarded. The kinematical variables have a pure geometrical meaning, the primary quantities are: displacements, velocities, accelerations, deformations (strains, shear strains), etc. In addition, one has to distinguish rigid body motions and deformations.

Let us introduce at first the strains.

Definition 9 (*Strain*) Given that strain results in the deformation of a body, it can be measured by calculating the change in length of a line or by the change in angle between two lines (where these lines are theoretical constructs within the deformed body). The change in length of a line is termed the stretch, absolute strain, or extension, and may be written as δl. Then the (relative) strain, ε, is given by

$$\varepsilon = \frac{\delta l}{l_0} = \frac{l - l_0}{l_0},$$

where ε is the strain in measured direction l_0 is the original length of the specimen l is the current length of the specimen.

Such a simple computing is possible if the strains in the specimen are homogeneous. If we have inhomogeneous strains, the computing should be modified. The position in the unloaded tension bar is given by x and the infinitesimal distance is $\mathrm{d}x$. In the loaded tension bar we have $R(x)$ and $\mathrm{d}R(x)$. Then

$$\mathrm{d}R = \frac{\mathrm{d}R(x)}{\mathrm{d}x}\mathrm{d}x = R'(x)\mathrm{d}x. \tag{1.77}$$

With **Definition** 9 and Eq. (1.77) we get

$$\varepsilon(x) = \frac{\mathrm{d}R - \mathrm{d}x}{\mathrm{d}x} = \frac{R'(x)\mathrm{d}x - \mathrm{d}x}{\mathrm{d}x}. = R'(x) - 1$$

Finally with the displacement $u(x)$,

$$R(x) = x + u(x) \Rightarrow R'(x) = 1 + u'(x) = 1 + \varepsilon(x).$$

The inhomogeneous strain in the tension bar is defined as

$$\varepsilon(x) = u'(x).$$

In the case of non-linear continuum mechanics the deformation cannot be computed in such a simple manner. However, even in the general case we can use the simple definitions of strain and shear strain used in strength of materials (Altenbach, 2020; Beer et al., 2009). We have to consider two configurations: the reference configuration with $t = t_0$ (mostly $t_0 = 0$), A_0, V_0 and the actual configuration with $t > t_0$, A, V. First, an origin is introduced. In the invariant form of representation, a coordinate system does not necessarily have to be specified for this. If $P(t_0$ and $P(t)$ are arbitrary points in the reference and the actual configuration, possible ways to change the configuration $P(t_0) \rightarrow P(t)$ are: translation as a rigid body, rotation as a rigid body, and deformation (changing volume and shape). The position of $P(t_0)$ with respect to the origin is given by the vector $\boldsymbol{x}$ and the position of $P(t)$ by the vector $\boldsymbol{X}$. The displacement vector is now

$$\boldsymbol{u} = \boldsymbol{X} - \boldsymbol{x}.$$

It is obvious that $\boldsymbol{u}$ is the shortest distance between the two positions. The real trajectory of changing the position can be much longer. In the geometrical linear theory the actual and the reference configurations coincide and the length of $\boldsymbol{u}$ is identical with the length of the trajectory.

Introducing infinitesimal neighborhoods $\mathrm{d}\boldsymbol{x}$ and $\mathrm{d}\boldsymbol{X}$ of the points $P(t_0)$ and $P(t)$, in the case simple material of the degree one (Noll, 1958), corresponding to the differential neighborhoods, the classical definition of strain yields

$$\varepsilon = \frac{|\mathrm{d}\boldsymbol{X}| - |\mathrm{d}\boldsymbol{x}|}{|\mathrm{d}\boldsymbol{x}|}. \tag{1.78}$$

The description of shear strains is related to the comparison $\angle(\mathrm{d}\boldsymbol{x}_1, \mathrm{d}\boldsymbol{x}_2)$ and $\angle(\mathrm{d}\boldsymbol{X}_1, \mathrm{d}\boldsymbol{X}_2)$. For the sake of simplicity, it is assumed $\angle(\mathrm{d}\boldsymbol{x}_1, \mathrm{d}\boldsymbol{x}_2) = \frac{\pi}{2}$ and the classical shear strain definition results in

$$\gamma = \frac{\pi}{2} - \angle(\mathrm{d}\boldsymbol{X}_1, \mathrm{d}\boldsymbol{X}_2). \tag{1.79}$$

The mathematical description of the mapping $\mathrm{d}\boldsymbol{x} \Rightarrow \mathrm{d}\boldsymbol{X}$ can be presented as follows. With the Cartesian coordinates x, y, z and $\boldsymbol{x} = \boldsymbol{x}(x, y, z)$, $\boldsymbol{X} = \boldsymbol{X}(x, y, z)$ we have

$$\mathrm{d}\boldsymbol{x} = \mathrm{d}x\, \boldsymbol{e}_x + \mathrm{d}y\, \boldsymbol{e}_y + \mathrm{d}z\, \boldsymbol{e}_z, \quad \mathrm{d}\boldsymbol{X} = \frac{\partial \boldsymbol{X}}{\partial x}\mathrm{d}x + \frac{\partial \boldsymbol{X}}{\partial y}\mathrm{d}y + \frac{\partial \boldsymbol{X}}{\partial z}\mathrm{d}z,$$

and with

$$\mathrm{d}x = \mathrm{d}\boldsymbol{x} \cdot \boldsymbol{e}_x, \quad \mathrm{d}y = \mathrm{d}\boldsymbol{x} \cdot \boldsymbol{e}_y, \quad \mathrm{d}z = \mathrm{d}\boldsymbol{x} \cdot \boldsymbol{e}_z$$

one gets

$$\mathrm{d}\boldsymbol{X} = \mathrm{d}\boldsymbol{x} \cdot \left[\boldsymbol{e}_x \otimes \frac{\partial \boldsymbol{X}}{\partial x} + \boldsymbol{e}_y \otimes \frac{\partial \boldsymbol{X}}{\partial y} + \boldsymbol{e}_z \otimes \frac{\partial \boldsymbol{X}}{\partial z}\right] = \mathrm{d}\boldsymbol{x} \cdot \boldsymbol{\nabla} \boldsymbol{X}. \tag{1.80}$$

$\boldsymbol{\nabla}$ denotes the Nabla (Hamilton) operator

$$\boldsymbol{\nabla}(...) \equiv \boldsymbol{e}_x \otimes \frac{\partial (...)}{\partial x} + \boldsymbol{e}_y \otimes \frac{\partial (...)}{\partial y} + \boldsymbol{e}_z \otimes \frac{\partial (...)}{\partial z}.$$

Equation (1.80) is the definition of the deformation gradient.

Definition 10 (*Deformation gradient*[13]) The deformation gradient tensor $\boldsymbol{F}$ is related to both the reference and current configuration, therefore it is a two-point tensor. It is defined as

$$\boldsymbol{F} = (\boldsymbol{\nabla}_{\boldsymbol{x}}\boldsymbol{X})^{\mathrm{T}}.$$

[13] Note that in some textbooks the definition is given without “transposed”.

The derivatives performed by the Nabla operator should be calculated with respect to $\boldsymbol{x}$. This is the pull forward operation (mapping of $\mathrm{d}\boldsymbol{x}$ onto $\mathrm{d}\boldsymbol{X}$)

$$\mathrm{d}\boldsymbol{X} = \boldsymbol{F} \cdot \mathrm{d}\boldsymbol{x} = \mathrm{d}\boldsymbol{x} \cdot \boldsymbol{F}^{\mathrm{T}}.$$

In Continuum Mechanics $\det \boldsymbol{F}$ is always positive ($\det \boldsymbol{F} > 0$) that means $\boldsymbol{F}$ is regular (non-singular) and $\boldsymbol{F}^{-1}$ can be computed. The reason for this property is given later. The push backward operation is defined by

$$\mathrm{d}\boldsymbol{x} = \boldsymbol{F}^{-1} \cdot \mathrm{d}\boldsymbol{X} = \mathrm{d}\boldsymbol{X} \cdot \boldsymbol{F}^{-1^{\mathrm{T}}}.$$

If $\boldsymbol{F}$ transfers $\mathrm{d}\boldsymbol{x}$ onto $\mathrm{d}\boldsymbol{X}$, the tensor $\boldsymbol{F}^{-1}$ transfers $\mathrm{d}\boldsymbol{X}$ onto $\mathrm{d}\boldsymbol{x}$.

Now, introducing three infinitesimal line elements $\mathrm{d}\boldsymbol{x}_x = \boldsymbol{X}_x \,\mathrm{d}x,\ \mathrm{d}\boldsymbol{x}_y = \boldsymbol{X}_y \,\mathrm{d}y$, $\mathrm{d}\boldsymbol{x}_z = \boldsymbol{X}_z \,\mathrm{d}z$ we define the infinitesimal volume element in the reference configuration

$$\mathrm{d}V_0 = \left(\mathrm{d}\boldsymbol{x}_x \times \mathrm{d}\boldsymbol{x}_y\right) \cdot \mathrm{d}\boldsymbol{x}_z = \left(\mathrm{d}\boldsymbol{X}_x \times \mathrm{d}\boldsymbol{X}_y\right) \cdot \mathrm{d}\boldsymbol{X}_z \,\mathrm{d}x \,\mathrm{d}y \,\mathrm{d}z.$$

Then we calculate the infinitesimal volume element in the actual configuration

$$\mathrm{d}V = \left(\mathrm{d}\boldsymbol{X}_x \times \mathrm{d}\boldsymbol{X}_y\right) \cdot \mathrm{d}\boldsymbol{X}_z = \left[(\mathbf{F} \cdot \mathrm{d}\boldsymbol{x}_x) \times \left(\mathbf{F} \cdot \mathrm{d}\boldsymbol{x}_y\right)\right] \cdot \mathbf{F} \cdot \mathrm{d}\boldsymbol{x}_z,$$

and after some mathematical manipulations (Altenbach, 2018) finally we get a relation between the infinitesimal volume elements in both configurations

$$\mathrm{d}V = \det \mathbf{F} \left(\mathrm{d}\boldsymbol{x}_x \times \mathrm{d}\boldsymbol{x}_y\right) \cdot \mathrm{d}\boldsymbol{x}_z = \det \mathbf{F} \mathrm{d}V_0.$$

Summarizing, we get

$$\det \boldsymbol{F} = \frac{\mathrm{d}V}{\mathrm{d}V_0} > 0.$$

Since $\mathrm{d}V$ and $\mathrm{d}V_0$ are positive ($\mathrm{d}V = 0$ or $\mathrm{d}V_0 = 0$ contradict the continuum definition) we have to distinguish three cases

- $\det \boldsymbol{F} > 1$ which means $\mathrm{d}V > \mathrm{d}V_0$ (extension or volume increase),
- $\det \boldsymbol{F} < 1$ which means $\mathrm{d}V < \mathrm{d}V_0$ (contraction or volume reduction), and
- $\det \boldsymbol{F} = 1$ which means $\mathrm{d}V = \mathrm{d}V_0$ (volume preserved).

The last case is the isochoric or incompressible behavior.

With the relation between $\boldsymbol{F}$ and $\boldsymbol{u}$ one gets from

$$\boldsymbol{X} = \boldsymbol{x} + \boldsymbol{u}$$

considering

$$\nabla_x \boldsymbol{x} = \boldsymbol{I}$$

finally

$$\nabla_x X = I + \nabla_x u.$$

In addition, the deformation gradient can be obtained

$$(\nabla X)^{\mathrm{T}} = I^{\mathrm{T}} + (\nabla u)^{\mathrm{T}} = I + (\nabla u)^{\mathrm{T}} = F.$$

The normal strains are defined as follows. Let us introduce a line element in the reference configuration

$$\mathrm{d}x = \mathrm{d}L m, \quad (|m| = 1)$$

and a line element in the actual configuration

$$\mathrm{d}X = \mathrm{d}l\widetilde{m}, \quad (|\widetilde{m}| = 1, \ \widetilde{m} \neq m).$$

Definition 11 (*Normal strain*) The normal strain can be presented as

$$\varepsilon_{mm} = \frac{\mathrm{d}l - \mathrm{d}L}{\mathrm{d}L} = \frac{\mathrm{d}l}{\mathrm{d}L} - 1$$

ε_{mm} is the normal strain in the neighborhood of P in direction m.

This definition corresponds to Eq. (1.78). The square length of the infinitesimal line element is

$$\mathrm{d}l^2 = \mathrm{d}X \cdot \mathrm{d}X = \mathrm{d}X \cdot F^{-1} \cdot F \cdot \mathrm{d}X = \mathrm{d}L^2 m \cdot F^{\mathrm{T}} \cdot F \cdot m.$$

Then

$$\left(\frac{\mathrm{d}l}{\mathrm{d}L}\right)^2 = m \cdot F^{\mathrm{T}} \cdot F \cdot m = (\varepsilon_{mm} + 1)^2 = \varepsilon_{mm}^2 + 2\varepsilon_{mm} + 1 \approx 2\varepsilon_{mm} + 1$$

and

$$\begin{aligned} \varepsilon_{mm} &= \frac{1}{2} m \cdot F^{\mathrm{T}} \cdot F \cdot m - \frac{1}{2} \\ &= \frac{1}{2} m \cdot F^{\mathrm{T}} \cdot F \cdot m - \frac{1}{2} m \cdot I \cdot m \\ &= m \cdot \frac{1}{2}\left(F^{\mathrm{T}} \cdot F - I\right) \cdot m \\ &= m \cdot G \cdot m. \end{aligned} \tag{1.81}$$

G is the Green-Lagrange strain tensor.

The shear can be presented by Eq. (1.79) with the introduction of two infinitesimal line elements in the reference configuration

$$\mathrm{d}x_n = \mathrm{d}L_n n, \ \mathrm{d}x_p = \mathrm{d}L_p p.$$

Two infinitesimal line elements in the actual configuration are

$$\mathrm{d}\boldsymbol{X}_{\tilde{n}} = \mathrm{d}l_{\tilde{n}}\,\widetilde{\boldsymbol{n}}, \ \mathrm{d}\boldsymbol{X}_{\tilde{p}} = \mathrm{d}l_{\tilde{p}}\,\widetilde{\boldsymbol{p}}.$$

The shear strain related to the plane with the directions $\boldsymbol{n}$ and $\boldsymbol{p}$ ($\boldsymbol{m}$, $\boldsymbol{p}$ and $\boldsymbol{p}$ are a orthonormal base) is

$$\gamma_{np} = \frac{\pi}{2} - \alpha_{np}.$$

Some calculations yield

$$\mathrm{d}\boldsymbol{X}_{\tilde{n}} \cdot \mathrm{d}\boldsymbol{X}_{\tilde{p}} = \mathrm{d}l_{\tilde{n}}\,\mathrm{d}l_{\tilde{p}}\, \cos\alpha_{np} = \mathrm{d}l_{\tilde{n}}\,\mathrm{d}l_{\tilde{p}}\, \sin\gamma_{np}$$

and

$$\mathrm{d}\boldsymbol{X}_{\tilde{n}} \cdot \mathrm{d}\boldsymbol{X}_{\tilde{p}} = (1+\varepsilon_{nn})\left(1+\varepsilon_{pp}\right)\sin\gamma_{np}\,\mathrm{d}L_n\,\mathrm{d}L_p.$$

With the assumption

$$\gamma_{np} \ll 1$$

follows

$$\sin\gamma_{np} \approx \gamma_{np},$$

and finally we get

$$(1+\varepsilon_{nn})\left(1+\varepsilon_{pp}\right)\sin\gamma_{np} \approx \left(1+\varepsilon_{nn}+\varepsilon_{pp}+\varepsilon_{nn}\varepsilon_{pp}\right)\gamma_{np} \approx \gamma_{np},$$

$$\mathrm{d}\boldsymbol{X}_{\tilde{n}} \cdot \mathrm{d}\boldsymbol{X}_{\tilde{p}} = \mathrm{d}\boldsymbol{x}_n \cdot \boldsymbol{F}^{\mathrm{T}} \cdot \boldsymbol{F} \cdot \mathrm{d}\boldsymbol{x}_p = \mathrm{d}L_n\mathrm{d}L_p\boldsymbol{n} \cdot \boldsymbol{F}^{\mathrm{T}} \cdot \boldsymbol{F} \cdot \boldsymbol{p} = \gamma_{np}\mathrm{d}L_n\mathrm{d}L_p$$

and

$$\gamma_{np} = \boldsymbol{n} \cdot \boldsymbol{F}^{\mathrm{T}} \cdot \boldsymbol{F} \cdot \boldsymbol{p}$$

with $\boldsymbol{F}^{\mathrm{T}} \cdot \boldsymbol{F} = 2\boldsymbol{G} + \boldsymbol{I}$ resulting in

$$\gamma_{np} = 2\boldsymbol{n} \cdot \boldsymbol{G} \cdot \boldsymbol{p}. \tag{1.82}$$

Equations (1.81) and (1.82) allow a interpretation of the components of the Green-Lagrange strain tensor

$$\boldsymbol{G} = \frac{1}{2}\left(\boldsymbol{F}^{\mathrm{T}} \cdot \boldsymbol{F} - \boldsymbol{I}\right).$$

With

$$\boldsymbol{F} = \boldsymbol{I} + (\nabla\boldsymbol{u})^{\mathrm{T}} \quad \text{and} \quad \boldsymbol{F}^{\mathrm{T}} = \boldsymbol{I} + \nabla\boldsymbol{u}$$

we get

$$G = \frac{1}{2}\left\{[I + \nabla u] \cdot \left[I + (\nabla u)^{\mathrm{T}}\right] - I\right\} = \frac{1}{2}\left[\nabla u + (\nabla u)^{\mathrm{T}} + \nabla u \cdot (\nabla u)^{\mathrm{T}}\right].$$

After linearization the Cauchy strain tensor follows

$$G \approx \frac{1}{2}\left[\nabla u + (\nabla u)^{\mathrm{T}}\right] = \varepsilon$$

with the components

- normal strains

$$\varepsilon_{xx} = \frac{\partial u}{\partial x}, \quad \varepsilon_{yy} = \frac{\partial u}{\partial y}, \quad \varepsilon_{zz} = \frac{\partial u}{\partial z}$$

- and shear strains

$$\gamma_{xy} = \frac{\partial u}{\partial y} + \frac{\partial u}{\partial x} = 2\varepsilon_{xy}, \quad \gamma_{xz} = \frac{\partial u}{\partial z} + \frac{\partial u}{\partial x} = 2\varepsilon_{xz}, \quad \gamma_{yz} = \frac{\partial u}{\partial z} + \frac{\partial u}{\partial y} = 2\varepsilon_{yz}.$$

They are given in relevant textbooks of strength of materials (Altenbach, 2020; Beer et al., 2009).

Remark 9 (*Symmetry of the Cauchy strain tensor*) If the stress tensor $\boldsymbol{\sigma}$ is a symmetric tensor, the corresponding (work-conjugated) tensor has only symmetrical components

$$\varepsilon = \varepsilon^{\mathrm{T}}.$$

With the polar decomposition theorem (Eremeyev et al., 2018) we obtain

$$\begin{aligned} F &= R \cdot U = V \cdot R, \\ C &= U^2 = F^{\mathrm{T}} \cdot F, \\ B &= V^2 = F \cdot F^{\mathrm{T}}. \end{aligned}$$

$\boldsymbol{R}$ is a rotation tensor with the properties $\boldsymbol{R}^{\mathrm{T}} \cdot \boldsymbol{R} = \boldsymbol{R}^{-1} \cdot \boldsymbol{R} = \boldsymbol{I}$, $\boldsymbol{U}$ is the symmetric right-stretch tensor and $\boldsymbol{V}$ the corresponding symmetric left-stretch tensor. $\boldsymbol{C}$ is the right Cauchy-Green deformation tensor, $\boldsymbol{B}$ is the left Cauchy-Green deformation tensor. Finally, we have for example

$$G = \frac{1}{2}\left(F^{\mathrm{T}} \cdot F - I\right) = \frac{1}{2}\left(U \cdot R^{\mathrm{T}} \cdot R \cdot U - I\right) = \frac{1}{2}(C - I).$$

Further detailed discussions are given in Lurie (2005), Altenbach (2018).

Mechanical Loadings. The following classification of the external mechanical loading can be introduced: natural models like body/mass/volume loading and surface/contact loading and, in addition, artificial loading models like line loading and single point loading. In all cases the loadings are forces and moments. The differences can be explained by dimensional analysis: for the force $[\mathbf{F}] = \mathrm{N}$ and for the moment

[M] = Nm. In this case, body loading - per volume, surface loading - per area and line loading - per line. The body force can presented as $\rho\,(\boldsymbol{X},t)\,\boldsymbol{k}\,(\boldsymbol{X},t)=\boldsymbol{k}^{\mathrm{V}}\,(\boldsymbol{X},t)$ and by analogy the body moment as $\rho\,(\boldsymbol{X},t)\,\boldsymbol{l}\,(V\boldsymbol{X},t)=\boldsymbol{l}^{\mathrm{V}}\,(\boldsymbol{X},t)$. Examples are

- the weight force $\rho\boldsymbol{k}=-\rho g\boldsymbol{e}_3$ with g as the gravitational acceleration and $\boldsymbol{e}_3$ as the direction (the weight force of a body is always directed towards the center of the celestial body),
- the centrifugal force $\rho\boldsymbol{k}=-\rho\boldsymbol{\omega}\times(\boldsymbol{\omega}\times\boldsymbol{X})$ with $\boldsymbol{\omega}$ as the angular velocity, and
- the potential force $\rho\boldsymbol{k}=-\rho\boldsymbol{\nabla}\Pi\,(\boldsymbol{X}t)$ with the potential Π.

Let us introduce the stresses. The stress vector is

$$\boldsymbol{t}=\lim_{\Delta A\to 0}\frac{\Delta\boldsymbol{f}}{\Delta A}.$$

$\Delta\boldsymbol{f}$ is a part of the surface load $\boldsymbol{f}$, acting on the part of the surface ΔA. By analogy we can introduce the couple stress vector

$$\boldsymbol{g}=\lim_{\Delta A\to 0}\frac{\Delta\boldsymbol{m}}{\Delta A}$$

with $\Delta\boldsymbol{m}$ as a part of the surface moment, acting on ΔA. The values at the point on the surface are obtained as the limit value.

The resultant force and th resultant moment can be computed

$$\boldsymbol{f}^{\mathrm{R}}=\int_V\rho\boldsymbol{k}\,\mathrm{d}V+\int_A\boldsymbol{t}\,\mathrm{d}A,\qquad \boldsymbol{m}_0^{\mathrm{R}}=\int_V\rho\,(\boldsymbol{l}+\boldsymbol{X}\times\boldsymbol{k})\,\mathrm{d}V+\int_A(\boldsymbol{g}+\boldsymbol{X}\times\boldsymbol{t})\,\mathrm{d}A.$$

$\boldsymbol{X}$ is the radius vector to the reference point 0 in the actual configuration. From Newton's axioms it follows that if $\boldsymbol{f}^{\mathrm{R}}=\boldsymbol{0}$ and $\boldsymbol{m}_0^{\mathrm{R}}=\boldsymbol{0}$ the continuum under consideration is at rest or moves at a steady velocity.

The stress tensor $\boldsymbol{T}$ is a measure of the stress state in a material point. The base of the stress tensor is defined by two vectors: $\boldsymbol{t}$ (the stress vector) and $\boldsymbol{n}$ - the normal to the surface through the point. The components of the stress vector are

$$\boldsymbol{t}=t_n\boldsymbol{n}+t_t\boldsymbol{e}_t=t_n\boldsymbol{n}+t_{t_1}\boldsymbol{e}_{t_1}+t_{t_2}\boldsymbol{e}_{t_2}$$

where $\boldsymbol{e}_{t_1}$ and $\boldsymbol{e}_{t_2}$ are arbitrary tangential directions in the surface, $\boldsymbol{n}\perp\boldsymbol{e}_{t_1}$, $\boldsymbol{n}\perp\boldsymbol{e}_{t_2}$, $\boldsymbol{e}_{t_1}\perp\boldsymbol{e}_{t_2}$ so that $\boldsymbol{n}$, $\boldsymbol{e}_{t_1}$ and $\boldsymbol{e}_{t_2}$ form an arbitrary orthonormal base. The Cauchy's lemma can be established by equilibrium considerations (Naumenko and Altenbach, 2016)

$$\boldsymbol{t}\,(\boldsymbol{X},\boldsymbol{n},t)=\boldsymbol{n}\cdot\boldsymbol{T}\,(\boldsymbol{X},t)\,.\tag{1.83}$$

Equilibrium Equations. Let us simplify our model taking into account only forces action. For the classical Cauchy continuum one gets

$$\int_V \rho \boldsymbol{k}\, \mathrm{d}V + \int_A \boldsymbol{t}\, \mathrm{d}A = \boldsymbol{0}, \qquad \text{and} \qquad \int_V (\boldsymbol{X} \times \rho \boldsymbol{k})\, \mathrm{d}V + \int_A (\boldsymbol{X} \times \boldsymbol{t})\, \mathrm{d}A = \boldsymbol{0}. \tag{1.84}$$

With the divergence theorem of Gauß and Ostrogradsky (Lurie, 2005; Altenbach, 2018) and Eq. (1.83)

$$\int_A \boldsymbol{t}\, \mathrm{d}A = \int_A \boldsymbol{n} \cdot \boldsymbol{T}\, \mathrm{d}A = \int_V \nabla \cdot \boldsymbol{T}\, \mathrm{d}A$$

finally we get

$$\int_V (\rho \boldsymbol{k} + \nabla \cdot \boldsymbol{T})\, \mathrm{d}V = \boldsymbol{0}$$

and if all fields are smooth enough the local form can be established

$$\nabla \cdot \boldsymbol{T} + \rho \boldsymbol{k} = \boldsymbol{0}.$$

In the case of non-smooth fields we refer to Šilhavý (1997), Casey (2011).

Eulerian Equations of Motion. Applying d'Alambert's principle (only forces, but inertia is considered)

$$\int_V \rho \boldsymbol{k}\, \mathrm{d}V + \int_A \boldsymbol{t}\, \mathrm{d}A - \int_V \rho \ddot{\boldsymbol{X}}\, \mathrm{d}V = \boldsymbol{0}$$

the global form of the first Eulerian equation of motion can be established. With the divergence theorem

$$\int_A \boldsymbol{t}\, \mathrm{d}A = \int_A \boldsymbol{n} \cdot \boldsymbol{T}\, \mathrm{d}A = \int_V \nabla \cdot \boldsymbol{T}\, \mathrm{d}A$$

finally we obtain

$$\int_V \left(\rho \boldsymbol{k} + \nabla \cdot \boldsymbol{T} - \rho \ddot{\boldsymbol{X}}\right) \mathrm{d}V = \boldsymbol{0}$$

and the local form

$$\nabla \cdot \boldsymbol{T} + \rho \boldsymbol{k} = \rho \ddot{\boldsymbol{X}}$$

is valid if all fields are smooth enough, otherwise jump conditions should be included.

The conclusion from Eq. $(1.84)_2$ in case of only forces acting is very simple: the stress tensor $\boldsymbol{T}$ is symmetric. If we have force and moment actions, one gets the second Eulerian law of motion as a independent relation (Truesdell, 1964).

Laws of Thermodynamics. Let us extend slighty our model taking into account thermo-mechanical actions. Then the laws of thermodynamics should be taken into account. In general, one should distinguish the equilibrium thermodynamics and the non-equilibrium thermodynamics. Here, we assume that equilibrium thermodynamics is valid.

The following four laws of thermodynamics can be formulated:

Theorem 1 *Zeroth law of thermodynamics If two systems are each in thermal equilibrium with a third, they are also in thermal equilibrium with each other.*

Theorem 2 *First law of thermodynamics In a process without transfer of matter, the change in internal energy of a thermodynamic system is equal to the energy gained by heat addition to a system and work (as energy) done by the surroundings on the system.*

Theorem 3 *Second law of thermodynamics Heat does not spontaneously flow from a colder body to a hotter.*

Theorem 4 *Third law of thermodynamics As the temperature of a system approaches absolute zero, all processes cease and the entropy of the system approaches a minimum value.*

In Rational Thermomechanics these four theorems are the basics. The first law is the energy balance which is an extension of the conversation of energy law. The second law defines the process direction allowing the split of physically admissible and physically not admissible constitutive equations.

Let us introduce state variables. In mechanical systems, the position coordinates and the velocities are state variables. This follows from the Newton's fundamental law. In thermodynamics we have the internal energy, the entropy, but also temperature, density, and volume. As usual they are assumed to be macroscopic, measurable, and independent. We have to distinguish extensive (additive) variables, which are proportional to the mass (for example, inner energy, which depends only on the kinematics and the temperature), and intensive variables, which are not proportional to the mass (for example, the density or the temperature). In the following, we introduce the following limitations: homogeneity (each material point with identical properties—they are constant with respect to the coordinates) and closed systems (no mass exchange with the surrounding - the mass preservation is valid).

General Balance Equation. Let $\Psi(\boldsymbol{X}, t)$ and $\Psi_0(\boldsymbol{x}, t)$ be specific scalar properties distributed in $\mathrm{d}V$ or $\mathrm{d}V_0$. The integration over all body points results in balance property $Y(t)$

$$Y(t) = \int\limits_V \Psi(\boldsymbol{X}, t)\mathrm{d}V = \int\limits_{V_0} \Psi_0(\boldsymbol{x}, t)\mathrm{d}V.$$

With $\mathrm{d}V = (\det\boldsymbol{F})\mathrm{d}V_0$ one gets $\Psi_0(\boldsymbol{x}, t) = (\det\boldsymbol{F})\Psi(\boldsymbol{X}, t)$ and we obtain the global balance equation for the actual configuration

Fig. 1.2 General balance equation (global formulation): Ψ - balance variable, Φ - action through the surface (flux), Ξ - action onto the volume (surface)

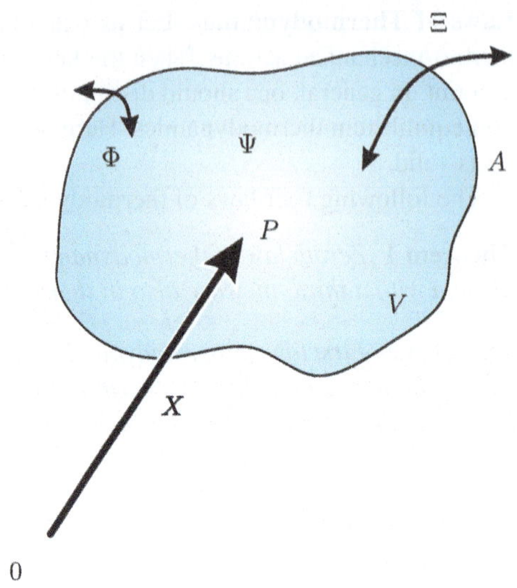

$$\frac{\mathrm{D}}{\mathrm{D}t}Y(t) = \frac{\mathrm{D}}{\mathrm{D}t}\int\limits_V \Psi(\boldsymbol{X},t)\mathrm{d}V = \int\limits_A \Phi(\boldsymbol{X},t)\mathrm{d}A + \int\limits_V \Xi(\boldsymbol{X},t)\mathrm{d}V$$

and the reference configuration

$$\frac{\mathrm{D}}{\mathrm{D}t}Y(t) = \frac{\mathrm{D}}{\mathrm{D}t}\int\limits_{V_0} \Psi_0(\boldsymbol{x},t)\mathrm{d}V = \int\limits_{A_0} \Phi_0(\boldsymbol{x},t)\mathrm{d}A + \int\limits_{V_0} \Xi_0(\boldsymbol{x},t)\mathrm{d}V.$$

The balance quantities can change their value by actions through the surface or by far distance actions. This is shown in Fig. 1.2.

The following comments concerning the general formulation can be made:

- Φ is the action through the surface A with the property of the surface A—the orientation $\boldsymbol{n}$

$$\Phi(\boldsymbol{X},t) \Rightarrow \Phi(\boldsymbol{X},\boldsymbol{n},t).$$

- Cauchy's theorem is valid

$$^{(n)}\boldsymbol{\Phi}(\boldsymbol{X},t) = \boldsymbol{n}^{(n+1)} \cdot \boldsymbol{\Phi}(\boldsymbol{X},t).$$

- actio = reactio is valid

$$\boldsymbol{\Phi}(\boldsymbol{n}) = -\boldsymbol{\Phi}(-\boldsymbol{n}).$$

- $\boldsymbol{\Psi}$, $\boldsymbol{\Xi}$ are tensor fields of the same rank n, $\boldsymbol{\Phi}$ is a tensor field of the rank $n+1$.
- The formulation with respect to the mass is equivalent to the formulation with respect to the volume

$$\frac{\mathrm{D}}{\mathrm{D}t}\int\limits_{m}\Psi(\boldsymbol{X},t)\,\mathrm{d}m=\frac{\mathrm{D}}{\mathrm{D}t}\int\limits_{V}\Psi(\boldsymbol{X},t)\rho(\boldsymbol{X},t)\,\mathrm{d}V.$$

- With the help of the Gauss-Ostrogradsky theorem we obtain

$$\int\limits_{A}\boldsymbol{n}\cdot\boldsymbol{\Phi}\,\mathrm{d}A-\int\limits_{V}\boldsymbol{\nabla}\cdot\boldsymbol{\Phi}\,\mathrm{d}V$$

- and the local form can be deduced

$$\frac{\mathrm{D}}{\mathrm{D}t}(\rho\Psi)=\boldsymbol{\nabla}\cdot\boldsymbol{\Phi}+\rho\Xi,$$

  if all fields are smooth enough.

Special Balances. Balance equations are general principles for all processes. These are not limited to thermo-mechanical problems. Other fields (electric, magnetic, chemical, etc.) can also be included. In this paper we limit ourselves to thermo-mechanical fields.

Let us introduce the following balance quantities

- mass,
- momentum,
- angular momentum,
- energy, and
- entropy

The special balance equations can be presented as follows:

- *Balance of Mass*
  As usual in solid mechanics this is the conservation of mass:

$$m=\int\limits_{V}\rho(P,t)\mathrm{d}V=\text{const.}$$

  The integral form can be expressed by

$$\frac{\mathrm{D}m}{\mathrm{D}t}=\frac{\mathrm{D}}{\mathrm{D}t}\int\limits_{V}\rho(\boldsymbol{X},t)\mathrm{d}V=\frac{\partial}{\partial t}\int\limits_{V_0}\rho_0(\boldsymbol{x})\mathrm{d}V=0$$

  and the local form is

$$\frac{\mathrm{D}}{\mathrm{D}t}(\mathrm{d}m)=\frac{\mathrm{D}}{\mathrm{D}t}(\rho\,\mathrm{d}V)=\frac{\partial}{\partial t}(\rho_0\,\mathrm{d}V_0)=0.$$

  In fluid mechanics this is the continuity equation

$$\frac{\mathrm{D}\rho}{\mathrm{D}t} + \rho \nabla \cdot \boldsymbol{v} = 0.$$

- *Balance of Linear Momentum*
  The integral form can be written down as

$$\frac{\mathrm{D}}{\mathrm{D}t} \int\limits_V \rho(\boldsymbol{X}, t)\boldsymbol{v}(\boldsymbol{X}, t)\mathrm{d}V = \int\limits_A \boldsymbol{t}(\boldsymbol{X}, t)\mathrm{d}A + \int\limits_V \rho(\boldsymbol{X}, t)\boldsymbol{k}(\boldsymbol{X}, t)\mathrm{d}V$$

  and in the case of smooth fields the local form can be deduced

$$\rho(\boldsymbol{X}, t)\frac{\mathrm{D}}{\mathrm{D}t}\boldsymbol{v}(\boldsymbol{X}, t) = \nabla \cdot \boldsymbol{T}(\boldsymbol{X}, t) + \rho(\boldsymbol{X}, t)\boldsymbol{k}(\boldsymbol{X}, t).$$

- *Balance of Moment of Momentum*
  The integral form can be expressed as

$$\frac{\mathrm{D}}{\mathrm{D}t} \int\limits_V [\boldsymbol{X} \times \rho(\boldsymbol{X}, t)\boldsymbol{v}(\boldsymbol{X}, t)]\mathrm{d}V = \int\limits_A [\boldsymbol{X} \times \boldsymbol{t}(\boldsymbol{X}, \boldsymbol{n}, t)]\mathrm{d}A + \int\limits_V [\boldsymbol{X} \times \rho(\boldsymbol{X}, t)\boldsymbol{k}(\boldsymbol{X}, t)]\mathrm{d}V.$$

  Considering the balance of momentum finally we get

$$\int\limits_V (\boldsymbol{I} \cdot \times \boldsymbol{T})\,\mathrm{d}V = \boldsymbol{0}$$

  or the local form

$$\boldsymbol{I} \cdot \times \boldsymbol{T} = \boldsymbol{0}.$$

  This is the symmetry of the stress tensor condition $\boldsymbol{T} = \boldsymbol{T}^{\mathrm{T}}$. Note that this condition is valid only in the case of the classical (Cauchy) continuum.
- *Balance of Energy: only Mechanical Actions*
  In this case the first law of thermodynamics in the integral form can be formulated as

$$\frac{\mathrm{D}}{\mathrm{D}t} \int\limits_V \left(\frac{1}{2}\boldsymbol{v} \cdot \boldsymbol{v} + u\right) \rho \mathrm{d}V = \int\limits_A \boldsymbol{t} \cdot \boldsymbol{v}\mathrm{d}A + \int\limits_V \boldsymbol{k} \cdot \boldsymbol{v}\rho\mathrm{d}V$$

  and the local form is

$$\rho\dot{u} = \boldsymbol{T} \cdot\cdot (\nabla \boldsymbol{v})^{\mathrm{T}} = \boldsymbol{T} \cdot\cdot \boldsymbol{D}.$$

  The velocity gradient $\nabla \boldsymbol{v}$ can always be split into a symmetrical and an antimetric part. Since the stress tensor is symmetrical here, only the symmetrical part of the velocity gradient remains. This part is denoted by $\boldsymbol{D}$.

- *Balance of Energy - First Law of Thermodynamics*
  The changes in time of the total energy W within the volume is equal to the heat flux Q and the power of all external loadings P_{ex}

$$\frac{\mathrm{D}}{\mathrm{D}t}W = P_{\text{ex}} + Q.$$

Here U is the inner energy

$$U = \int_m u \,\mathrm{d}m = \int_V \rho u \,\mathrm{d}V$$

and the total energy $W = U + K$ with K as the kinetic energy

$$K = \frac{1}{2}\int_V \boldsymbol{v}\cdot\boldsymbol{v}\rho\,\mathrm{d}V.$$

If $\boldsymbol{t}$ denotes the surface traction and the power of all external loadings is

$$P_{\text{ex}} = \int_A \boldsymbol{t}\cdot\boldsymbol{v}\,\mathrm{d}A + \int_V \boldsymbol{k}\cdot\boldsymbol{v}\rho\,\mathrm{d}V$$

with the heat flux

$$Q = \int_V \rho r\,\mathrm{d}V - \int_A \boldsymbol{n}\cdot\boldsymbol{h}\,\mathrm{d}A$$

and r as inner heat sources and $\boldsymbol{h}$ as the heat flux vector. Finally, one gets

$$\frac{\mathrm{D}}{\mathrm{D}t}\int_V\left(u + \frac{1}{2}\boldsymbol{v}\cdot\boldsymbol{v}\right)\rho\,\mathrm{d}V = \int_A \boldsymbol{t}\cdot\boldsymbol{v}\,\mathrm{d}A + \int_V \boldsymbol{k}\cdot\boldsymbol{v}\rho\,\mathrm{d}V - \int_A \boldsymbol{n}\cdot\boldsymbol{h}\,\mathrm{d}A + \int_V \rho r\,\mathrm{d}V.$$

Let us make some mathematical manipulations (Altenbach, 2018)

$$\frac{\mathrm{D}}{\mathrm{D}t}\int_V (\ldots) = \int_V \frac{\mathrm{D}}{\mathrm{D}t}(\ldots),$$

$$\frac{\mathrm{D}}{\mathrm{D}t}\left(\frac{1}{2}\boldsymbol{v}\cdot\boldsymbol{v}\right) = \frac{1}{2}\left(\dot{\boldsymbol{v}}\cdot\boldsymbol{v} + \boldsymbol{v}\cdot\dot{\boldsymbol{v}}\right), = \dot{\boldsymbol{v}}\cdot\boldsymbol{v}$$

$$\int_A \boldsymbol{n}\cdot(\boldsymbol{T}\cdot\boldsymbol{v} - \boldsymbol{h})\,\mathrm{d}A = \int_V \left[\nabla\cdot(\boldsymbol{T}\cdot\boldsymbol{v}) - \nabla\cdot\boldsymbol{h}\right]\mathrm{d}V$$

and

$$\nabla \cdot (\boldsymbol{T} \cdot \boldsymbol{v}) = (\nabla \cdot \boldsymbol{T}) \cdot \boldsymbol{v} + \boldsymbol{T} \cdot\cdot (\nabla \boldsymbol{v})^{\mathrm{T}} = (\nabla \cdot \boldsymbol{T}) \cdot \boldsymbol{v} + \boldsymbol{T} \cdot\cdot \boldsymbol{D}$$

If all fields are smooth, the local form can be established. The underlined terms can be deleted since the balance of linear momentum is valid

$$\int\limits_V \left(\frac{\mathrm{D}u}{\mathrm{D}t} + \underline{\dot{\boldsymbol{v}} \cdot \boldsymbol{v}}\right) \rho \mathrm{d}V = \int\limits_V (\boldsymbol{T} \cdot\cdot \boldsymbol{D} - \nabla \cdot \boldsymbol{h} + \rho r)\,\mathrm{d}V + \int\limits_V \underline{[(\nabla \cdot \boldsymbol{T}) \cdot \boldsymbol{v} + \rho \boldsymbol{k} \cdot \boldsymbol{v}]}\mathrm{d}V.$$

In this case from

$$\int\limits_V (\rho \dot{u} - \boldsymbol{T} \cdot\cdot \boldsymbol{D} + \nabla \cdot \boldsymbol{h} - \rho r)\,\mathrm{d}V = 0$$

the local form follows

$$\rho u = \boldsymbol{T} \cdot\cdot \boldsymbol{D} - \nabla \cdot \boldsymbol{h} + \rho r. \tag{1.85}$$

- *Second Law of Thermodynamics*
  The integral formulation is

$$\frac{\mathrm{D}}{\mathrm{D}t} \int\limits_V \rho s\,\mathrm{d}V \geq \int\limits_V \frac{r}{\Theta} \rho \mathrm{d}V - \int\limits_A \frac{\boldsymbol{n} \cdot \boldsymbol{h}}{\Theta}\,\mathrm{d}A.$$

The specif entropy is denoted by s and Θ is the absolute temperature. The change in time of the entropy within the volume under consideration is not smaller then the rate of the outer entropy flux. After the transformation

$$\int\limits_A \frac{\boldsymbol{n} \cdot \boldsymbol{h}}{\Theta}\,\mathrm{d}A = \int\limits_V \nabla \cdot \frac{\boldsymbol{h}}{\Theta}\,\mathrm{d}V = \int\limits_V \left(\frac{\nabla \cdot \boldsymbol{h}}{\Theta} - \frac{\boldsymbol{h} \cdot \nabla\Theta}{\Theta^2}\right)\mathrm{d}V$$

in the case of smooth fields we obtain the local form

$$\rho\Theta\dot{s} \geq \rho r - \nabla \cdot \boldsymbol{h} + \frac{1}{\Theta}\boldsymbol{h} \cdot \nabla\Theta$$

and with

$$\frac{1}{\Theta}\boldsymbol{h} \cdot \nabla\Theta = \boldsymbol{h} \cdot \nabla \ln\Theta$$

we get

$$\rho\Theta\dot{s} - \underline{(\rho r - \nabla \cdot \boldsymbol{h})} - \boldsymbol{h} \cdot \nabla \ln\Theta \geq 0.$$

The underlined term can be substituted with the help of the local form of the first law of thermodynamics (1.85)

$$\rho\Theta\dot{s} + \boldsymbol{T} \cdot\cdot\, \boldsymbol{D} - \rho\dot{u} - \boldsymbol{h} \cdot \nabla \ln\Theta \geq 0.$$

With

$$\rho\Theta\dot{s} = \rho\,(\Theta s)^{\cdot} - \rho s\dot{\Theta}$$

follows

$$\rho\,(\Theta s - u)^{\cdot} - \rho s\dot{\Theta} + \boldsymbol{T} \cdot\cdot\, \boldsymbol{D} - \boldsymbol{h} \cdot \nabla \ln\Theta \geq 0.$$

Substituting the Helmholtz' free energy

$$f = u - \Theta s$$

we can introduce the specific dissipation function

$$\boldsymbol{T} \cdot\cdot\, \boldsymbol{D} - \rho\left(\dot{f} + s\dot{\Theta}\right) = \Phi \geq 0$$

and the second law can be written as

$$\Phi - \boldsymbol{h} \cdot \nabla \ln\Theta \geq 0$$

Φ is positive definite. In the case of no dissipation ($\Phi = 0$), we obtain

$$\boldsymbol{h} \cdot \nabla \ln\Theta = \frac{\boldsymbol{h}}{\Theta} \cdot \nabla\Theta \leq 0.$$

Since the absolute temperature is $\Theta > 0$, the inequality has the following solutions:

- $\boldsymbol{h} = \boldsymbol{0}$—adiabatic process,
- $\nabla\Theta = \boldsymbol{0}$—homogeneous temperature field, or
- the angle between $\boldsymbol{h}$ and $\nabla\Theta$ is more then 90°.

1.3.5 Latest Developments

Among the latest developments we mention here only one-dimensional and two-dimensional continuum theories, micropolar theories, and nanomechanics. There are much more, for example, gradient theories, but we refer only to a recent publication (Bertram and Forest, 2020).

One-dimensional and Two-dimensional Continuum Theories. Since the end of the 1950th (after the paper of Ericksen and Truesdell 1958) an increasing number of papers dedicated to the direct approach in the theory of rods and shells and applications were published (see, for example Altenbach et al. 2010; Altenbach and Eremeyev 2013b, 2017 and the literature therein). The basic models are a deformable line and a deformable surface. In both cases, each material point has six degrees of freedom (three translations and three rotations). However, models with a lower number of degrees of freedom can be deduced as special cases. In addition, two basic definitions can be introduced.

Definition 12 (*Simple Rod*) A simple rod is a one-dimensional continuum in which the interaction between neighboring parts is due to forces and moments.

Definition 13 (*Simple Shell*) A simple shell is a two-dimensional continuum in which the interaction between neighboring parts is due to forces and moments.

It is obvious that both models correspond to the Cosserat model (Cosserat and Cosserat, 1909).

Micropolar Continuum Theories. The Cosserat continuum is partly named micropolar continuum. This approach is widely discussed in the literature, for example, in Eringen (1999), Eringen (2001), Nowacki (1986), Maugin and Metrikine (2010), Altenbach et al. (2011), Markert (2011), Altenbach and Eremeyev (2013b), Eremeyev et al. (2013). The Cosserat continuum is focussed on the fact that the continuum translations and rotations can be defined independently. One consequence is that momentum and angular momentum balances are also independent (see Truesdell 1964) and the force and moment actions in the continuum can be introduced independently as in rigid body dynamics or structural mechanics. In addition to the ordinary (force) stresses in the theory of micropolar continuum, couple stresses are introduced. Arguments for introducing moment stresses or for ignoring them are given e.g. in Nowacki (1986). By accounting couple stresses we have the possibility to describe more complex media, for example, micro-inhomogeneous materials, foams, cellular solids, lattices, masonries, particle assemblies, magnetic rheological fluids, liquid crystals, etc. (Eremeyev et al., 2013).

Remark 10 (Schaefer 1967) "...everyone, who is thinking about the foundations of Continuum Mechanics, will attend the world of images of the COSSERAT brothers."

Nanomechanics. In recent years, the trend towards even smaller structural elements has increased. Consequently, the question arose whether continuum mechanics could be applied to nanostructures. There are different opinions, but following the basic idea that a split between the bulk and the surface part in the continuum theory can be made, the continuum mechanics can be extended to nanostructures:

- one should consider that the surface part has more influence then in the classical theory, and
- a combination of three-dimensional and two-dimensional theories can be realized.

A suitable theory was suggested, for example, in Altenbach and Eremeyev (2013c). Considering such approach, the following problems are solved

- increase/decrease of the plate stiffness parameters in dependence of the thickness parameter,
- in particular, the bending stiffness is bigger for the shells with surface stresses than for shells without surface elasticity,
- "growing" plates,
- viscoelastic plates: in-plane and out-of-plane stiffness parameters are influenced by the surface effects, however, no influence on the transverse stiffness was obtained, and
- in dependence on the material parameters, the rod stiffness can decrease with an increasing number of pores or can monotonic increase.

More details are presented in Altenbach and Eremeyev (2009), Altenbach and Eremeyev (2010), Altenbach et al. (2012a), Altenbach et al. (2012b), Altenbach and Eremeyev (2013a), Altenbach and Eremeyev (2013c), Altenbach and Eremeyev (2014), Eremeyev and Altenbach (2014), Altenbach and Eremeyev (2015).

1.3.6 Conclusions

As it was shown the continuum mechanics have some important advantages. We have unique formalism for solids and fluids, for various dimensions, and for different scales. At the same time, several approaches (deductive, inductive, rheological models) can be applied for the description of complex material behavior. A brief introduction to the method of rheological models is given in the next section and in the chapter of this book written by Thomas Seifert some applications are presented. However, there are also big challenges concerning the estimation of the material parameters in the constitutive equations, the introduction of higher gradients, the numerical treatment of problems, and the introduction of equivalent stresses.

1.4 Rheological Models

1.4.1 Some Remarks

The method of rheological models was developed in the beginning of the 20th century. The term *rheology* comes from ancient Greek, meaning the theory ($\lambda o\gamma o\alpha$) of flow ($\rho\epsilon\omega$).

Definition 14 (*Rheology*, Schowalter 1978) Rheology is a branch of physics (mechanics), and it is the science that deals with the deformation and flow of materials, both solids and liquids.

Rheological modeling can be applied to fluids (liquids or gases), but also to "soft" solids or solids under conditions of plastic flow (moderate temperature, loads beyond the yield stress) or creep (elevated temperature, loads below the yield stress). Maybe the starting point for scientific studies of rheological questions was in the 17th century. Newton defined the viscosity for liquids. His viscous law was a proportional dependency between shear stress and the shear strain rate. This was similar to Hooke's law for linear elasticity. A brief historical survey is presented in Tanner (1985), Doraiswamy (2002). The following brief introduction into the method of rheological models are presented with more details in Altenbach and Eisenträger (2021).

Several classes and subclasses of rheological models can be introduced:

- ideal materials with the subclasses
  - rigid or Euclidian solids (no deformation),
  - elastic solids (linear and non-linear),
  - inviscid or Pascalian fluids (pressure in the fluid is the same in all directions or no resistance to flow),
  - Newtonian fluids (proportional dependency between shear stress and the shear strain rate),
- linear viscoelasticity (as a combination of elastic and viscous models),
- generalized Newtonian materials,
- non-linear viscosity, etc.

Many scientists contribute results to rheology. Today, we have many rheological models named after these researchers, like the Hookean model, the Newtonian model, the Prandtl model, the Kelvin-Voigt model, the Maxwell model, the Schwedoff model, the Bingham model, the Burgers model, amongst others (Reiner, 1960).

The term "rheology" for the science was only established in the late 1920s by Bingham and Reiner (Giesekus, 1994). It should be mentioned that Bingham was a chemist and Reiner—a civil engineer. On August 29th, 1929, Bingham founded the Society of Rheology in Columbus, Ohio with others (Doraiswamy, 2002). Today rheology itself is used in many areas of technology and science, e.g., in materials science, in geology, but also in food technology. In addition, rheology and continuum mechanics are in a close interaction. The equations describing the material behavior (the individual response to loadings) in continuum mechanics can be established by combining rheological models. It is straightforward to combine rheological elements by connecting them in parallel or in series to obtain more complex models. Furthermore, the physical admissibility of complex rheological models is guaranteed as long as the individual rheological elements are physically admissible. Further simplifications result from the following axioms of rheology (Reiner, 1960):

(1) Under the action of hydrostatic pressure, all materials behave in the same manner as perfectly elastic bodies.
(2) Each matter features all rheological properties, but in different degrees. Note that the main rheological properties include elasticity, viscosity, plasticity, etc.
(3) There is a hierarchy of ideal bodies such that the rheological equation of the simpler body, i.e., a body lower in the hierarchy, can be derived by setting one

or other of the constants of the rheological equation of the less simple body, i.e., a body higher in the hierarchy, equal to zero.

Let us note that one-dimensional rheological models can be generalized to three dimensions in many cases.

1.4.2 Simplest Three-dimensional Rheological Models

One-dimensional models are presented on different levels in the literature. Here we mention only a few of them: Reiner (1960), Tanner (1985), Krawietz (1986), Giesekus (1994), Palmov (1998), Palmov (2014).[14]

Let us give a short overview with respect to Palmov's method of rheological modeling. Let us assume isotropy for the material behaviour. In addition, we consider the split of the stress tensor $\boldsymbol{\sigma}$ into its hydrostatic part $\sigma_{\mathrm{m}}\boldsymbol{I}$ and deviatoric part $\boldsymbol{s} = \boldsymbol{\sigma} - \sigma_{\mathrm{m}}\boldsymbol{I}$ with the second-order unit tensor $\boldsymbol{I}$ and the hydrostatic stress $\sigma_{\mathrm{m}} = \frac{1}{3}\mathrm{tr}\,\boldsymbol{\sigma}$. This split is unique. A similar split can be suggested for the strain tensor $\boldsymbol{\varepsilon}$, whose deviatoric part is denoted by $\boldsymbol{e} = \boldsymbol{\varepsilon} - \frac{1}{3}\epsilon\boldsymbol{I}$ with the volumetric strain $\epsilon = \mathrm{tr}\,\boldsymbol{\varepsilon}$. The constitutive equations of rheological element α for the stress deviator $\boldsymbol{s}_\alpha$, the free energy F_α, and the entropy S_α for the rheological element α can be presented as functions of the temperature Θ, the temperature gradient $\nabla\Theta$, the volumetric strain ϵ, and the strain deviator $\boldsymbol{e}$ can be given as follows

$$\begin{aligned} \boldsymbol{s}_\alpha &= \boldsymbol{s}_\alpha(\Theta, \nabla\Theta, \vartheta, \boldsymbol{e}), \\ F_\alpha &= F_\alpha(\Theta, \nabla\Theta, \boldsymbol{e}), \\ S_\alpha &= S_\alpha(\Theta, \nabla\Theta, \boldsymbol{e}). \end{aligned} \tag{1.86}$$

The following connections can be established

- in parallel

$$\begin{aligned} \boldsymbol{s} &= \boldsymbol{s}_1 + \boldsymbol{s}_2, & \boldsymbol{e} &= \boldsymbol{e}_1 = \boldsymbol{e}_2, \\ F &= F_1 + F_2, & S &= S_1 + S_2 \end{aligned} \tag{1.87}$$

- in series

$$\begin{aligned} \boldsymbol{s} &= \boldsymbol{s}_1 = \boldsymbol{s}_2, & \boldsymbol{e} &= \boldsymbol{e}_1 + \boldsymbol{e}_2, \\ F &= F_1 + F_2, & S &= S_1 + S_2 \end{aligned} \tag{1.88}$$

The following basic elements can be established

- Elasticity - Hookean element

$$\boldsymbol{s} = 2\mu\boldsymbol{e},$$

- Viscosity - Newtonian element

$$\boldsymbol{s} = 2\nu\dot{\boldsymbol{e}},$$

[14] Palmov (1998) is the translation of the original Russian version from 1976.

- Plasticity - St. Venant element

$$\begin{cases} N(\sigma) < \sigma_{\mathrm{Y}} \ \dot{\boldsymbol{e}} = \boldsymbol{0}, \\ N(\sigma) = \sigma_{\mathrm{Y}} \ \dot{\boldsymbol{e}} = \dfrac{\boldsymbol{s}}{\lambda}. \end{cases}$$

- In addition, a purely elastic constitutive equation for the hydrostatic part can be introduced

$$\sigma = K\vartheta$$

Complex models result as combinations

- Viscoelastic models
  - Kelvin-Voigt model as elastic and viscous elements in parallel
  - Maxwell model as elastic and viscous elements in series
  - Poynting model as elastic and Maxwell elements in parallel
  - Generalized Kelvin-Voigt model—n Kelvin-Voigt elements in series
  - Generalized Maxwell model—n Maxwell elements in parallel
- Plastic models
  - Prandtl model as elastic and plastic model in series
  - Bingham model as viscous and plastic elements in parallel with elastic in series
  - Generalized Prandtl model as n Prandtl models in parallel

This simple approach is related to some open questions:

- One of the basic assumptions is the isotropy. The split of the stress tensor and the strain tensor into volumetric and deviatoric parts is unique. How can we formulate the anisotropic constitutive equations?
- The volumetric part in the constitutive equations is assumed to be purely elastic. There are experimental data that this is not always correct (see, for example, Bridgman 1949).
- Rheological modeling can be used the Langrangian or the Eulerian description. What is better?

Here, we need further research efforts to give proper answers. For further reading concerning large strains w.r.t. Palmov's approach we refer to Palmov (1997). In addition, in many textbooks on continuum mechanics, it is written that the Langrangian description should be used for solids, while the Eulerian approach should be preferred for fluids. However, this is questionable since the plastic flow (plastic behaviour of a material) is similar to the behaviour of fluids (see Bruhns 2020).

1.4.3 Simplest Two-dimensional Rheological Models

In Altenbach (1985) Palmov's approach was applied to two-dimensional continua. In this paper, the following items were established:

- governing equations with respect to the direct plate theory (see, for example, Altenbach 1984),
- basic elements, and
- the statement that plasticity cannot be formulated in the same way as done for elasticity and viscosity.

Within the framework of a Zhilin-type theory (Zhilin, 1976), let us introduce a Cosserat plate. The main variables are

- the stress resultants, i.e., the transverse force vector $\boldsymbol{F}$ and the moment tensor $\boldsymbol{M}$,
- the strains, i.e., the transverse shear strain vector $\boldsymbol{\gamma}$ and the tensor of the bending and torsional strains $\boldsymbol{\kappa}$, and
- an energetic variable, i.e., the free energy H.

A new variable can be introduced (Palmow and Altenbach, 1982):

$$\boldsymbol{G} = \boldsymbol{M} \times \boldsymbol{n}, \tag{1.89}$$

whereby $\boldsymbol{G}$ is referred to as "polar moment tensor" Aßmus et al. (2017). It can also be obtained based on the stress tensor:

$$\boldsymbol{G} = \langle \boldsymbol{a}_1 \cdot \boldsymbol{\sigma} z \cdot \boldsymbol{a}_1 \rangle, \tag{1.90}$$

where $\boldsymbol{\sigma}$ is the classical symmetric stress tensor, $\boldsymbol{a}_1 = \boldsymbol{e}_\alpha \boldsymbol{e}_\alpha$ is the first metric tensor (note that we consider the two-dimensional orthonormal coordinate system $\boldsymbol{e}_\alpha,\ \alpha = \{1, 2\}$, and make use of Einstein's sum convention), z is the coordinate in the transverse direction of the plate, and the brackets $\langle \ldots \rangle$ denote the integration over the thickness of the plate-like body. It is obvious that $\boldsymbol{G} = \boldsymbol{G}^{\mathrm{T}}$. For the introduced moment tensors, the following relation is valid:

$$\boldsymbol{M}^{\mathrm{T}} \cdot\cdot\, \boldsymbol{\kappa} = \boldsymbol{M} \cdot\cdot\, \boldsymbol{\kappa}^{\mathrm{T}} = \boldsymbol{G} \cdot\cdot\, \boldsymbol{\mu} \tag{1.91}$$

with $\boldsymbol{\mu} = \boldsymbol{\kappa} \times \boldsymbol{n}$. Note that the variable $\boldsymbol{\mu}$ is introduced in analogy to the transformation of the moment tensor $\boldsymbol{M}$ into the polar moment tensor $\boldsymbol{G}$, cf. Eq. (1.89).

The simplest basic two-dimensional rheological elements are presented in Altenbach (1985).

1.4.4 Advanced Rheological Models

Advanced rheological models are characterized by application of traditional (mostly linear) models, but also highly non-linear models. Nice one-dimensional examples are presented in the contribution of Thomas Seifert in this book. In addition, based on Naumenko et al. (2011a), Naumenko et al. (2011b) phenomenological constitutive equations that describe inelastic behavior at elevated temperature were developed. To characterize hardening, recovery, and softening processes a fraction model with creep-hard and creep-soft constituents is introduced. In the case of heat resistent steels the volume fraction of the creep-hard constituent is assumed to decrease towards a saturation value. Such an approach describes well the primary creep as a result of stress redistribution between constituents and tertiary creep as a result of softening (decrease of the volume fraction). To describe the whole creep curve a damage parameter in the sense of continuum damage mechanics is introduced. The material parameters and the response functions in the model were calibrated against published experimental curves for X20CrMoV12-1 steel. To verify the model predictions with experimental creep curve under stress change conditions and the stress-strain curve under constant strain rate are compared. The consideration of both softening and damage processes is necessary to characterize the long term strength in a wide stress range. The model can be generalized to the case of multi-axial stress state. The details can be taken from Eisenträger et al. (2017), Eisenträger et al. (2018), Eisenträger et al. (2018), Eisenträger et al. (2019). Finally, it can be shown that such approach is applicable for other materials (as an example the creep behavior of thermoplastics Altenbach et al. 2015).

References

Altenbach, H. (1984). Die Grundgleichungen einer linearen Theorie für dünne, elastische Platten und Scheiben mit inhomogenen Materialeigenschaften in Dickenrichtung. *Technische Mechanik, 5*(2), 51–58.

Altenbach, H. (1985). Zur Theorie der Cosserat-Platten. *Technische Mechanik, 6*(2), 43–50.

Altenbach, H. (2002). Creep analysis of thin-walled structures. *ZAMM - Zeitschrift für Angewandte Mathematik und Mechanik, 82*(8), 507–533. https://doi.org/10.1002/1521-4001(200208)82:8<507::AID-ZAMM507>3.0.CO;2-Y

Altenbach, H. (2018). *Kontinuumsmechanik - Eine elementare Einführung in die materialunabhängigen und materialabhängigen Gleichungen*, 4th edn. Springer. https://doi.org/10.1007/978-3-662-57504-8

Altenbach, H. (2020). *Holzmann/Meyer/Schumpich Technische Mechanik Festigkeitslehre*. Vieweg, Wiesbaden: Springer. 14. verb. u. erw. Aufl. https://doi.org/10.1007/978-3-658-32023-2

Altenbach, H., & Eisenträger, J. (2020). Introduction to creep mechanics. In H. Altenbach, & A. Öchsner (Eds.), *Encyclopedia of Continuum Mechanics* (pp. 1337–1344). Berlin: Springer. https://doi.org/10.1007/978-3-662-55771-6_155

Altenbach, H., & Eisenträger, J. (2021). Rheological modeling in solid mechanics from the beginning up to now. *Lecture Notes of TICMI, 22*, 13–29.

Altenbach, H., & Eremeyev, V. (Eds.). (2017). *Shell-Like Structures - Advanced Theories and Applications, CISM International Centre for Mechanical Sciences* (Vol. 572). Cham: Springer. https://doi.org/10.1007/978-3-319-42277-0

Altenbach, H., & Eremeyev, V. A. (2009). On the linear theory of micropolar plates. *ZAMM - Journal of Applied Mathematics and Mechanics / Zeitschrift für Angewandte Mathematik und Mechanik, 89*(4), 242–256. https://doi.org/10.1002/zamm.200800207

Altenbach, H., & Eremeyev, V. A. (2010). On the effective stiffness of plates made of hyperelastic materials with initial stresses. *International Journal of Non-Linear Mechanics, 45*(10), 976–981. https://doi.org/10.1016/j.ijnonlinmec.2010.04.007

Altenbach, H., & Eremeyev, V. A. (2013a). On the constitutive equations of viscoelastic micropolar plates and shells. *Proceedings of Indian National Science Academy Part A, 79*(4, SI), 513–518. https://doi.org/10.16943/ptinsa/2013/v79i4/47984

Altenbach, H., & Eremeyev, V. A. (2014). Vibration analysis of non-linear 6-parameter prestressed shells. *Meccanica, 49*(8), 1751–1761. https://doi.org/10.1007/s11012-013-9845-1

Altenbach, H., & Eremeyev, V. A. (2015). On the constitutive equations of viscoelastic micropolar plates and shells of differential type. *Mathematics and Mechanics of Complex Systems, 3*(3), 273–283. https://doi.org/10.2140/memocs.2015.3.273

Altenbach, H., & Eremeyev, V. A. (Eds.). (2013). *Generalized Continua - from the Theory to Engineering Applications, CISM International Centre for Mechanical Sciences* (Vol. 541). Wien: Springer. https://doi.org/10.1007/978-3-7091-1371-4

Altenbach, H., & Eremeyev, V. A. (2013c). On the continuum mechanics approach in modeling nanosized structural elements. In A. Öchsner, & A. Shokuhfar (Eds.), *New Frontiers of Nanoparticles and Nanocomposite Materials*, Advanced Structured Materials (Vol. 4, pp. 351–371). Berlin: Springer. https://doi.org/10.1007/8611_2012_67

Altenbach, H., & Knape, K. (2020). On the main directions in creep mechanics of metallic materials. *Proceedings of the National Academy of Sciences of Armenia: Mechanics, 73*(3), 24–43. https://doi.org/10.33018/73.3.2

Altenbach, H., & Kruch, S. (Eds.). (2013). *Advanced Materials Modelling for Structures*, Advanced Structured Materials (Vol. 19). Heidelberg: Springer. https://doi.org/10.1007/978-3-642-35167-9

Altenbach, H., & Naumenko, K. (1997). Creep bending of thin-walled shells and plates by consideration of finite deflections. *Computational Mechanics, 19*(6), 490–495. https://doi.org/10.1007/s004660050197

Altenbach, H., & Naumenko, K. (2002). Shear correction factors in creep-damage analysis of beams, plates and shells. *JSME International Journal Series A Solid Mechanics and Material Engineering, 45*(1), 77–83. https://doi.org/10.1299/jsmea.45.77

Altenbach, H., & Naumenko, K. (2007). Long term creep analysis of pipe bends in a steam transfer line at elevated temperature. *Key Engineering Materials, 340–341*, 795–802. https://doi.org/10.4028/www.scientific.net/KEM.340-341.795

Altenbach, H., & Skrzypek, J. J. (Eds.). (1999). *Creep and Damage in Materials and Structures. CISM International Centre for Mechanical Sciences Courses and Lectures (Vol. 399)*. Vienna: Springer.

Altenbach, H., Altenbach, J., & Naumenko, K. (1997). On the prediction of creep damage by bending of thin-walled structures. *Mechanics of Time-Dependent Materials, 1*(2), 181–193. https://doi.org/10.1023/A:1009794001209

Altenbach, H., Breslavsky, D., Morachkovsky, O., & Naumenko, K. (2000). Cyclic creep-damage in thin-walled structures. *Journal of Strain Analysis for Engineering Design, 35*(1), 1–12. https://doi.org/10.1243/0309324001513964

Altenbach, H., Kolarov, G., Morachkovsky, O. K., & Naumenko, K. (2000). On the accuracy of creep-damage predictions in thinwalled structures using the finite element method. *Computational Mechanics, 25*(1), 87–98. https://doi.org/10.1007/s004660050018

Altenbach, H., Huang, C., & Naumenko, K. (2001). Modelling of the creep-damage under reversed stress states by considering damage activation and deactivation. *Technische Mechanik, 21*(4), 273–282.

Altenbach, H., Kushnevsky, V., & Naumenko, K. (2001). On the use of solid- and shell-type finite elements in creep-damage predictions of thinwalled structures. *Archive of Applied Mechanics, 71*(2), 164–181. https://doi.org/10.1007/s004190000132

Altenbach, H., Huang, C., & Naumenko, K. (2002). Creep-damage predictions in thin-walled structures by use of isotropic and anisotropic damage models. *The Journal of Strain Analysis for Engineering Design, 37*(3), 265–275. https://doi.org/10.1243/0309324021515023

Altenbach, H., Gorash, Y., & Naumenko, K. (2008). Creep analysis for a wide stress range based on relaxation data. *International Journal of Modern Physics B, 22*(31 & 32), 5413–5418. https://doi.org/10.1142/S0217979208050589

Altenbach, H., Gorash, Y., & Naumenko, K. (2008). Steady-state creep of a pressurized thick cylinder in both the linear and the power law ranges. *Acta Mechanica, 195*(1–4), 263–274. https://doi.org/10.1007/s00707-007-0546-5

Altenbach, H., Maugin, G. A., & Erofeev, V. (Eds.). (2011). *Mechanics of Generalized Continua*, Advanced Structured Materials (Vol. 7). Springer. https://doi.org/10.1007/978-3-642-19219-7

Altenbach, H., Eremeyev, V. A., Ivanova, E. A., & Morozov, N. F. (2012a). Bending of a three-layer plate with near-zero transverse shear stiffness (in Russ.). *Fizicheskaya Mezomekhanika (Physical Mesomechanics), 15*(6), 15–19.

Altenbach, H., Eremeyev, V. A., & Morozov, N. F. (2012). Surface viscoelasticity and effective properties of thin-walled structures at the nanoscale. *International Journal of Engineering Science, 59*, 83–89. https://doi.org/10.1016/j.ijengsci.2012.03.004

Altenbach, H., Bîrsan, M., & Eremeyev, V. A. (2013). Cosserat-type rods. In H. Altenbach, & V. A. Eremeyev (Eds.) *Generalized Continua from the Theory to Engineering Applications*. CISM International Centre for Mechanical Sciences (Vol. 541, pp. 178–248). Vienna, Wien: Springer. https://doi.org/10.1007/978-3-7091-1371-4_4

Altenbach, H., Bolchoun, A., & Kolupaev, V. A. (2014). Phenomenological yield and failure criteria. In H. Altenbach, & A. Öchsner (Eds.), *Plasticity of Pressure-Sensitive Materials* (pp. 49–152). Berlin, Heidelberg: Springer. https://doi.org/10.1007/978-3-642-40945-5_2

Altenbach, H., Girchenko, A., Kutschke, A., & Naumenko, K. (2015). Creep behavior modeling of polyoxymethylene (POM) applying rheological models. In H. Altenbach, & M. Brünig (Eds.), *Inelastic Behavior of Materials and Structures Under Monotonic and Cyclic Loading*, Advanced Structured Materials (Vol. 57, pp. 1–15). Cham: Springer. https://doi.org/10.1007/978-3-319-14660-7_1

Altenbach, V., & Öchsner, A. (Eds.). (2020). *Encyclopedia of Continuum Mechanics* (Vols. 1–3). Berlin, Heidelberg: Springer. https://doi.org/10.1007/978-3-662-55770-9

Altenbach, J., Altenbach, H., & Naumenko, K. (2004). Edge effects in moderately thick plates under creep-damage conditions. *Technische Mechanik, 24*(3), 254–263.

Altenbach, J., Altenbach, H., & Eremeyev, V. A. (2010). On generalized Cosserat-type theories of plates and shells: a short review and bibliography. *Archive of Applied Mechanics, 80*(1), 73–92. https://doi.org/10.1007/s00419-009-0365-3

Apostol, T. M. (1967). *Calculus, Volume 1: One-Variable Calculus with an Introduction to Linear Algebra*. New York: Wiley.

Arutyunyan, N. K. (1966). *Some Problems in the Theory of Creep*. Oxford: Pergamon Press.

Aßmus, M., Eisenträger, J., & Altenbach, H. (2017). Projector representation of isotropic linear elastic material laws for directed surfaces. *ZAMM - Zeitschrift für Angewandte Mathematik und Mechanik, 97*(12), 1625–1634. https://doi.org/10.1002/zamm.201700122

Bailey, R. W. (1935). The utilization of creep test data in engineering design. *Proceedings of the Institution of Mechanical Engineers, 131*(1), 131–349. https://doi.org/10.1243/PIME_PROC_1935_131_012_02

Beer, F. P., Johnston, E. R., Jr., Dewolf, J. T., & Mazurek, D. F. (2009). *Mechanics of Materials* (6th ed.). New York: McGraw-Hill.

Bertram, A., & Forest, S. (Eds.). (2020). *Mechanics of Strain Gradient Materials, CISM International Centre for Mechanical Sciences* (Vol. 600). Cham: Springer. https://doi.org/10.1007/978-3-030-43830-2

Betten, J. (1992). Applications of tensor functions in continuum damage mechanics. *International Journal of Damage Mechanics, 1*(1), 47–59. https://doi.org/10.1177/105678959200100103

Betten, v,. (2001). *Kontinuumsmechanik - Elastisches und inelastisches Verhalten isotroper und anisotroper Stoffe* (2nd ed.). Berlin: Springer. https://doi.org/10.1007/978-3-642-56562-5

Bridgman, P. W. (1949). *The Physics of High Pressure*. London: G. Bell and Sons.

Bruhns, O. T. (2020). History of plasticity. In H. Altenbach & A. Öchsner (Eds.), *Encyclopedia of Continuum Mechanics* (pp. 1129–1190). Berlin, Heidelberg: Springer. https://doi.org/10.1007/978-3-662-53605-6_281-1

Cane, B. J. (1981). Creep fracture of dispersion strengthened low alloy ferritic steels. *Acta Metallurgica, 29*, 1581–1591.

Casey, J. (2011). On the derivation of jump conditions in continuum mechanics. *International Journal of Structural Changes in Solids, 3*(2), 61–84.

Casey, J., & Jahedmotlagh, H. (1984). The strength-differential effect in plasticity. *International Journal of Solids and Structures, 20*(4), 377–393. https://doi.org/10.1016/0020-7683(84)90047-7

Chaboche, J.-L. (1993). Development of continuum damage mechanics for elastic solids sustaining anisotropic and unilateral damage. *International Journal of Damage Mechanics, 2*(4), 311–329. https://doi.org/10.1177/105678959300200401

Chaboche, J. L. (1987). Continuum damage mechanics: Present state and future trends. *Nuclear Engineering and Design, 105*(1), 19–33. https://doi.org/10.1016/0029-5493(87)90225-1

Chaboche, J. L. (1988). Continuum Damage Mechanics: Part I - General Concepts. *Journal of Applied Mechanics, 55*(1), 59–64. https://doi.org/10.1115/1.3173661

Chaboche, J. L. (1988). Continuum Damage Mechanics: Part II - Damage Growth, Crack Initiation, and Crack Growth. *Journal of Applied Mechanics, 55*(1), 65–72. https://doi.org/10.1115/1.3173662

Cotrell, A. H. (1967). *An Introduction to Metallurgy*. London: Edward Arnold.

Cosserat, E., & Cosserat, F. (1909). *Théorie des Corps Déformables*. Paris: A. Hermann et Fils.

da Costa Andrade, E. N. (1910). On the viscous flow of metals, and allied phenomena. *Proceedings of the Royal Society of London. Series A, Containing Papers of a Mathematical and Physical Character, LXXXIV*(567), 1–12. https://doi.org/10.1098/rspa.1910.0050

da Costa Andrade, E. N., & Porter, A. W. (1914). The flow in metals under large constant stresses. *Proceedings of the Royal Society of London. Series A, Containing Papers of a Mathematical and Physical Character, XC*(619), 329–342. https://doi.org/10.1098/rspa.1914.0056

de Baas, A. F. (Ed.) (2017). *What Makes a Material Function? Let Me Compute the Ways ...*. Luxembourg: Publications Office of the European Union. https://doi.org/10.2777/417118. 6th version.

Dill, E. H. (2006). *Continuum Mechanics: Elasticity, Plasticity, Viscoelasticity*. Boca Raton: CRC Press. https://doi.org/10.1201/9781420009828

Doraiswamy, D. (2002). The origins of rheology: A short historical excursion. *Rheology Bulletin, 71*(1), 7–17.

Dyson, B. F., & Oosgerby, S. (1993). Modelling and analysis of creep deformation and fracture in a 1Cr 0.5Mo femtic steel. NPL Report DMM(A)116, National Physical Laboratory.

Eisenträger, J., Naumenko, K., & Altenbach, H. (2018). Numerical implementation of a phase mixture model for rate-dependent inelasticity of tempered martensitic steels. *Acta Mechanica, 229*(7), 3051–3068. https://doi.org/10.1007/s00707-018-2151-1

Eisenträger, J., Naumenko, K., & Altenbach, H. (2018). Calibration of a phase mixture model for hardening and softening regimes in tempered martensitic steel over wide stress and temperature ranges. *The Journal of Strain Analysis for Engineering Design, 53*(3), 156–177. https://doi.org/10.1177/0309324718755956

Eisenträger, J., & Altenbach, H. (2020). Creep in heat-resistant steels at elevated temperatures. In H. Altenbach, & A. Öchsner (Eds.), *State of the Art and Future Trends in Material Modeling*, Advanced Structured Materials (Vol. 100, pp. 79–112). Cham: Springer International Publishing. https://doi.org/10.1007/978-3-030-30355-6_4

Eisenträger, J., Naumenko, K., Altenbach, H., & Gariboldi, E. (2017). Analysis of temperature and strain rate dependencies of softening regime for tempered martensitic steel. *The Journal of Strain Analysis for Engineering Design, 52*(4), 226–238. https://doi.org/10.1177/0309324717699746

Eisenträger, J., Naumenko, K., & Altenbach, K. (2019). Numerical analysis of a steam turbine rotor subjected to thermo-mechanical cyclic loads. *Technische Mechnik, 19*(3), 261–281. https://doi.org/10.24352/UB.OVGU-2019-024

Eremeyev, V. A., & Altenbach, H. (2014). Equilibrium of a second-gradient fluid and an elastic solid with surface stresses. *Meccanica, 49*(11), 2635–2643. https://doi.org/10.1007/s11012-013-9851-3

Eremeyev, V. A., Lebedev, L. P., & Altenbach, H. (2013). *Foundations of Micropolar Mechanics*. Springer Briefs in Applied Sciences and Technology - Continuum Mechanics. Heidelberg, New York, Dordrecht, London: Springer. https://doi.org/10.1007/978-3-642-28353-6

Eremeyev, V. A., Cloud, M. J., & Lebedev, L. P. (2018). *Applications of Tensor Analysis in Continuum Mechanics*. Singapore: World Scientific. https://doi.org/10.1142/10959

Ericksen, J. L., & Truesdell, C. (1958). Exact theory of stress and strain in rods and shells. *Archive for Rational Mechanics and Analysis, 1*(1), 295–323. https://doi.org/10.1007/BF00298012

Eringen, A. C. (1999). *Microcontinuum Field Theory, Foundations and Solids* (Vol. II). New York: Springer. https://doi.org/10.1007/978-1-4612-0555-5

Eringen, A. C. (2001). *Microcontinuum Field Theory, Fluent Media* (Vol. II). New York: Springer. https://doi.org/10.1007/978-1-4419-3192-4

Feynman, R., Leighton, R., & Sands, M. (1964, 1965). *The Feynman Lectures on Physics* (Vol. 1–3). Boston: Addison-Wesley.

Frost, H. J., & Ashby, M. F. (1982). *Deformation-mechanism Maps - the Plasticity and Creep of Metals and Ceramics*. Oxford: Pergamon Press.

Giesekus, H. (1994). *Phänomenologische Rheologie. Eine Einführung*. Berlin, Heidelberg, New York: Springer. https://doi.org/10.1007/978-3-642-57953-0

Gorash, Y., Altenbach, H., & Lvov, G. (2012). Modelling of high-temperature inelastic behaviour of the austenitic steel aisi type 316 using a continuum damage mechanics approach. *The Journal of Strain Analysis for Engineering Design, 47*(4), 229–243. https://doi.org/10.1177/0309324712440764

Green, A. E., Naghdi, P. M., & Wenner, M. L. (1974). On the theory of rods. I. Derivations from the three-dimensional equations. *Proceedings of the Royal Society of London. A. Mathematical and Physical Sciences, 337*(1611), 451–483. https://doi.org/10.1098/rspa.1974.0061

Haupt, P.(2002). *Continuum Mechanics and Theory of Materials*. Advanced Texts in Physics, 2nd edn. Berlin-Heidelberg: Springer. https://doi.org/10.1007/978-3-662-04775-0

Hayhurst, D. R. (1972). Creep rupture under multi-axial states of stress. *Journal of the Mechanics and Physics of Solids, 20*(6), 381–382. https://doi.org/10.1016/0022-5096(72)90015-4

Hayhurst, D. R. (1994). The use of continuum damage mechanics in creep analysis for design. *The Journal of Strain Analysis for Engineering Design, 29*(3), 233–241. https://doi.org/10.1243/03093247V293233

Hoff, N. (Ed.). (1962). *Creep in Structures*. Berlin: Springer.

Hoff, N. J. (1953). The Necking and the Rupture of Rods Subjected to Constant Tensile Loads. *Journal of Applied Mechanics, 20*(1), 105–108. https://doi.org/10.1115/1.4010601

Hoff, N. J. (1957). Buckling at high temperature. *The Journal of the Royal Aeronautical Society, 61*(563), 756–774. https://doi.org/10.1017/S0368393100133735

Hult, J. (Ed.). (1972). *Creep in Structures*. Berlin/Heidelberg/New York: Springer.

Hutter, K., & Jöhnk, K. (2010). *Continuum Methods of Physical Modeling - Continuum Mechanics, Dimensional Analysis, Turbulence*. Berlin: Springer. https://doi.org/10.1007/978-3-662-06402-3

Kachanov, L. M. (1958). O vremeni razrusheniya v usloviyakh polzuchesti (On the time to rupture under creep conditions, in Russ.). *Izvestiya AN SSSR. Otdelenie Tekhnicheskikh Nauk, 8*, 26–31.

Kachanov, L. M. (1974). *Osnovy mekhaniki razrusheniia (Fundamentals of Fracture Mechanics, in Russ.)*. Moscow: Nauka.

Kolupaev, V. A. (2018). *Equivalent Stress Concept for Limit State Analysis, Advanced Structured Materials* (Vol. 68). Cham: Springer. https://doi.org/10.1007/978-3-319-73049-3

Kolupaev, V. A., Bolchoun, A., & Altenbach, H. (2009). Aktuelle Trends beim Einsatz von Festigkeitshypothesen. *Konstruktion, 5*, 59–66.

Kolupaev, V. A., Bolchoun, A., & Altenbach, H. (2013). Yield criteria for incompressible materials in the shear stress space. In A. Öchsner, & H. Altenbach (Eds.), *Experimental and Numerical Investigation of Advanced Materials and Structures*, Advanced Structured Materials (Vol. 41, pp. 107–119). Cham: Springer. https://doi.org/10.1007/978-3-319-00506-5_6

Kowalewski, Z. L., Hayhurst, D. R., & Dyson, B. F. (1994). Mechanisms-based creep constitutive equations for an aluminium alloy. *J. Strain Analysis, 29*, 309–316.

Krajcinovic, D. (1995). Continuum damage mechanics: When and how? *International Journal of Damage Mechanics, 4*(3), 217–229. https://doi.org/10.1177/105678959500400302

Krajcinovic, D. (1996). *Damage Mechanics, North-Holland Series in Applied Mathematics and Mechanics* (Vol. 41). Amsterdam: North-Holland. https://doi.org/10.1016/S0167-5931(06)80062-8

Krajcinovic, D., & Lemaitre, J. (Eds.). (1987). *Continuum Damage Mechanics - Theory and Application, CISM International Centre for Mechanical Sciences Courses and Lectures (Vol. 295)*. Vienna: Springer.

Krawietz, A. (1986). *Materialtheorie*. Berlin, Heidelberg, New York, Tokyo: Springer. https://doi.org/10.1007/978-3-642-82512-5

Lacy, T. E., McDowell, D. L., Willice, P. A., & Talreja, R. (1997). On representation of damage evolution in continuum damage mechanics. *International Journal of Damage Mechanics, 6*(1), 62–95. https://doi.org/10.1177/105678959700600106

Lai, W. M., Rubin, D., & Krempl, E. (2010). *Introduction to Continuum Mechanics* (4th ed.). Oxford: Butterworth-Heinemann.

Leckie, F. A., & Hayhurst, D. R. (1977). Constitutive equations for creep rupture. *Acta Metallurgica, 25*(9), 1059–1070. https://doi.org/10.1016/0001-6160(77)90135-3

Lemaitre, J. (1984). How to use damage mechanics. *Nuclear Engineering and Design, 80*(2), 233–245. https://doi.org/10.1016/0029-5493(84)90169-9. 4th Special Issue on Smirt-7.

Lemaitre, J., & Chaboche, J.-L. (1990). *Mechanics of Solid Materials*. Cambridge: Cambridge University Press. https://doi.org/10.1017/CBO9781139167970

Lemaitre, J., & Desmorat, R. (2005). *Engineering Damage Mechanics: Ductile, Creep, Fatigue and Brittle Failures*. Berlin, Heidelberg, New York: Springer.

Lemaitre, J., & Dufailly, J. (1987). Damage Measurements. *Engineering Fracture Mechanics, 28*(5), 643–661. https://doi.org/10.1016/0013-7944(87)90059-2. Special Issue in Honor of Professor Takeo Yokobori.

Lifshitz, I. M., & Slyozov, V. V. (1961). The kinematics of precipitation from supersaturated solid solutions. *J. Phys. Chem. Solids, 19*, 35–50.

Lurie, A. I. (2005). *Theory of Elasticity. Foundations of Engineering Mechanics*. Berlin: Springer.

Lü, Y.-B., & Chen, X.-F. (1989). The order of a damage tensor. *Applied Mathematics and Mechanics, 10*(3), 251–258. https://doi.org/10.1007/BF02014619

Manzhirov, A. V., Altenbach, H., & Gupta, N. (Eds.) *Procedia IUTAM - IUTAM Symposium on Growing Solids, 23-27 June 2015, Moscow, Russia* (Vol. 23). IUTAM, Elsevier.

Markert, B. (Ed.) (2011). *Advances in Extended and Multifield Theories for Continua, Lecture Notes in Applied and Computational Mechanics* (Vol. 59). Springer. https://doi.org/10.1007/978-3-642-22738-7

Maugin, G. A., & Metrikine, A. V. (Eds.) *Mechanics of Generalized Continua: One Hundred Years After the Cosserats*, Advances in Mechanics and Mathematics (Vol. 21). New York: Springer. https://doi.org/10.1007/978-1-4419-5695-8

Menzel, A., & Sprave, L. (2020). Continuum damage mechanics - modelling and simulation. In J. Merodio, & R. Ogden (Eds.), *Constitutive Modelling of Solid Continua*, Solid Mechanics and Its Applications (Vol. 262). Cham. https://doi.org/10.1007/978-3-030-31547-4_8

Murakami, S. (2012). *Continuum Damage Mechanics: A Continuum Mechanics Approach to the Analysis of Damage and Fracture, Solid Mechanics and Its Applications* (Vol. 185). Dordrecht, Heidelberg, London, New York: Springer. https://doi.org/10.1007/978-94-007-2666-6

Murakami, S., & Ohno, N. (Eds.). (2001). *IUTAM Symposium on Creep in Structures, Solid Mechanics and Its Applications (Vol. 86)*. Dordrecht: Kluwer.

Nadai, A. (1938). The influence of time upon creep. the hyperbolic sine creep law. In J. M. Lessells, J. P. den Hartog, G. B. Karelitz, R. E. Peterson, & H. M. Westergard (Eds.), *Contributions to the Mechanics of Solids dedicated to Stephen Timoshenko by his friends on the occasion of his sixties birthday anniversary* (pp. 155–170). New York: Mcmillian.

Naue, G. (1984). Kontinuumsbegriff und Erhaltungssätze in der Mechanik seit Leonhard Euler. *Technische Mechanik, 5*(4-), 62–66.

Naumenko, K., & Altenbach, H. (2005). A phenomenological model for anisotropic creep in a multipass weld metal. *Archive of Applied Mechanics, 74*(11), 808–819. https://doi.org/10.1007/s00419-005-0409-2

Naumenko, K., & Altenbach, H. (2016). *Modeling High Temperature Materials Behavior for Structural Analysis. Part I: Continuum Mechanics Foundations and Constitutive Models*, Advanced Structured Materials (Vol. 28). Cham: Springer. https://doi.org/10.1007/978-3-319-31629-1

Naumenko, K., & Altenbach., H. (2020) *Modeling High Temperature Materials Behavior for Structural Analysis. Part II: Solution Procedures and Structural Analysis Examples*, Advanced Structured Materials (Vol. 112). Cham: Springer. https://doi.org/10.1007/978-3-030-20381-8

Naumenko, K., & Gariboldi, E. (2014). A phase mixture model for anisotropic creep of forged Al-Cu-Mg-Si alloy. *Materials Science and Engineering: A, 618*, 368–376. https://doi.org/10.1016/j.msea.2014.09.012

Naumenko, K., & Kostenko, Y. (2009). Structural analysis of a power plant component using a stress-range-dependent creep-damage constitutive model. *Materials Science and Engineering: A, 510*, 169–174.

Naumenko, K., Altenbach, H., & Gorash, Y. (2009). Creep analysis with a stress range dependent constitutive model. *Archive of Applied Mechanics, 79*(6–7), 619–630. https://doi.org/10.1007/s00419-008-0287-5

Naumenko, K., Altenbach, H., & Kutschke, A. (2011). A combined model for hardening, softening, and damage processes in advanced heat resistant steels at elevated temperature. *International Journal of Damage Mechanics, 20*(4), 578–597. https://doi.org/10.1177/1056789510386851

Naumenko, K., Altenbach, H., & Kutschke, A. (2011b). Inelastic analysis versus simplified rules for stress concentration fields under variable loading and high temperature. *Materials Research Innovations, 15*(Supl. 1), s205–s208. https://doi.org/10.1179/143307511X12858957673275

Naumenko, K., Gariboldi, E., & Nizinkovskyi, R. (2020). Stress-regime-dependence of inelastic anisotropy in forged age-hardening aluminium alloys at elevated temperature: Constitutive modeling, identification and validation. *Mechanics of Materials, 141*, 103262. https://doi.org/10.1016/j.mechmat.2019.103262

Noll, W. (1958). A mathematical theory of the mechanical behavior of continuous media. *Archive for Rational Mechanics and Analysis, 2*(1), 197–226. https://doi.org/10.1007/BF00277929

Norton, F. H. (1929). *Creep of Steel at High Tmperatures*. New York: McGraw-Hill.

Nowacki, W. (1986). *Theory of Asymmetric Elasticity*. Oxford: Pergamon.

Odqvist, F. K. G. (1935). Theory of creep under combined stresses. In C. B. Biezeno, & J. M. Burgers (Eds.) *Proceedings of the Fourth International Congress for Applied Mechanics* (p. 228). Cambridge.

Odqvist, F. K. G. (1974). *Mathematical Theory of Creep and Creep Rupture*. Oxford: Oxford University Press.

Odqvist, F. K. G. (1981). Historical survey of the development of creep mechanics from its beginnings in the last century to 1970. In A. R. S. Ponter, & D. R. Hayhurst (Eds.) *Creep in Structures*,

International Union of Theoretical and Applied Mechanics (pp. 1–12). Berlin: Springer. https://doi.org/10.1007/978-3-642-81598-0_1
Odqvist, F. K. G., & Hult, J. (1962). *Kriechfestigkeit metallischer Werkstoffe*. Berlin: Springer. https://doi.org/10.1007/978-3-642-52432-5
Olsen-Kettle, L. (2019). Bridging the macro to mesoscale: Evaluating the fourth-order anisotropic damage tensor parameters from ultrasonic measurements of an isotropic solid under triaxial stress loading. *International Journal of Damage Mechanics, 28*(2), 219–232. https://doi.org/10.1177/1056789518757293
Palmov, V. A. (1997). Large strains in viscoelastoplasticity. *Acta Mechanica, 125*(1), 129–139. https://doi.org/10.1007/BF01177303
Palmov, V. A. (1998). *Vibrations of Elasto-Plastic Bodies. Foundations of Engineering Mechanics. Heidelberg: Springer.* https://doi.org/10.1007/978-3-540-69636-0
Palmov, V. A. (2014). *Nonlinear Mechanics of Deformable Bodies*. St. Peterburg: Polytechnic University Publisher.
Palmow, W. A., & Altenbach, H. (1982). Über eine Cosseratsche Theorie für elastische Platten. *Technische Mechanik, 3*(3), 5–9.
Perrin, I. J., & Hayhurst, D. R. (1996). Creep constitutive equations for a 0.5Cr-0.5Mo-0.25V ferritic steel in the temperature range 600-675°C. *The Journal of Strain Analysis for Engineering Design, 31*(4), 299–314. https://doi.org/10.1243/03093247V314299
Ponter, A. R. S., & Hayhurst, D. R. (1981). *Creep in Structures*. International Union of Theoretical and Applied Mechanics. Berlin: Springer. https://doi.org/10.1007/978-3-642-81598-0
Prager, W. (1945). Strain hardening under combined stresses. *Journal of Applied Physics, 16*(12), 837–840. https://doi.org/10.1063/1.1707548
Prygorniev, O. (2015). *Statistical analysis of stress and deformation state in polycristalline aggregates with a large numbers of grains*. Ph.D. thesis, Faculty of Mechanical Engineering, Otto-von-Guericke-Universität, Magdeburg.
Rabotnov, Yu. N. (1959). O mechanizme dlitel'nogo razrusheniya (A mechanism of the long term fracture, in Russ.). In *Voprosy prochnosti materialov i konstruktsii* (pp. 5–7). Moscow: Publ. AN SSSR.
Rabotnov, Yu. N. (1969). *Creep Problems in Structural Members, North-Holland Series in Applied Mathematics and Mechanics* (Vol. 7). Amsterdam: North-Holland.
Rabotnov, Yu. N. (1980). *Elements of Hereditary Solid Meanics*. Moscow: Mir.
Reiner, M. (1943). *Ten lectures on theoretical rheology*. London: H.K. Lewis and Co.
Reiner, M. (1945). A mathematical theory of dilatancy. *American Journal of Mathematics, 67*(3), 350–362. https://doi.org/10.2307/2371950
Reiner, M. (1960). *Deformation, Strain and Flow: An Elementary Introduction to Rheology*. London: H.K. Lewis.
Rodin, G. J., & Parks, D. M. (1986). Constitutive models of a power-law matrix containing aligned penny-shaped cracks. *Mechanics of Materials, 5*(3), 221–228. https://doi.org/10.1016/0167-6636(86)90019-0
Schaefer, H. (1967). Das Cosserat Kontinuum. *ZAMM - Journal of Applied Mathematics and Mechanics / Zeitschrift für Angewandte Mathematik und Mechanik, 47*(8), 485–498. https://doi.org/10.1002/zamm.19670470802
Schowalter, W. R. (1978). *Mechanics of Non-Newtonian Fluids*. Oxford: Pergamon.
Sdobyrev, V. P. (1959). Criterion for the long term strength of some heat-resistant alloys at a multiaxial loading (in Russ.: Kriterij dlitelnoj prochnosti dlja nekotorykh zharoprochnykh splavov pri slozhnom naprjazhennom sostojanii). *Izvestija Akademii Nauk SSSR, Otdelenie Tekhnicheskikh Nauk, Mekhanika i Mashinostroenie, 6*, 93–99.
Skrzypek, J., & Ganczarski, A. (1999). Modelling of Material Damage and Failure of Structures: Theory and Applications. *Foundations of Engineering Mechanics*. Berlin: Springer. https://doi.org/10.1007/978-3-540-69637-7
Soderberg, C. R. (1938). Plasticity and creep in machine design. In J. M. Lessells, J. P. den Hartog, G. B. Karelitz, R. E. Peterson, & H. M. Westergard (Eds.), *Contributions to the Mechanics of*

Solids Dedicated to Stephen Timoshenko by His Friends on the Occasion of His Sixties Birthday Anniversary (pp. 197–210). New York: Mcmillian.

Sosnin, O. V. (1973). Energy version of the theory of creep and long-term (creep) strength. creep and rupture of nonstrengthening materials. I. *Strength of Materials, 5*(5), 564–568. https://doi.org/10.1007/BF00762312

Sosnin, O. V., Gorev, B. V., & Nikitenko, A. F. (1976). Energy variant of theory of creep. *Strength of Materials, 8*(5), 1255–1260. https://doi.org/10.1007/BF01528744

Stodola, A. (1933). Die Kriecherscheinungen, ein neuer technisch wichtiger Aufgabenkreis der Elastizitätstheorie. *ZAMM - Journal of Applied Mathematics and Mechanics / Zeitschrift für Angewandte Mathematik und Mechanik, 13*(2), 143–146. https://doi.org/10.1002/zamm.19330130221

Tanner, R. I. (1985). *Engineering Rheology*. Oxford: Clarendon.

Truesdell, C. (1964). Die Entwicklung des Drallsatzes. *ZAMM - Journal of Applied Mathematics and Mechanics / Zeitschrift für Angewandte Mathematik und Mechanik, 44*(4–5), 149–158. https://doi.org/10.1002/zamm.19640440402

Truesdell, C., & Noll, W. (2004). *The Non-linear Field Theories of Mechanics*. Berlin: Springer. https://doi.org/10.1007/978-3-662-10388-3

Truesdell, C., & Toupin., R. A. (1960). The Classical Field Theories. In S. Flügge (Ed.), *Handbuch der Physik* (Vol. III/l). Berlin: Springer. https://doi.org/10.1007/978-3-642-45943-6_2

Trunin, I. I. (1965). Criteria of creep strength under complex stress (in Russ.). *Prikladnaya Mekhanika,* **1**(7), 77–83.

Voyiadjis, G. Z., & Kattan, P. I. (2009). A comparative study of damage variables in continuum damage mechanics. *International Journal of Damage Mechanics, 18*(4), 315–340. https://doi.org/10.1177/1056789508097546

Šilhavý, M. (1997). *The Mechanics and Thermodynamics of Continuous Media*. Heidelberg: Springer.

Zheng, Q.-S., & Betten, J. (1995). On the tensor function representations of 2nd-order and 4th-order tensors. Part I. *ZAMM - Journal of Applied Mathematics and Mechanics/Zeitschrift für Angewandte Mathematik und Mechanik, 75*(4), 269–281. https://doi.org/10.1002/zamm.19950750410

Zhilin, P. A. (1976). Mechanics of deformable directed surfaces. *International Journal of Solids and Structures, 12*(9), 635–648. https://doi.org/10.1016/0020-7683(76)90010-X

Życzkowski, M. (1991). Creep in Structures. *International Union of Theoretical and Applied Mechanics. Berlin: Springer.* https://doi.org/10.1007/978-3-642-84455-3

Chapter 2
Anisotropic Plasticity During Non-proportional Loading

Frédéric Barlat and Seong-Yong Yoon

Abstract Modeling of the elasto-plastic behavior of isotropic and anisotropic metals for applications to forming process simulations is discussed. In particular, the macroscopic flow theory of plasticity combined with the concept of isotropic hardening, in which a single monotonic stress-strain curve serves as a reference, is briefly reviewed. Selected non-proportional loading test procedures are described and the main deviations of the material behavior compared to an isotropic hardening response are discussed on the basis of underlying mechanisms of deformation at lower scale. The failure of isotropic hardening to accurately capture the behavior of a material subjected to non-linear strain paths demonstrates the need for more advanced hardening theories. Thus, theories based on kinematic hardening, possibly combined with distortional plasticity concepts, are succinctly reviewed. A pressure-dependent, distortional-only, plasticity approach recently proposed is discussed in more details and its relevance is illustrated with the prediction of stress-strain curves of advanced high strength steel sheets deformed along non-linear strain paths. A finite element (FE) implementation of this distortional plasticity model is outlined, with special attention to the formulation of the stress integration algorithm and the elasto-plastic tangent tensor. Application examples on several steel sheet samples subjected to various strain path changes are given for validation purpose. Simulations are conducted with a stand-alone (SA) code containing the constitutive equations only and a FE code with only one element. Comparisons between these predictions and experimental results demonstrate the accuracy of the model and the excellent performance of the FE implementation. Applications on advanced high strength steel (AHSS) sheet demonstrate why the pressure-dependency in the model is an important feature.

F. Barlat (✉) · S.-Y. Yoon
Graduate Institute of Ferrous and Energy Materials Technology, Pohang University of Science and Technology, Pohang, Republic of Korea
e-mail: f.barlat@postech.ac.kr

H. Altenbach and A. Ganczarski (eds.), *Advanced Theories for Deformation, Damage and Failure in Materials*, CISM International Centre for Mechanical Sciences 605, https://doi.org/10.1007/978-3-031-04354-3_2

2.1 Introduction

Nowadays, numerical simulations are indispensable in many applications involving material processing, product manufacturing, and structural performance assessment of metals, composites and other materials. Among the number of physical and numerical input data necessary to achieve reliable simulations, a proper description of the material behavior, the constitutive model, is very important in this context. For metal processing, manufacturing and performance, an accurate elasto-plastic description of the stress-strain relationship is of highest importance. Foremost, the development of elasto-plastic constitutive equations for metals requires scientific rigor allowing the fulfillment of physical and mathematical constraints. It is important to provide a precise description of the material stress-strain response with reliable interpolation and extrapolation capabilities. In addition, it is also imperative to achieve a robust implementation through algorithms that allow a smooth transfer of the accuracy of the mathematical model to numerical simulations. Finally, in view of its relevance to technological processes, a constitutive model requires enough flexibility to facilitate its use in engineering applications.

The objective of this chapter is to demonstrate how the requirements of rigor and flexibility are combined to deliver accurate models with enough practicality to conduct forming simulations of products with complex shapes such as those illustrated in Fig. 2.1. In Sect. 2.2, elements of solid mechanics are provided to define the fundamental variables necessary to develop constitutive equations. In Sects. 2.3 an 2.4, the isotropic and anisotropic elasto-plastic behavior of metallic materials based on isotropic hardening, is concisely reviewed. In Sect. 2.5, the critical experiments and analyses necessary to characterize the anisotropic hardening behavior occurring during non-proportional loading are summarized. In Sect. 2.6, a few advanced plasticity models that account for these effects are introduced. In Sect. 2.7, a recently proposed pressure-dependent, distortional-only, so-called HAH_{20} plasticity approach is discussed in more details. Finally, in Sect. 2.8, the formulation and the algorithms developed for the implementation of HAH_{20} in finite element codes is explained. Application examples on advanced high strength steel (AHSS) sheets deformed with various non-linear strain paths are shown to validate the HAH_{20} model and its FE implementation.

2.2 Stress States

2.2.1 *Uniaxial*

A tension specimen, with initial gauge length l_0 and cross-section area A_0 is deformed in monotonic loading by the application of a load P under quasi-static conditions. The current length is l and the cross-section area A. The engineering strain is $e = \Delta l / l_0$ where $\Delta l = l - l_0$ and the engineering stress is $\sigma_E = P / A_0$. At low engineering

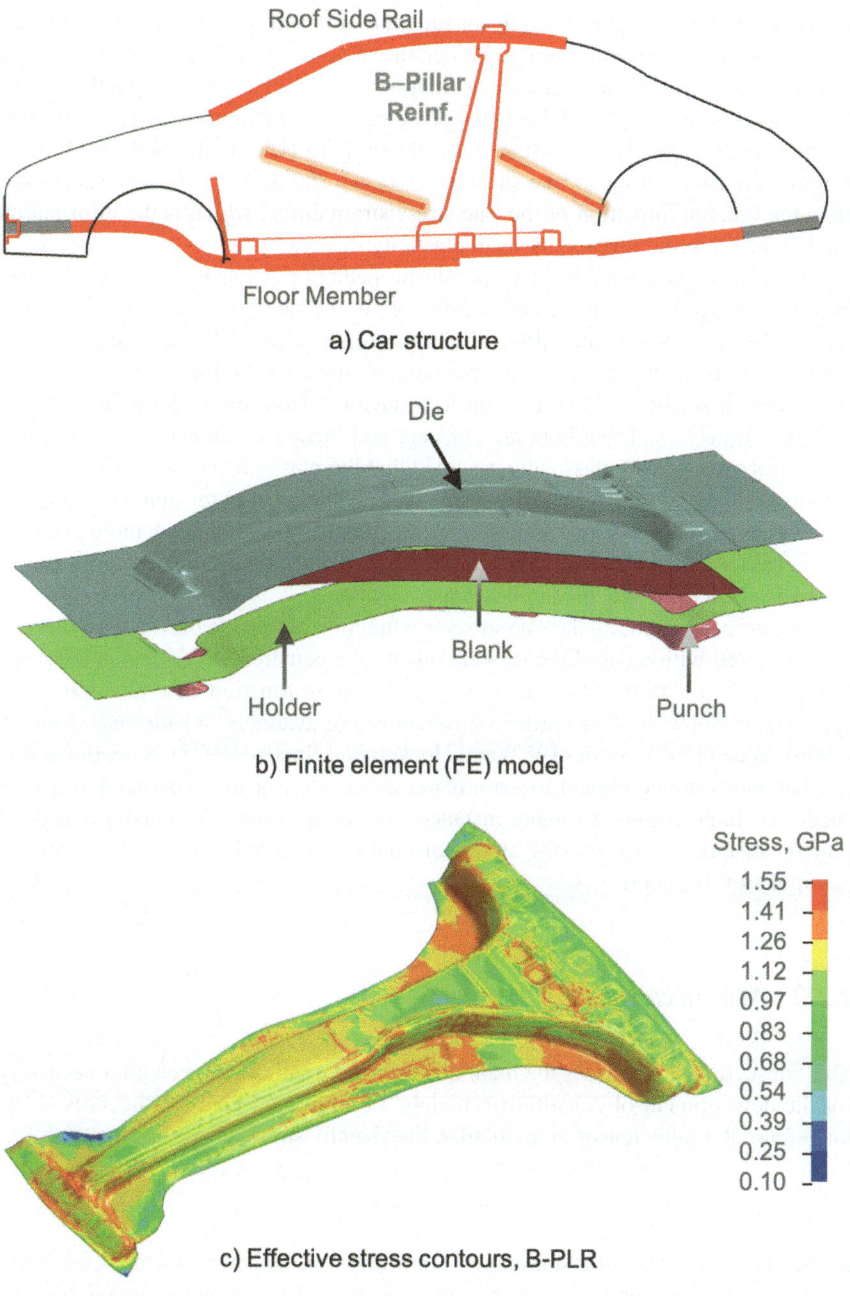

Fig. 2.1 **a** Sketch of: **a** car structure with B-pillar (from POSCO EV concept) reinforcement; **b** finite element (FE) model of B-pillar forming and; **c** FE prediction of formed B-pillar showing levels of residual stresses at process end before unloading

stress, the material response is a linear function of the strain with Young's modulus E as the slope. As the load becomes higher, the stress reaches the yield strength (YS), designated by σ_y, at which plastic deformation initiates. The slope of the flow curve decreases progressively until the engineering stress reaches a maximum value called the tensile strength (TS) or ultimate tensile strength (UTS), and denoted σ_u. The longitudinal true strain and true stress are $\varepsilon = \ln(l/l_0)$ and $\sigma = P/A$, respectively, allowing the determination of the true stress-strain curve, which is the information needed for scientific investigations in plasticity.

Classical assumptions are that the elastic behavior is linear and reversible but plasticity is highly non-linear and irreversible in the thermodynamic sense. For an elastic-plastic material, the additive decomposition allows the subtraction of the elastic strain from the total strain to obtain the plastic contribution $\varepsilon^p = \varepsilon = \varepsilon^t - \varepsilon^e$. Very often, it is assumed that the elastic behavior is isotropic with modulus E and Poisson's ratio ν, and that both are constant and identical in tension and compression. Furthermore, it is classically assumed that the plastic behavior is isotropic and incompressible, and the true stress-strain curve is identical in tension and compression. Of course, all the above assumptions are idealizations of a much more complex real behavior but necessary to develop useful analyses. Additional features maybe added as needed for specific applications.

The curve representing the true stress as a function of the true plastic strain may be approximated with a one-dimensional constitutive equation with a few coefficients such as K and n in the Hollomon (1945) hardening function $\sigma = K\varepsilon^n$. Different approximations with other hardening functions are available, for instance, Ludwik (1909), Voce (1948), Swift (1952) and Hockett and Sherby (1975). Sometimes several equations are combined to get a better description of the experimental data at small and large strains. In many instances, the strain range in uniaxial tension is low, for instance less than 10% strain for dual-phase DP980 steel and an accurate extrapolation is doubtful.

2.2.2 *Multiaxial*

This subsection summarizes the main quantities of continuum mechanics necessary for the development of constitutive models. In multiaxial loading, the stress is no longer a scalar but a tensor, in particular, the Cauchy stress tensor

$$\boldsymbol{\sigma} = \sigma_{pq}\mathbf{e}_p\mathbf{e}_q \tag{2.1}$$

In the above equation, the summation convention on repeated indices is assumed. σ_{pq} is a component and $\mathbf{e}_p\mathbf{e}_q$, often denoted $\mathbf{e}_p \otimes \mathbf{e}_q$, is the open (tensor) product of two base vectors of the set $\mathcal{B} \equiv (\mathbf{e}_1, \mathbf{e}_2, \mathbf{e}_3)$. The stress tensor may be represented in $\mathcal{B}$ as a matrix with nine components.

$$[\sigma] = \begin{bmatrix} \sigma_{11} & \sigma_{12} & \sigma_{13} \\ \sigma_{21} & \sigma_{22} & \sigma_{23} \\ \sigma_{31} & \sigma_{32} & \sigma_{33} \end{bmatrix}_{\mathcal{B}} = \left[\sigma_{pq}\right]_{\mathcal{B}} \tag{2.2}$$

Because of equilibrium considerations, the stress tensor is symmetric, that is, equal to the transpose of itself $\sigma_{pq}^T = \sigma_{qp} = \sigma_{pq}$. It is useful to decompose the stress tensor in the sum of a deviator (**dev**) and a spherical (**sph**) tensor

$$\boldsymbol{\sigma} = \mathbf{dev}(\boldsymbol{\sigma}) + \mathbf{sph}(\boldsymbol{\sigma}) = \boldsymbol{\sigma}' + \sigma_{\mathrm{m}}\mathbf{I} \tag{2.3}$$

where $\sigma_{\mathrm{m}} = \mathrm{tr}\boldsymbol{\sigma}\ /3$ is the mean stress and $\mathbf{I}$ the identity tensor.

Similarly, the deformation is characterized by its rate of change that produces an infinitesimal strain tensor $d\varepsilon = \dot{\varepsilon}dt$ during an infinitesimal time dt. In fact, these variables, $d\varepsilon$ or $\dot{\varepsilon}$ may be used interchangeably in elasto-plasticity. During a small but finite time increment δt, a small deformation tensor may be defined and possibly used for elastic deformations of metals.

The deformation at a continuum point occurs only when the gradient $\nabla\mathbf{v}$ of the velocity field $\mathbf{v} = v_p\mathbf{e}_p$

$$\nabla\mathbf{v} = \frac{\partial v_p}{\partial x_q}\mathbf{e}_p\mathbf{e}_q \tag{2.4}$$

is not nil at all points and does not correspond to an elastic rotation. In order to eliminate rotations, the rate of deformation tensor is defined as $\dot{\boldsymbol{\varepsilon}} = \dot{\varepsilon}_{pq}\mathbf{e}_p\mathbf{e}_q$ with

$$\dot{\varepsilon}_{pq}=\frac{1}{2}\left(\frac{\partial v_p}{\partial x_q}+\frac{\partial v_q}{\partial x_p}\right) \tag{2.5}$$

This expression shows that the rate of deformation tensor is symmetric. It may be represented using a matrix

$$[\dot{\boldsymbol{\varepsilon}}] = \begin{bmatrix} \dot{\varepsilon}_{11} & \dot{\varepsilon}_{12} & \dot{\varepsilon}_{13} \\ \dot{\varepsilon}_{21} & \dot{\varepsilon}_{22} & \dot{\varepsilon}_{23} \\ \dot{\varepsilon}_{31} & \dot{\varepsilon}_{32} & \dot{\varepsilon}_{33} \end{bmatrix}_{\mathcal{B}} = \left[\dot{\varepsilon}_{pq}\right]_{\mathcal{B}} \tag{2.6}$$

The rate of deformation, as any 2nd order tensor, may be decomposed as the sum of a deviator and a spherical tensor, i.e.

$$\dot{\boldsymbol{\varepsilon}} = \mathbf{dev}(\dot{\boldsymbol{\varepsilon}}) + \mathbf{sph}(\dot{\boldsymbol{\varepsilon}}) = \dot{\boldsymbol{\varepsilon}}' + \frac{1}{3}\mathrm{tr}(\dot{\boldsymbol{\varepsilon}}) \tag{2.7}$$

Expressed with the stress and the rate of deformation tensor, the power of deformation per unit volume is

$$\boldsymbol{\sigma} : \dot{\boldsymbol{\varepsilon}} = \sigma_{pq}\dot{\varepsilon}_{pq}=\boldsymbol{\sigma}' : \dot{\boldsymbol{\varepsilon}}' + \frac{1}{3}\mathrm{tr}(\dot{\boldsymbol{\varepsilon}})\,\mathrm{tr}(\boldsymbol{\sigma}) \tag{2.8}$$

where the symbol : corresponds to the double dot product. The field variables reviewed in this subsection are sufficient to develop elasto-plastic constitutive equations for metals.

2.3 Elasto-Plasticity

2.3.1 Elasticity

In order to solve elasto-plastic problems, a constitutive model, i.e., the relationships between stresses and strains, is needed. The elasto-plastic behavior of metals is usually well described with the Cauchy stress tensor $\boldsymbol{\sigma}$ and the rate of deformation tensor $\dot{\boldsymbol{\varepsilon}}$ (or the corresponding increment $d\boldsymbol{\varepsilon}$). For an isotropic material, the relationship between stress and strains must be developed based on tensor invariants. For instance, the simplest form of the generalized Hooke's isotropic law for linear elasticity is expressed in invariant form as

$$\begin{cases} \dot{\boldsymbol{\varepsilon}}^{e\prime} = \dot{\boldsymbol{\sigma}}'/2\mu \\ \text{tr}\ \dot{\boldsymbol{\varepsilon}}^{e} = \text{tr}\ \dot{\boldsymbol{\sigma}}/3\kappa \end{cases} \tag{2.9}$$

where $2\mu = E/(1+\nu)$ and $2\kappa = E/(1-2\nu)$ are the elastic shear and bulk moduli, respectively. This is arguably the simple form of Hooke's law.

2.3.2 Plasticity

Although, engineering materials are all anisotropic because of prior thermo-mechanical or other types of processing, they are assumed to be isotropic in classical flow theory of plasticity. At first approximation, plasticity is considered to occurs without volume change and is not influenced by the hydrostatic pressure. When a material is loaded in the plastic range, unloaded and reloaded in the reverse direction, the flow stress at reloading is assumed to be equal to that just before unloading, which means that the material hardening is isotropic although this is not in accordance with experimental evidence as indicated, for instance, by the Bauschinger effect. Temperature and strain rate have a strong influence on plasticity but these effects are not discussed here. In multiaxial loading, the classical theory of plasticity is defined using a yield condition, a flow rule and a hardening function.

2.3.3 Yield Condition

The yield condition indicates whether the material is elastic or plastic. For an isotropic material, plastic deformation occurs in uniaxial tension if the yield condition,

$$\Phi(\sigma) = \sigma - \sigma_y = 0, \tag{2.10}$$

is satisfied. For multiaxial loading, the yield condition is simply defined as

$$\Phi(\boldsymbol{\sigma}) = \bar{\phi}\left(\sigma_{pq}\right) - \sigma_y = 0 \tag{2.11}$$

Here, the yield function $\overline{\phi}\left(\sigma_{pq}\right)$ is homogeneous of first degree, i.e., an equivalent stress that lumps all the stress tensor components in an isotropic scalar function supposed to behave like an uniaxial tension stress. The yield function may be defined based on different physical interpretations. For instance, Tresca assumes that plasticity occurs when the maximum shear stress reaches a critical value. Von Mises assumes that plasticity initiates when the distortional elastic energy per unit volume reaches a critical value, namely, a given function of the yield stress in uniaxial tension. Many other definitions of the effective stress are possible for an isotropic material as long as they are expressed with stress tensor invariants such as the principal values. For instance, as proposed by Hershey (1954), the following equation

$$\overline{\phi} = \left(\frac{|\sigma_1 - \sigma_2|^a + |\sigma_2 - \sigma_3|^a + |\sigma_3 - \sigma_1|^a}{2}\right)^{\frac{1}{a}} \tag{2.12}$$

is a non-quadratic yield function with the exponent a larger than 1 to guarantee convexity. This formulation has no straight-forward physical interpretation but has some meaning based on crystal plasticity simulations for cubic metals when recommend values of a are 6 or 8 (Logan and Hosford 1980; Barlat et al. 2003a).

Because the hydrostatic stress (pressure) has classically no influence on yielding, the yield condition leads to a cylindrical surface in stress space. It is common to represent the cross-section of the cylindrical yield surface, in the deviatoric (or π-) plane. In this plane, von Mises yield condition is a circle and Tresca a regular hexagon with straight sides and sharp vertices. Hershey yield condition exhibits areas with relatively flat sides and others with rounded vertices that become sharper when the exponent tends towards 1 or ∞. A plane stress cross-section where one of the principal stresses is nil is also a common representation of a yield surface.

Crystal plasticity considerations indicate that a yield condition (yield surface) is likely to exhibit areas with low and high curvatures even for an isotropic polycrystal, very similar to those predicted with Hershey's. In addition, the yield condition should be convex according to the analysis of Bishop and Hill (1951), which restricts the number of possible mathematical expressions for the continuum scale formulation.

2.3.4 Flow Rule

The flow rule indicates the relationship between stresses and strains. Since plasticity is highly non-linear, the plastic rate of deformation (or infinitesimal plastic strain increment) is considered in the constitutive equations. Levy-Mises postulated that it is colinear to the stress deviator. This means that there is no volume change as observed experimentally. Thus,

$$\begin{cases} \dot{\boldsymbol{\varepsilon}}' = \dot{\lambda}\boldsymbol{\sigma}' \\ \text{tr } \dot{\boldsymbol{\varepsilon}} = 0 \end{cases} \tag{2.13}$$

These equations are similar, although not identical, to Hooke's law in elasticity, with $\kappa = 0$. Because of the additive decomposition, the total deviatoric and volumetric rates of deformation are easily obtained from Eqs. (2.9) and (2.13).

More generally, the crystal plasticity theory applied to polycrystals (Bishop and Hill 1951) indicates that the yield condition is a potential for the strain increment, that is

$$\dot{\boldsymbol{\varepsilon}} = \dot{\lambda}\frac{\partial\bar{\phi}}{\partial\boldsymbol{\sigma}} \quad \text{or} \quad \dot{\varepsilon}_{pq} = \dot{\lambda}\frac{\partial\bar{\phi}}{\partial\sigma_{pq}} \tag{2.14}$$

The effective strain is defined as the plastic work equivalent of the effective stress

$$\bar{\phi}\,\dot{\bar{\varepsilon}} = \sigma_{pq}\dot{\varepsilon}_{pq} = \boldsymbol{\sigma}:\dot{\boldsymbol{\varepsilon}} \tag{2.15}$$

Substituting the associated flow rule and using Euler theorem on homogeneous functions of first degree indicates that $\dot{\lambda} = \dot{\bar{\varepsilon}}$.

2.3.5 Hardening

In the previous section, it was shown that the uniaxial tension test allows the determination of a true stress-true strain curve. This flow curve is now used as a reference hardening function $\sigma_r(\bar{\varepsilon})$ for the plastic behavior of the solid in multiaxial stress states by substituting the yield stress by $\sigma_r(\bar{\varepsilon})$ in the yield condition Eq. (2.11), namely

$$\bar{\phi}\left(\sigma_{pq}\right) - \sigma_r(\bar{\varepsilon}) = 0 \tag{2.16}$$

The reference curve is not necessarily a uniaxial flow curve though. In fact, the bulge test produces a balanced biaxial stress state at the pole of the specimen from which a flow curve can be extracted. Such curve is considered as a better alternative compared to uniaxial tension because it is usually defined over a much larger strain range. Therefore, whenever possible, this flow curve should be used.

Since a single parameter $\sigma_r(\bar{\varepsilon})$ controls the size of the yield condition, the yield surface expands only. This is called isotropic hardening. Other forms of hardening are discussed later. Therefore, in classical flow theory of plasticity, it is only necessary to define the yield condition, that is, two scalar functions, namely, the effective stress and the hardening function. A few examples of these functions were mentioned above. Because the material is isotropic, all the previous relationships may be, in principle, defined only with the principal components, i.e., using only stress and strain components with one index, i.e., σ_k. These components may be transformed in a different set of reference vectors if necessary.

2.4 Anisotropy

For elasto-plasticity, the additive decomposition of the total strain facilitates a separate treatment of elasticity and plasticity.

2.4.1 Elasticity

For elasticity, Hooke's law is generalized as

$$\varepsilon_P^e = C_{PQ}^e \sigma_Q \tag{2.17}$$

where C_{PQ}^e are the compliance coefficients. For orthotropic symmetry with nine independent coefficients instead of only two for isotropy, Hooke's law in matrix form writes as

$$\begin{bmatrix} \varepsilon_{11}^e \\ \varepsilon_{22}^e \\ \varepsilon_{33}^e \\ 2\varepsilon_{23}^e \\ 2\varepsilon_{31}^e \\ 2\varepsilon_{12}^e \end{bmatrix}_{\mathcal{B}} = \begin{bmatrix} C_{11}^e & C_{12}^e & C_{13}^e & 0 & 0 & 0 \\ C_{21}^e & C_{22}^e & C_{23}^e & 0 & 0 & 0 \\ C_{31}^e & C_{32}^e & C_{33}^e & 0 & 0 & 0 \\ 0 & 0 & 0 & C_{44}^e & 0 & 0 \\ 0 & 0 & 0 & 0 & C_{55}^e & 0 \\ 0 & 0 & 0 & 0 & 0 & C_{66}^e \end{bmatrix}_{\mathcal{B}} \begin{bmatrix} \sigma_{11} \\ \sigma_{22} \\ \sigma_{33} \\ \sigma_{23} \\ \sigma_{31} \\ \sigma_{12} \end{bmatrix}_{\mathcal{B}} \tag{2.18}$$

Note that this relationship is valid only if written in a set of base vectors attached to the symmetry axes of the material, for instance, rolling ($\mathbf{e}_x$), transverse ($\mathbf{e}_y$) and normal ($\mathbf{e}_z$) directions, respectively, for a sheet. This is a difference with isotropic materials in which Hooke's law may be used for stress and strains expressed in any set of base vectors. The matrix $[\mathbf{C}^e]$ represents the 4th order compliance tensor $\mathbf{C}^e$. Tensor transformations are necessary if the stresses are not known in the set of reference vectors $\mathcal{B} \equiv (\mathbf{e}_x, \mathbf{e}_y, \mathbf{e}_z)$. The indices P and Q vary between 1 and 6. $C_{PQ} = C_{QP}$ are the compliance coefficients. Conversely, the stiffness coefficients S_{PQ} are used for the inverse relationship

$$\sigma_Q = S_{QP}^e \varepsilon_P^e \tag{2.19}$$

2.4.2 Plasticity

Many approaches to plastic anisotropy are available as reviewed by Banabic et al. (2020). The following is the most efficient in the authors' view. The constitutive equations of plasticity for isotropic materials are valid for anisotropic materials as well with an anisotropic effective stress $\bar{\phi}$ but the same reference hardening function. The main difference is that, in order to make the effective stress anisotropic, the principal stresses are replaced by those of one, or several, linearly transformed stress deviator $\tilde{\boldsymbol{\sigma}}'$, i.e., for an orthotropic material

$$\begin{bmatrix} \tilde{\sigma}'_{11} \\ \tilde{\sigma}'_{22} \\ \tilde{\sigma}'_{33} \\ \tilde{\sigma}'_{23} \\ \tilde{\sigma}'_{31} \\ \tilde{\sigma}'_{12} \end{bmatrix}_{\mathcal{B}_x} = \begin{bmatrix} \tilde{c}_{11} & \tilde{c}_{12} & \tilde{c}_{13} & 0 & 0 & 0 \\ \tilde{c}_{21} & \tilde{c}_{22} & \tilde{c}_{23} & 0 & 0 & 0 \\ \tilde{c}_{31} & \tilde{c}_{32} & \tilde{c}_{33} & 0 & 0 & 0 \\ 0 & 0 & 0 & \tilde{c}_{44} & 0 & 0 \\ 0 & 0 & 0 & 0 & \tilde{c}_{55} & 0 \\ 0 & 0 & 0 & 0 & 0 & \tilde{c}_{66} \end{bmatrix}_{\mathcal{B}_x} \begin{bmatrix} \sigma'_{11} \\ \sigma'_{22} \\ \sigma'_{33} \\ \sigma'_{23} \\ \sigma'_{31} \\ \sigma'_{12} \end{bmatrix}_{\mathcal{B}_x} \tag{2.20}$$

Unlike for elasticity, both columns in the above relationship are stresses, i.e., transformed and deviatoric. In this case, the coefficients are not compliance or stiffness but anisotropy. Another main difference is that, as in the elastic case, the linear transformation is only valid when all the quantities are expressed in the reference base vector set attached to the anisotropy axes. The use of tensor transformation formulae is necessary each time the stresses are not expressed in this set of base vectors. The matrix in Eq. (2.20) reflects the symmetries of the material and, therefore, has a similar structure as the compliance (or stiffness) matrix in elasticity, although the symmetry $c_{PQ} = c_{QP}$ is not necessary. However, because the trace of a deviator is equal to zero, all the coefficients are not independent.

Many yield conditions for isotropic materials have been proposed over the last decades. In particular, any isotropic yield condition, written as a function of the principal values become anisotropic with the linear transformation method explained above. Although not demonstrated here, it is even possible to write (Hill, 1948) classical anisotropic yield function under the linear transformation framework.

Non-quadratic models with a sufficient number of coefficients for anisotropic materials have been successful to describe a number of metals with different crystal structures (Barlat and Lian 1989; Barlat et al. 1991, 2003a, 2005). This also includes materials with pressure-independent strength-differential effects such as in Cazacu and Barlat (2001, 2003, 2004). Several yield conditions were proposed with this approach using one or two linear transformations. In particular, the so-called Yld2000-2d (Barlat et al. 2003a) for plane stress and Yld2004-18p (Barlat et al. 2005) for general stress cases are generalizations of Hershey (1954) non-quadratic yield function.

Both Yld2000-2d and Yld2004-18p employ two linear transformations of the stress deviator, namely,

$$\tilde{\boldsymbol{\sigma}}' = \tilde{\mathbf{C}}:\boldsymbol{\sigma}' \quad \text{and} \quad \tilde{\tilde{\boldsymbol{\sigma}}}' = \tilde{\tilde{\mathbf{C}}}:\boldsymbol{\sigma}' \tag{2.21}$$

The plane stress yield condition Yld2000-2d is defined using the principal values $\tilde{\sigma}'_p$ and $\tilde{\tilde{\sigma}}'_q$ of the linearly transformed tensors $\tilde{\boldsymbol{\sigma}}'$ and $\tilde{\tilde{\boldsymbol{\sigma}}}'$ in the effective stress $\bar{\phi}$ as follows

$$\bar{\phi} = \left(\frac{\phi_1 + \phi_2}{2}\right)^{\frac{1}{a}} = \sigma_r \tag{2.22}$$

with

$$\phi_1 = |\tilde{\sigma}_1 - \tilde{\sigma}_2|^a \quad \text{and} \quad \phi_2 = \left|2\tilde{\tilde{\sigma}}_1 + \tilde{\tilde{\sigma}}_2\right|^a + \left|\tilde{\tilde{\sigma}}_1 + 2\tilde{\tilde{\sigma}}_2\right|^a \tag{2.23}$$

A similar approach was adopted for a general stress case with Yld2004-18p for which the yield condition is defined as

$$\bar{\phi} = \left(\frac{1}{4}\sum_{p,q}^{1,3}\left|\tilde{\sigma}_p - \tilde{\tilde{\sigma}}_q\right|^a\right)^{\frac{1}{a}} = \sigma_r \tag{2.24}$$

The coefficients of Yld2000-2d or Yld2004-18p are calculated based on uniaxial flow stresses and r-values in three different directions. The r-value characterizes the plastic flow anisotropy in uniaxial tension. If the longitudinal axis is $\mathbf{e}_1$, the r-value is defined as the ratio of the width-to-thickness strain rate (rate of deformation)

$$r = \frac{\dot{\varepsilon}_{22}}{\dot{\varepsilon}_{33}} \tag{2.25}$$

The r-value is equal to one for an isotropic material, irrespective of the tensile direction. In order to completely defines the yield function coefficients, two additional data are needed, for instance, the balanced biaxial flow stress and strain ratio $r_b = \dot{\varepsilon}_{yy} \,/\, \dot{\varepsilon}_{xx}$.

2.4.3 Hill's Yield Condition

For comparison purpose, the classical Hill (1948) anisotropic yield condition is considered. This is a generalization of the isotropic von Mises formulation with additional coefficients F, G, H, L, M and N

$$\begin{aligned}\bar{\phi}^2 = F\left(\sigma_{xx} - \sigma_{yy}\right)^2 + G\left(\sigma_{zz} - \sigma_{xx}\right)^2 + F\left(\sigma_{yy} - \sigma_{zz}\right)^2 \\ +L\sigma_{yz}^2 + M\sigma_{zx}^2 + M\sigma_{xy}^2 = \sigma_r^2\end{aligned} \tag{2.26}$$

The coefficients of this yield condition are calculated either based on flow stresses or r-values but not both because, for plane stress, only four independent coefficients are available, compared to eight and 16 for Yld2000-2d and Yld2004-18p, respectively.

2.4.4 *Application*

As an application of the anisotropic yield conditions described above, the variations of the normalized flow stress and r-value as a function of the tensile angle with respect to the RD are calculated for a 2090-T3 Al-Li alloy sheet sample. They are represented with two sets of coefficients in Fig. 2.2 for Hill's yield condition and only one set in Fig. 2.3 for Yld2000-2d and Yld2004-18p. Figure 2.2a corresponds to Hill's yield condition with coefficients calculated using four flow stresses (stress-based coefficients) and Fig. 2.2b with three r-values and one flow stress (r-value-based coefficients). The first figure, Fig. 2.2a, indicates that Hill's yield condition cannot predict the anisotropy of the r-value properly. Conversely, Fig. 2.2b demonstrates that Hill's r-value-based yield condition cannot predict the anisotropy of the yield stress. In fact, the difference is particularly large in Fig. 2.2b. For instance, uniaxial tension in the TD (90) shows an experimental normalized flow stress of about 0.9 while the predicted value is about 1.5!

In contrast, with Yld2000-2d and Yld2004-18p in Figs. 2.3a and b, respectively, both predicted flow stresses and r-values are in good agreement with the experiments, especially, with Yld2004-18p. This is because the number of independent coefficients of Yld2004-18p is larger (12 in plane stress case) than those of Yld2000-2d (8) and Hill (4).

2.5 Non-linear Strain Path

2.5.1 *Deformation History*

During the deformation of a solid, each point is subjected to a specific strain history. In monotonic test, the loading is proportional or the strain path is linear. Although technically different, either terminology is assumed to be interchangeable throughout this chapter. A load change may be a mere deviation from the original strain path, or a complete reversal. In the forming of complex geometries, many types of non-linear strain paths, as shown schematically in Fig. 2.4, may occur. For instance, when a sheet blank slides over a sharp tool radius, a bending-unbending sequence is likely to induce load reversal. Wi (2021) represented the simulated deformation histories of several points initially in a flat blank shaped as the B-pillar of a car structure by stamping (Fig. 2.1).

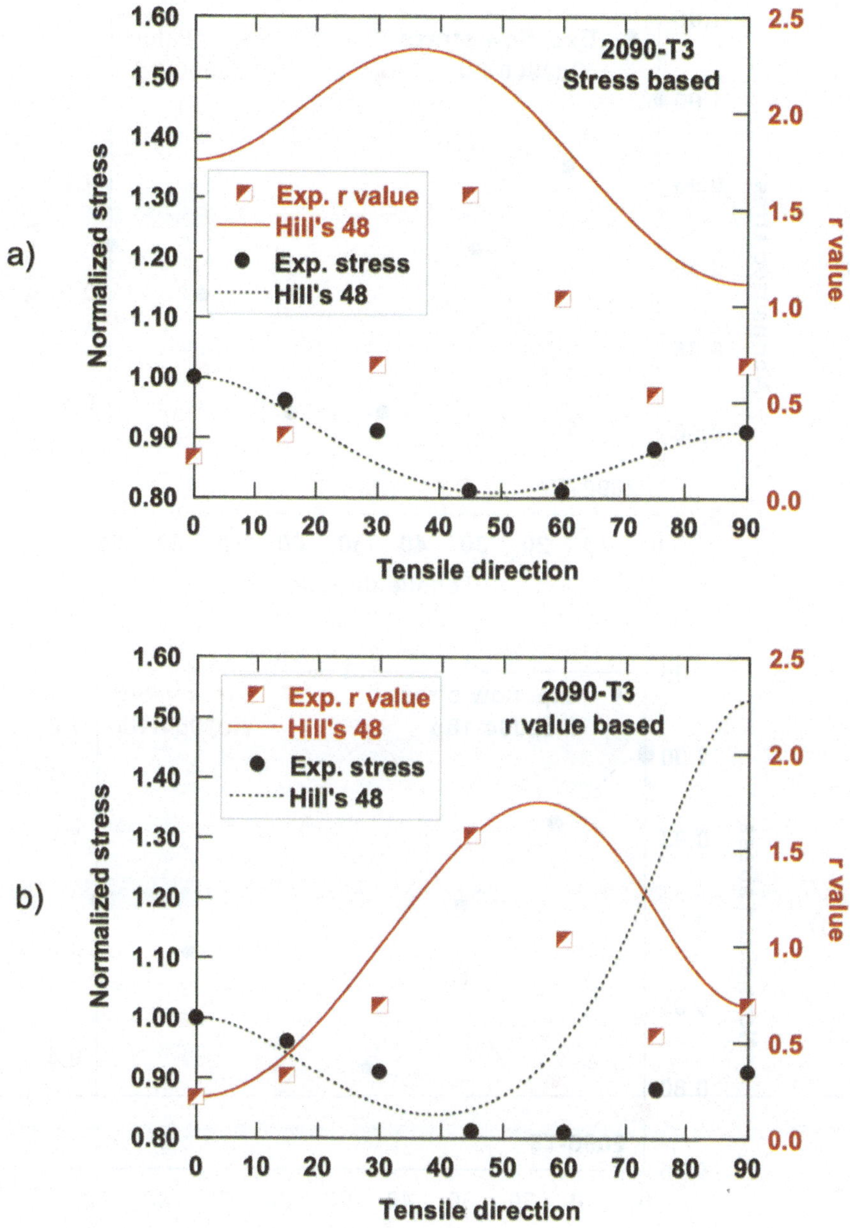

Fig. 2.2 Experimental and predicted (Hill 1948) flow stress (left scale axis) and r-value (right scale axis) anisotropy in 2090-T3 Al-Li sheet: **a** stress-based coefficients, **b** r-value-based coefficients

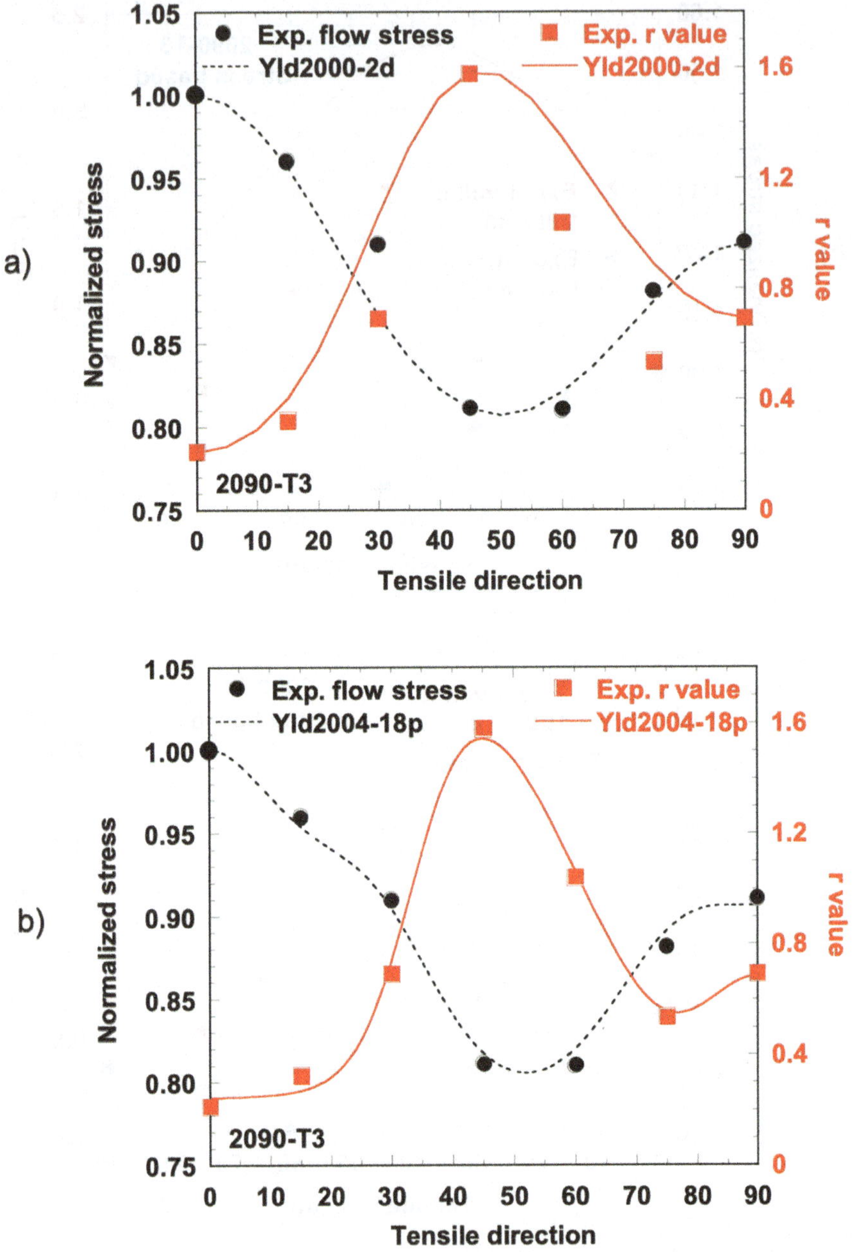

Fig. 2.3 Experimental flow stress (left scale axis) and r-value (right scale axis) anisotropy in 2090-T3 Al-Li sheet and non-quadratic yield function predictions: **a** Yld2000-2d, **b** Yld2004-18p

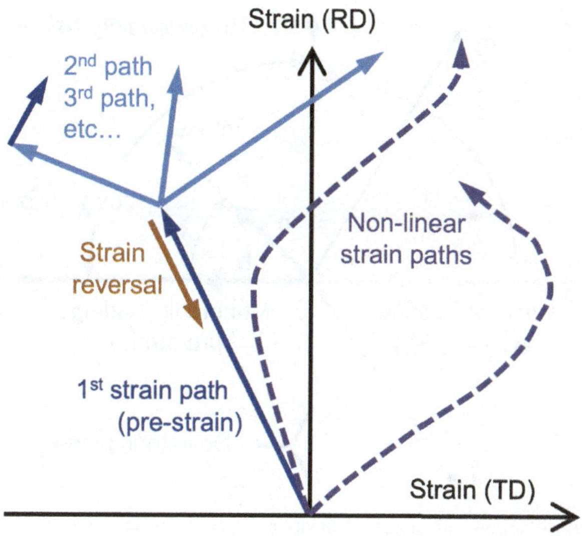

Fig. 2.4 Schematics of continuous or abrupt non-linear strain path changes at one point represented in the RD versus TD strain plane

These deformation histories represented by the major surface strain as a function of the minor surface strain demonstrate that each point is subjected to its own tortuous strain path. It is observed that the trajectories are generally not linear but curved. In addition, there are clear signs of drastic strain path changes, including reversals. In order to produce data for the characterization of strain path changes, reliable experiments are needed. This is a complex task because stresses and strains must be determined independently. Therefore, most of the experiments in the literature consist of a single strain path change, namely, a first path called pre-strain step followed by a second path called call subsequent step. Two-step tests are usually conducted using uniaxial tension or rolling as a pre-strain, because they result in a flat deformed area from which a new specimen can be machined easily. Therefore, the subsequent test may be any test applicable to flat sheets. For reverse loading, tension-compression and forward-reverse shear tests may be conducted with one or multiple reversals applied to a single specimen.

Figure 2.5 represents a single strain path change in the deviatoric stress plane. The pre-strain (pre-load) is marked as monotonic loading $\boldsymbol{\sigma}_1'$.

Load reversal occurs when the sign of the stress is changed $\boldsymbol{\sigma}_1' \rightarrow -\boldsymbol{\sigma}_1'$. Pure cross-loading is described by a new stress deviator orthogonal to the initial stress deviator in the deviatoric plane, i.e., ($\boldsymbol{\sigma}_1' \rightarrow \boldsymbol{\sigma}_C'$). Figure 2.5 indicates that an arbitrary single strain path change can be considered as: (1) monotonic loading (a pre-strain) combined with pure cross-loading for $\chi \leq \pi/2$ or; (2) a reversal combined with pure cross-loading for $\chi \geq \pi/2$.

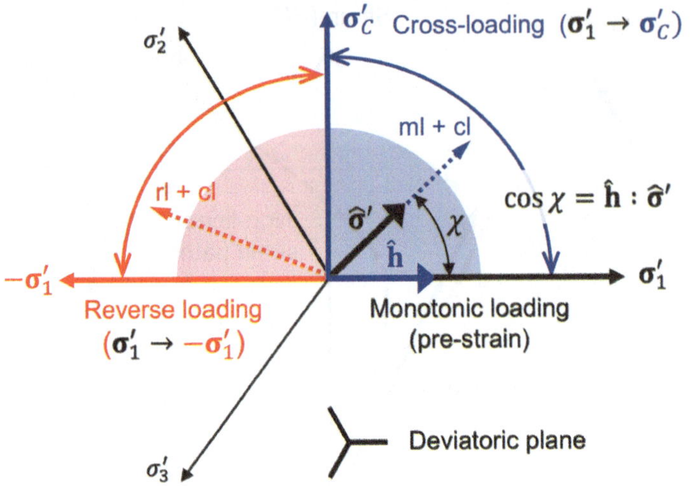

Fig. 2.5 Schematic of strain path change in deviatoric plane between two linear loading segments. Adapted from Lee et al. (2020a)

Classically, tests such as uniaxial tension are conducted under monotonic loading or linear strain path conditions, and numerical simulations are performed under the assumption of isotropic hardening. However, this is usually not what happens in real conditions. One issue with strain path changes is that the stress-strain response is not that of monotonic loading, as schematically illustrated in Fig. 2.6. In this figure, the solid line represents monotonic loading and the dot-and-dash line unloading at a given pre-strain. Reverse loading results in the Bauschinger effect and permanent softening, as indicated by the dash line, and cross-loading in either latent hardening or cross-loading contraction, as indicated by the dotted lines. These effects, in turn, affects the robustness of a forming process and the performance of a product. Therefore, it is important to numerically optimize a process based on the most accurate description of the elasto-plastic behavior of the material.

2.5.2 *Experimental*

Reverse Loading

Experiments capable of producing pure reversals are usually conducted with simple shear, torsion or uniaxial tests. For uniaxial tension-compression of a sheet, a special comb-shaped die, as proposed by Kuwabara (2013), is necessary to prevent buckling but several load reversals may be conducted with a single specimen. For forward-reverse simple shear, a special device designed for advanced high strength steels is mounted on a tension-compression machine (Choi et al. 2015) allowing the

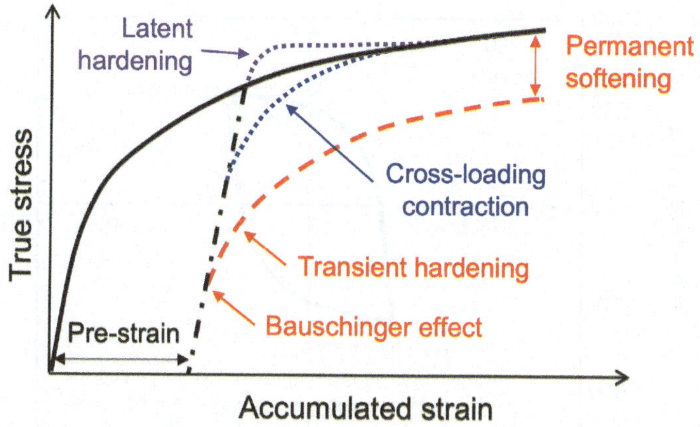

Fig. 2.6 Schematic hardening curves after different strain path changes with monotonic curve (solid line), reverse loading (dash line) and cross-loading (dotted lines). Adapted from Lee et al. (2020b)

testing of several load reversals as well. In this test, deformation occurs in the narrow zone of a specimen with a high length-to-width aspect ratio to minimize the end effects. Although tension-compression tests require strain gauges or extensometers, the deformation in a shear test may be measured using a digital image correlation (DIC) system.

Irrespective of the type of test employed, a common observation with reversal tests is the occurrence of the Bauschinger effect, that is, the reloading yield stress after reversal is much lower than the flow stress at unloading. During plastic reloading, a transient stage of strain hardening with high rate is also observed until it recovers the rate of the monotonic curve. However, the flow stress does usually not recover the level of the monotonic stress-strain curve, a phenomenon called permanent softening. As an example, tension-compression-tension (T-C-T) stress-strain curves with two different strain amplitudes for a dual-phase DP780 steel sheet are shown in Fig. 2.7. The Bauschinger effect is clearly visible and permanent softening is assessed by comparing the stress-strain curves with the isotropic hardening prediction for the case of the 6% strain amplitude. The latter in successive tension and compression segments may surprisingly lead to a high accumulated strain as shown in Fig. 2.7, much higher than that corresponding to uniform elongation of about 10% for DP780 steel sheet. The reason is that, the high strain hardening rate accompanying the Bauschinger effect after each reversal prevents plastic flow localization.

Strength-Differential Effect

This means that the stress-strain curve in compression and tension are different as demonstrated by Spitzig et al. (1975, 1976) with martensitic steels, and illustrated in Fig. 2.8. The SD effect is quantitatively defined as

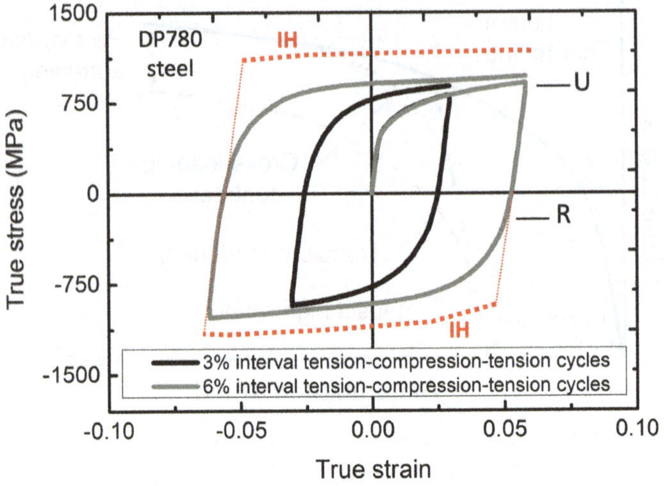

Fig. 2.7 Tension-compression-tension cycles (two strain ranges) on sheet (IH: Isotropic hardening; U: Plastic unloading; R: Plastic reloading). Adapted from Lee et al. (2020b)

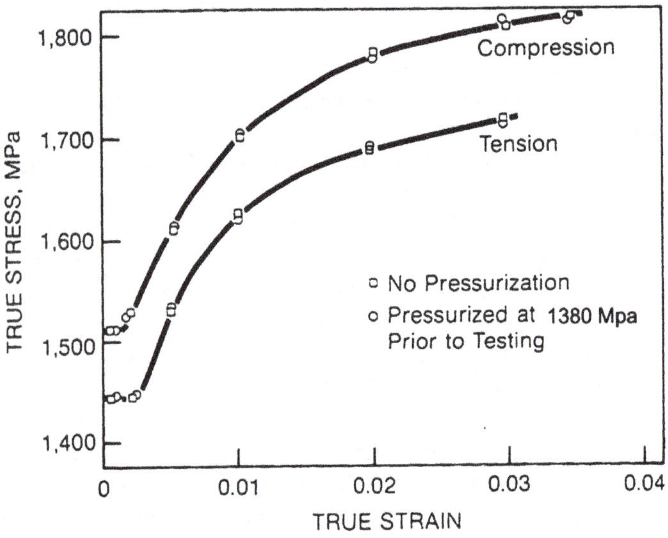

Fig. 2.8 Uniaxial tension and compression flow curves of 4330 high strength steel. Reproduced from Spitzig et al. (1975)

$$SD = 2\,\frac{|\sigma_c| - |\sigma_t|}{|\sigma_c| + |\sigma_t|} \tag{2.27}$$

where σ_c and σ_t are the compressive and tensile flow stresses at the same accumulated strain. For steels $|\sigma_c| \geq |\sigma_t|$ and, for strengths over 1 GPa, the SD effect is significant (Spitzig et al. 1975, 1976; Spitzig and Richmond 1984). Although this effect is not due to a strain path change, it likely alters the amount of apparent permanent softening that occurs during tension-compression testing, particularly for high strength materials in which the SD effect is high.

Cross-Loading

Cross-loading is obtained from two loading segments, a pre-strain with reloading in a different direction or in a different stress state (Zaman et al. 2018; Wi 2021). In order to produce cross-loading, two-step uniaxial tension tests are relatively easy to perform. The first loading step is conducted in a given direction and the subsequent step may be conducted in an arbitrary direction making an angle with the pre-strain. Depending on this angle, the characteristics of the strain path changes are different, for instance,

- 0°, monotonic loading
- ∼55°, pure cross-loading
- 90°, cross- and reverse loading combination

The angle of 55° above corresponds to a case for which the stress deviators of the two segments are orthogonal, as schematically illustrated by Fig. 2.5. After the pre-strain, smaller size specimens extracted from the gauge section (Fig. 2.9) are subsequently deformed leading to the stress-strain curves in Fig. 2.10 in which two types of material responses are observed.

In Fig. 2.10a, an extra deep drawing quality steel (EDDQ) sheet sample is pre-strained about 9% in uniaxial tension in the RD and reloaded in the DD (45° from RD), leading to a strain path change close to pure cross-loading. After reloading, the material is mostly elastic until it plastically overshoots the monotonic stress-strain curve due to latent hardening. Figure 2.10b shows a dual-phase steel (DP780) sheet sample deformed with the same sequence as EDDQ with two pre-strains of about 4 and 9%. This figure indicates that, after reloading, yielding occurs at a stress lower than that expected assuming isotropic hardening. This suggests that during the pre-strain, the yield surface contracts in a direction orthogonal to the loading direction in stress space, a phenomenon called cross-loading contraction. Unlike latent hardening, the contraction occurs during the monotonic pre-strain tensile loading but its effect on the stress-strain curve is observed only after unloading and reloading in a different direction.

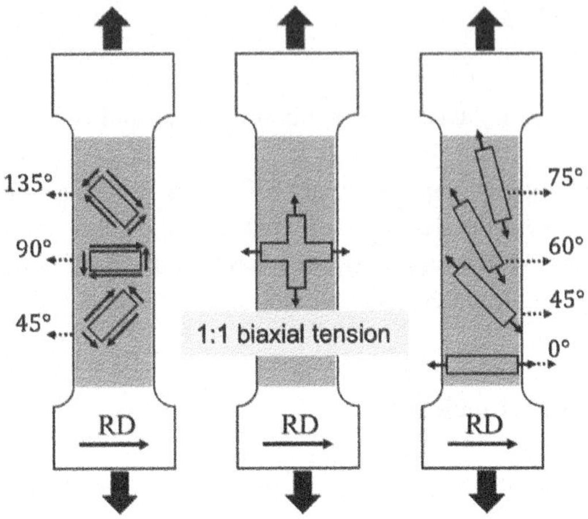

Fig. 2.9 Non-linear strain path experiment using a flat pre-strained sheet. Adapted from Wi (2021)

Double Cross-Loading

Double cross-loading with three linear strain path segments and two changes have been conducted as well (Vincze et al. 2013). However, the experiments are difficult to conduct and not often performed. This information is useful, though, to develop models because in order to predict the material response during an arbitrary deformation history, a minimum of two changes is necessary. For instance, after cross-loading contraction or latent hardening, many assumptions are possible for further evolution of the yield condition.

Yield Surface

Yield surface measurements such as for biaxial stress states are also of interest to assess strain path effects on strain hardening. The yield surfaces of a number of materials were measured by Tozawa in the late 1960 and 1970s using biaxial compression tests on stacked sheet specimen. As already mentioned above, the influence of the hydrostatic stress for low to medium strength metals may be neglected as indicated by Spitzig and Richmond (1984) allowing the determination of a yield surface in the entire deviatoric plane. Biaxial compression tests of stacked sheet specimens, schematically depicted in Fig. 2.11, allow the measurement of the entire yield locus by assuming yielding is independent of pressure.

Some yield surface results measured at different offset strains (strain after reloading) for brass pre-strained in rolling are reproduced from Tozawa (1978) in Fig. 2.11 demonstrating that the yield condition is definitely subjected to distortion during the pre-strain although part of the change could also be attributed to a translation. In Tozawa experiments for an Al-stabilized steel pre-strained in uniaxial tension (Fig. 2.12a), specimens were prepared for subsequent deformation in biaxial com-

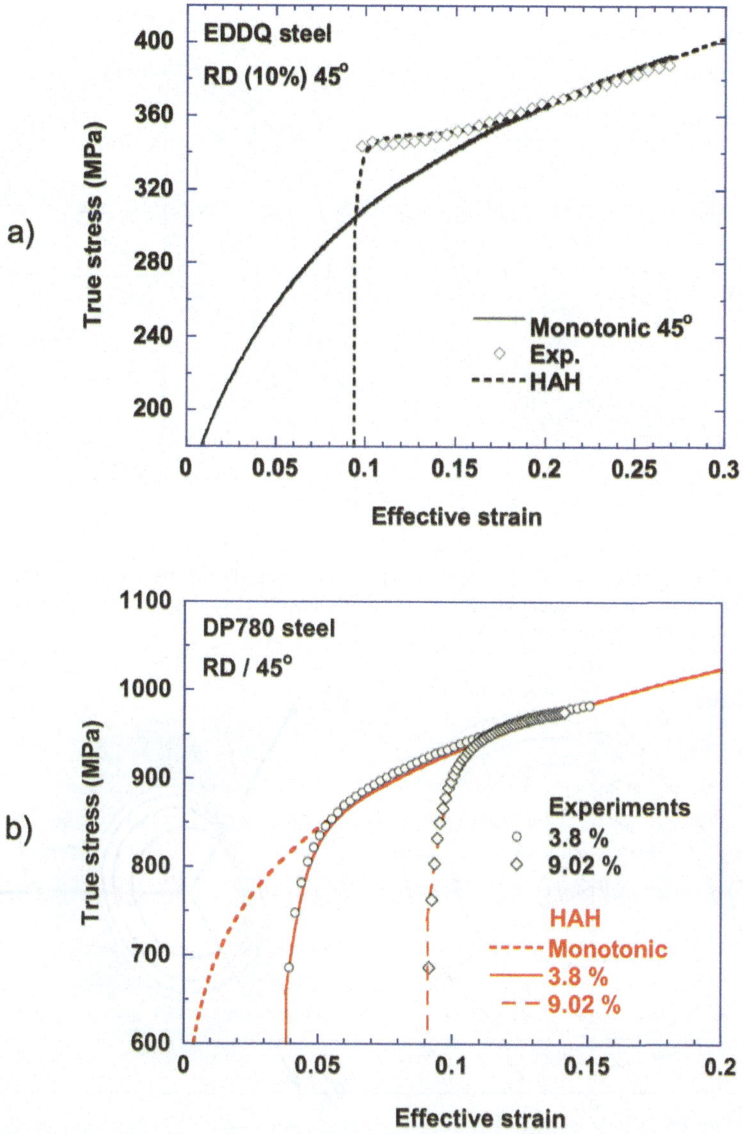

Fig. 2.10 Non-linear strain path examples consisting of two-step tension tests in different directions (RD followed by DD); **a** EDDQ steel, **b** DP870 steel. Reproduced from Ha et al. (2013)

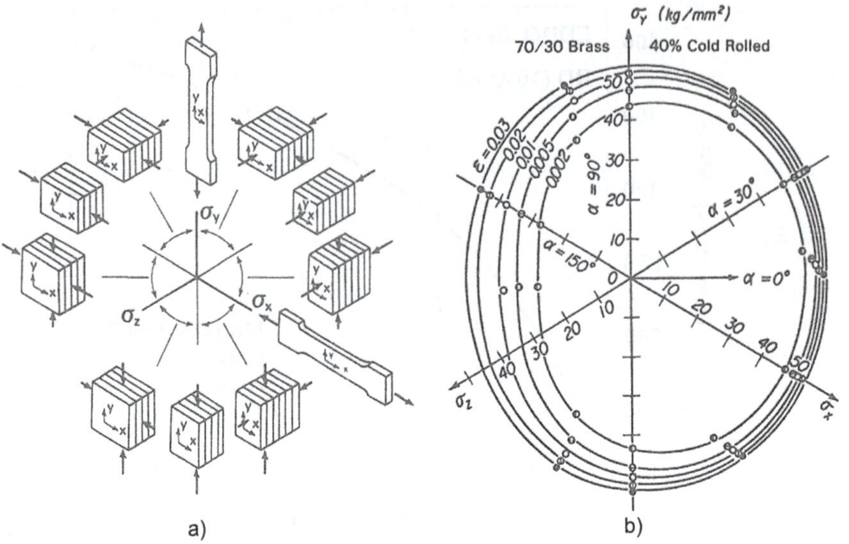

Fig. 2.11 **a** Schematic of biaxial compression testing of stacked sheet specimen for yield surface measurement and; **b** subsequent yield surfaces at different offset strains for 70/30 brass after 40% cold rolling reduction. Reproduced from Tozawa (1978)

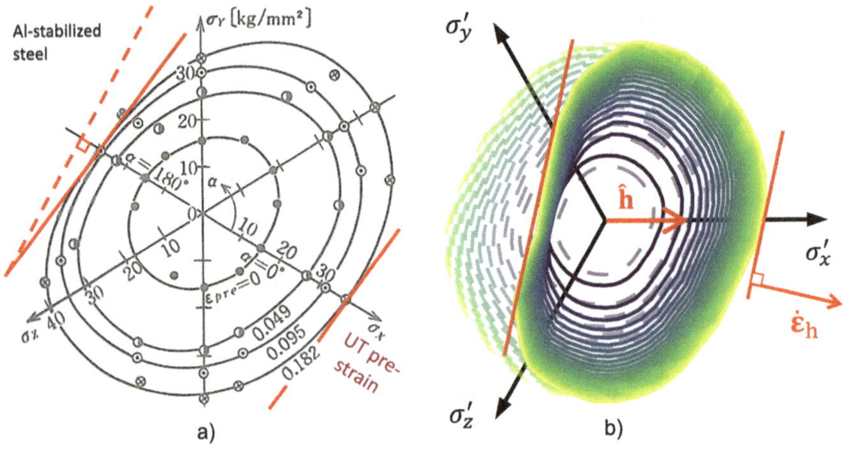

Fig. 2.12 Cross-loading in: **a** Low carbon steel, experimental results adapted from Tozawa (1978) and; **b** crystal plasticity combined with dislocation interaction modeling results adapted from Jeong et al. (2017)

pression to probe the yield surface with a 0.2% offset strain. This figure shows that a distortion and, possibly, a translation of the yield surface occur. In addition, the tangent of the yield surface at the loading point and its opposite (straight solid lines in Fig. 2.12a) are essentially parallel.

2.5.3 Interpretation

Each material is unique and develop specific mechanisms giving rise to the deviation of the flow stress with respect to the monotonic curve. However, it is useful to provide a few generic mechanisms that help in the understanding of the material behavior and the development of macroscopic plasticity models.

Dislocation Interactions

It is well recognized that during plastic deformation, the dislocation density increases, which is the main mechanism of strain hardening. An estimate of the strength of a crystal or a polycrystal is provided by Taylor equation

$$\sigma = \sigma_0 + \varsigma \mu b M \sqrt{\rho} \tag{2.28}$$

where σ_0 is the lattice friction, $\varsigma \approx 1$, μ the shear modulus, b the Bürgers vector, M a polycrystal factor close to 3 and ρ the dislocation density. Therefore, the stress-strain response of a material may fluctuate, depending on the dislocation interactions and the resulting density. For instance, during forward loading, ρ and σ increase due to dislocation accumulation at obstacles. During reversal, ρ and σ tends to decreases due to the recombination of dislocations. This leads to a permanent softening as shown by Hasegawa et al. (1975); Hasegawa and Yakou (1980) because this annihilation process is irreversible. During cross-loading, the slip may require a higher stress to overcome the dislocations on the previously active slip systems.

In single phase materials, during the pre-strain, dislocations create cell patterns that are sometimes clearly visible with transmission electron microscopy (TEM). After reloading, the dislocation microstructure induced by the pre-strain becomes unstable. It appears to dissolve, giving place to a new dislocation structure typical of the new strain path. For instance, for a commercially pure Al alloy, relatively equiaxed cells are produced in uniaxial tension and parallel dislocation walls in simple shear (Lopes et al. 2003; Barlat et al. 2003b). For low carbon steel, uniaxial tension and simple shear produce walls (Rauch et al. 2011).

Depending on the material and the subsequent load, different effects may occur in the microstructure. The interaction of mobile dislocations with the walls or cells built during the pre-strain strongly affects the stability of the flow curve. For EDDQ steel reloaded at an effective strain of about 0.09, the reloading flow stress overshoots the monotonic stress-strain curve as shown by Ha et al. (2013) in Fig. 2.10. This behavior is the result of latent hardening, i.e. dislocations on the new slip systems glide against dislocation walls or other structures formed during the pre-strain and require a higher stress to overcome.

The results of a modified crystal plasticity model based on Schmid's law and directional dislocation density accumulation (Kitayama et al. 2013) indicates that distortion is a good assumption for ensuing reversal, Fig. 2.12b. In addition, as observed in this figure, the tangent of the yield surface at the loading point and its opposite are parallel, which is consistent with the experiment of Fig. 2.12a. Although it is not possible to exclude kinematic hardening, all these observations justify the development of a purely distortional hardening model.

In multi-phase materials, such as dual-phase steel that consists of a soft ferrite matrix with a distribution of hard martensite particles, or aged-hardening Al alloys with a dispersion of non-shearable precipitates, dislocation pile-ups form at numerous obstacles. This leads to the occurrence of a back-stress. For instance, for DP780 steel sheet reloaded at effective strains of about 0.04 and 0.09 in Fig. 2.10, yielding occurs at a lower level compared with the flow stress just before unloading. This is possibly due to the development of a short range residual stress field due to the presence of hard martensitic particles. Therefore, for both pre-strains cross-loading contraction occurs during the first deformation step (Ha et al. 2013).

Again, each material is specific, and many other phenomena are likely to occur. However, generally speaking, circumstances that lead to a disruption of the dislocation behavior with a drop in dislocation density or a sudden change resulting in more obstacles to overcome lead to hardening fluctuations.

Dislocation-Pressure Interaction

Despite the fact that several explanations of the SD effect were examined, a main reason for steels and likely many metals is the influence of pressure on the yield condition. In the literature and textbooks, plasticity is assumed to be independent of pressure. This is generally a good assumption for low and medium strength materials, but not for high strength materials such as AHSS. Despite this pressure influence, it has been observed that the volume change is negligible (Table 2.1), suggesting a non-associated flow rule. Therefore, as inferred by Richmond and Spitzig (1980), the yield condition (YC) and the plastic potential (PP), respectively, may be written as follows

$$\begin{cases} \mathrm{YC}: \ \bar{\Sigma}(\boldsymbol{\sigma}) = \bar{\sigma}(\boldsymbol{\sigma}') + \alpha c(\bar{\varepsilon})\,\mathrm{tr}(\boldsymbol{\sigma}) = c(\bar{\varepsilon}) \\ \mathrm{PP}: \ \bar{\sigma}(\boldsymbol{\sigma}') = c(\bar{\varepsilon}) \end{cases} \tag{2.29}$$

In these equations, $\bar{\Sigma}$ and $\bar{\sigma}$ are the effective stresses of the pressure-dependent and pressure-independent models, respectively. $c(\bar{\varepsilon})$ is a parameter that represents the material strength (stress-strain curve) and α the pressure coefficient. Numerical values of the coefficients c at yield and α are listed in Table 2.1 for a number of steels and a commercially pure aluminum alloy. As an example, a load-reversal (tension-compression) curve for a DP780 steel sheet sample with several cycles and increasing amplitude is shown in Fig. 2.13. The curve starts to saturate in tension and compression. The saturation in compression is marked with a straight dash line. Using a symmetry with respect to the origin of the curve, another dash line is represented in the tension side. However, the tensile side of the stress cycle does not saturate

Table 2.1 Constitutive coefficients c and α with predicted and experimental S-D effect and volumetric strain for γ a number of pressure-sensitive materials. Adapted from Spitzig and Richmond (1984)

Material	c	α	S-D(%)		γ	
	(MPa)	(Tpa^{-1})	Observed	Predicted	Observed	Predicted*
Fe single crystal	35	23	–	0.2	–	0.002
Spheroidized 1045 steel	470	19	–	1.8	–	0.02
Fe-Ti-Mn single crystal	570	23	–	2.6	–	0.04
HY 80 steel	606	13	1.5	1.6	0.002	0.02
Maraging steel (unaged)	1005	17	3.5	3.4	0.001	0.05
4310 steel	1066	23	5.5	5.0	0.004	0.07
4330 steel	1480	20	6.0	5.8	0.004	0.07
Maraging steel (aged)	1833	20	7.0	7.4	0.007	0.11
1100 aluminium	25	56	0	0.3	0.0005	0.004

*Associated flow rule

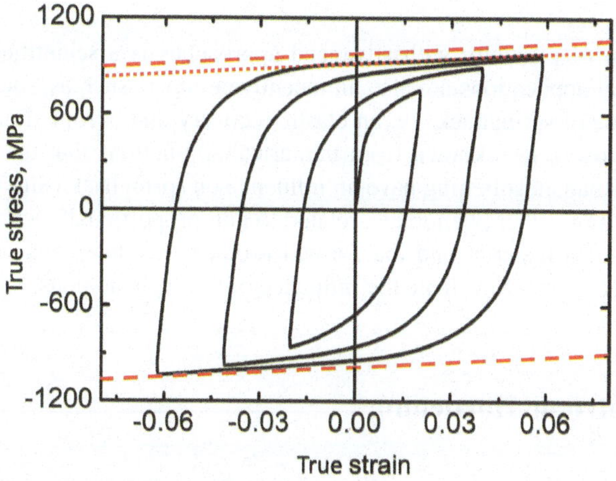

Fig. 2.13 Stress-strain curve for DP780 steel sheet sample with several sequential cycles of increasing amplitude. Dash lines are symmetric with respect to origin of figure. Adapted from Lee et al. (2020b)

towards this dash line but to a lower level represented by the dotted line. This effect is due to the influence of the hydrostatic stress on the flow stress.

Dislocation motion is controlled by the elastic properties of a material. Since it is known that pressure influences the elastic shear modulus (Jung 1981), it is therefore expected to affect plastic properties as confirmed by Spitzig et al. (1975, 1976). Several authors have investigated the influence of pressure on the shear modulus using dislocation approaches. For instance, the pressure coefficient α of the continuum theory described above may be approximated as (Jung 1981)

$$\alpha = \frac{2}{3\mu_0}\frac{d\mu}{dp} \tag{2.30}$$

leading to $\alpha = 20$ and 50 TPa^{-1} for steel and aluminum, respectively. These results are close to the experimental values listed in Table 2.1. Another estimate of the coefficient α for Al was calculated by Bulatov et al. (1999) using molecular static based on the embedded atom method. The authors found $\alpha = 48$ and 53 TPa^{-1} for screw and edge dislocations, respectively, which is a reasonable estimate compared with the experimental results listed in Table 2.1.

2.5.4 *Discussion*

The extension of plasticity with advanced approaches is a scientific endeavor in itself. Practical applications are also an incentive to do so such as, for instance, the important issue of springback, i.e. an elastic recovery that occurs after the forming loads are removed. It is known from numerical simulations that the Bauschinger effect and permanent softening have an influence on springback (Sun and Wagoner 2013). In addition, other phenomena related to non-proportional loading may affect the response of a material and the formed product. For these reasons, advanced models are necessary to evaluate the influence of these phenomena.

2.6 Anisotropic Hardening

The plasticity models assuming isotropic hardening are reasonable for cubic metals in monotonic loading. They are also safe to use in many applications involving strain path changes where high accuracy is not required. However, experimental evidence discussed in the previous section show that the yield surface does not merely expands during pre-strain due to the Bauschinger effect and other phenomena. After reversal, the reloading flow stress Y_R is lower than the unloading stress U_R as shown in Fig. 2.7. Associated phenomena, in particular, a high hardening rate after plastic reloading occurs, as well as permanent softening. Thus, models different from isotropic hardening are necessary to describe this complex behavior.

For uniaxial loading with reversals, the stress-strain curve is compared with the isotropic hardening response in order to assess the effect of a pre-strain. For multiaxial loading, the yield surface must be compared with an isotropic expansion. Several scenarios have been proposed to model the deviation from isotropic hardening, such as kinematic (yield surface translation), combined isotropic-kinematic, or distortional. Since the measurement of a yield surface is a very tedious work, one hardening scenario is often assumed arbitrarily, and the model is calibrated using a few experimental data.

2.6.1 Kinematic Hardening

Typically, kinematic hardening, with a back-stress as a tensorial state variable, has been used to capture the Bauschinger effect. The so-called back-stress **X** defines the center of the yield surface that translates, thus breaking the center of symmetry in stress space. In the earlier models, the evolution of the back-stress led to linear hardening (Prager 1949; Ziegler 1959) but were not really applicable to materials such as metals. Later, Armstrong and Frederick (1966); Chaboche (1977), proposed non-linear kinematic hardening models in which the yield condition of an initially isotropic material is defined as

$$\bar{\sigma}\left(\boldsymbol{\sigma}-\mathbf{X}\right)-\sigma_r\left(\bar{\varepsilon}\right)=0 \tag{2.31}$$

where $\bar{\sigma}$ is the Mises effective stress, σ_r is the reference hardening function, and $\bar{\varepsilon}$ the Mises effective strain. It is customary to write the hardening function as the sum of the yield stress and $R\left(\bar{\varepsilon}\right)$, the added hardening stress due to plastic deformation, i.e.

$$\sigma_r\left(\bar{\varepsilon}\right)=\sigma_y+R\left(\bar{\varepsilon}\right) \tag{2.32}$$

The non-linear back-stress evolution equation takes the form

$$\dot{\mathbf{X}}=\frac{2}{3}C\dot{\boldsymbol{\varepsilon}}-\gamma\mathbf{X}\dot{\bar{\varepsilon}} \tag{2.33}$$

where C and γ are two constant coefficients. Finally, the classical associated flow rule, indicating the normality between the strain rate and the yield surface

$$\dot{\boldsymbol{\varepsilon}}=\dot{\lambda}\frac{\partial\bar{\sigma}(\boldsymbol{\sigma}-\mathbf{X})}{\partial\boldsymbol{\sigma}} \tag{2.34}$$

is generally recognized as a good approximation.

Kinematic hardening has been mostly developed to capture the material behavior after load reversal, which includes all the phenomena associated with the Bauschinger effect. In particular, this model provides good results for cyclic plasticity. Chaboche (2008) provides an excellent review on the topic.

Such a kinematic hardening approach was refined by Teodosiu and Hu (1998) to account for additional effects that occur during arbitrary strain path changes such as latent hardening. The yield condition is defined as

$$\bar{\sigma} = \sqrt{(\boldsymbol{\sigma} - \mathbf{X}) : \mathbf{M} : (\boldsymbol{\sigma} - \mathbf{X})} = \sigma_y + f\,|\mathbf{S}| \tag{2.35}$$

in which **M** contains the anisotropy coefficients describing the initial state of the materials. The hardening curve is a function f of the norm of a fourth order tensor **S**, i.e., $|\mathbf{S}| = \sqrt{\mathbf{S} : \mathbf{S}}$, representing the strength of planar persistent dislocation structures. **S** depends on two terms, which involves the 2nd order tensor **P** associated with the polarity of persistent dislocation sheets. This model was developed based on macroscopic and microscopic observations of mild steel deformed along non-linear strain paths. Since mechanical and electron microscopy experiments were conducted, the model is based on the physical interpretation of dislocation patterning during plastic deformation.

A detailed description of the model is out of the scope of the present article. Simply speaking, plastic deformation develops dislocation structures during the pre-strain and the subsequent material response depends on the reloading direction with respect to the orientation of these structures. The model is particularly well suited to low carbon steels as it provides a detailed description of mechanisms observed in this material (Peeters et al. 2001a, b, 2002). In addition to **M**, the model includes 10 coefficients that require calibration.

2.6.2 *Mutli-surface Kinematic Hardening*

A multi-surface yield condition was proposed by Mróz (1967). Each surface translates inside the surface of immediate larger size and is associated with a given constant hardening modulus. Therefore, any stress-strain curve is approximated by a sequence of linear hardening segments. Two-surface models were proposed by Krieg (1975); Dafalias and Popov (1976), and lead to a smooth stress-strain response except, of course, when the strain path changes abruptly.

Among the numerous kinematic hardening models proposed in the literature, the formulation proposed by Yoshida and Uemori (2002), denoted YU, has received much attention in the forming community. The YU model is based on two surfaces with the inner surface called the yield surface and the outer the bounding surface. The yield surface (YF) translates while the bounding surface (BS) translates and expands

$$\begin{cases} \text{YS}: \; f = \bar{\phi}(\boldsymbol{\sigma} - \boldsymbol{\alpha}) - Y = 0 \\ \text{BS}: \; F = \Phi(\boldsymbol{\sigma} - \boldsymbol{\beta}) - (B + R) = 0 \end{cases} \tag{2.36}$$

where Y is the yield stress. B and $B + R$ are the initial and current sizes or the bounding surface. Two back-stresses $\boldsymbol{\alpha}$ and $\boldsymbol{\beta}$ with distinct evolution equations are necessary to describe the material behavior. Strain hardening is affected by the two rates of translation (YS and BS), by the BS expansion and also by the contact between yield and bounding surfaces.

2.6.3 *Distortional Hardening*

In order to improve the predictive accuracy, kinematic hardening has been supplemented by yield surface distortion such as in Ortiz and Popov (1983); Voyiadjis and Foroozesh (1990); François (2001); Feigenbaum and Dafalias (2007) to name a few. Kinematic, distortional and other forms of hardening have also been proposed by Kurtyka and Życzkowski (1996). More recently, a distortional plasticity approach was introduced in which the yield surface expands and distorts, but does not translate in stress space (Barlat et al. 2011). The most recent version of this model is described in more details in the next section.

2.7 Homogeneous Anisotropic Hardening Model

2.7.1 *Background*

The distortional-only plasticity model family, denoted HAH in its original version, was developed to describe the material behavior subjected to non-linear strain paths. A revised version of this approach, called HAH_{20}, was published recently (Barlat et al. 2020) to improve the predictive capability of the previous version, HAH_{14} (Barlat et al. 2014), and to introduce the influence of the hydrostatic stress as previously discussed. Relevant articles regarding the HAH family of distortional plasticity models may be found in He et al. (2013); Qin et al. (2017a, b, 2018, 2019) and many recent articles by the authors.

HAH_{20} is introduced semi-quantitatively in this section with more emphasis on the description of the yield condition. Although the state variable evolution equations are listed in the Appendix, the reader is referred to the original article (Barlat et al. 2020) for a better and practical understanding of the model. Based on experimental evidences and crystal plasticity simulation results presented in previous sections, it is assumed that three modes of deformation are sufficient to describe the material behavior, namely,

- Monotonic loading
- Stress reversal
- Cross-loading

Any arbitrary strain path change is assumed to be a combination of these three basic modes as represented in Fig. 2.5.

2.7.2 *Yield Condition*

The HAH distortional plasticity framework (Barlat et al. 2011, 2014, 2020) is similar as that explained in Sect. 2.4 on anisotropic plasticity, with an anisotropic yield condition, the associated flow rule, and the plastic work-based effective strain. The main difference is that, for isotropic hardening, only one state variable σ_r describes the state of the material. In the HAH model, additional state variables x_k are introduced to describe the distortion. The evolution equations of these variables are described in the corresponding article (Barlat et al. 2020) and also summarized in the Appendix.

The HAH yield condition is

$$\bar{\sigma}\left(\boldsymbol{\sigma}', x_k\right) = \left(\bar{\xi}^q + \phi_h\right)^{\frac{1}{q}} = \sigma_R\left(\bar{\varepsilon}\right) \tag{2.37}$$

where, $\bar{\sigma}$ is the HAH$_{20}$ effective stress and $\bar{\xi}$ the yield function similar to $\bar{\phi}$ in Eq. (2.16). The reference flow stress σ_R is similar to σ_r but account for permanent softening, q is a coefficient, and ϕ_h controls the reverse loading distortion. Two state variables g_C and g_L are available to transform $\bar{\phi}$ into $\bar{\xi}$ in order to account for the cross-loading effects. Figure 2.14a shows the yield surface under isotropic hardening represented by the dark solid line, which does not change during deformation because the yield surface is normalized. In this figure, the yield surface with the dash line corresponds to cross-loading contraction that occurs during the pre-strain. As a result, if cross-loading is applied, yielding occurs earlier compared to isotropic hardening. The dotted line in Fig. 2.14b represents the yield surface after a small amount of deformation during reloading, indicating that it overshoots the isotropic hardening curve, a phenomenon called latent hardening.

The truncation function ϕ_h is defined as

$$\phi_h = f_-^q\left|\hat{\mathbf{h}}:\boldsymbol{\sigma}' - \left|\hat{\mathbf{h}}:\boldsymbol{\sigma}'\right|\right|^q + f_+^q\left|\hat{\mathbf{h}}:\boldsymbol{\sigma}' + \left|\hat{\mathbf{h}}:\boldsymbol{\sigma}'\right|\right|^q \tag{2.38}$$

or, alternatively

$$\phi_\varepsilon = f_-^q\left|\frac{\hat{\boldsymbol{\varepsilon}}_h:\boldsymbol{\sigma}' - \left|\hat{\boldsymbol{\varepsilon}}_h:\boldsymbol{\sigma}'\right|}{\hat{\boldsymbol{\varepsilon}}_h:\hat{\mathbf{h}}}\right|^q + f_+^q\left|\frac{\hat{\boldsymbol{\varepsilon}}_h:\boldsymbol{\sigma}' + \left|\hat{\boldsymbol{\varepsilon}}_h:\boldsymbol{\sigma}'\right|}{\hat{\boldsymbol{\varepsilon}}_h:\hat{\mathbf{h}}}\right|^q \tag{2.39}$$

The first definition, ϕ_h, is simpler to use and appropriate for isotropic materials. The second, ϕ_ε, is more suitable for anisotropic materials because of the property regarding the slopes of the yield surface in Fig. 2.12. In Eqs. (2.38) and (2.39), f_- and f_+ are two scalar state variables controlling the Bauschinger effect. $\hat{\mathbf{h}}$ is a

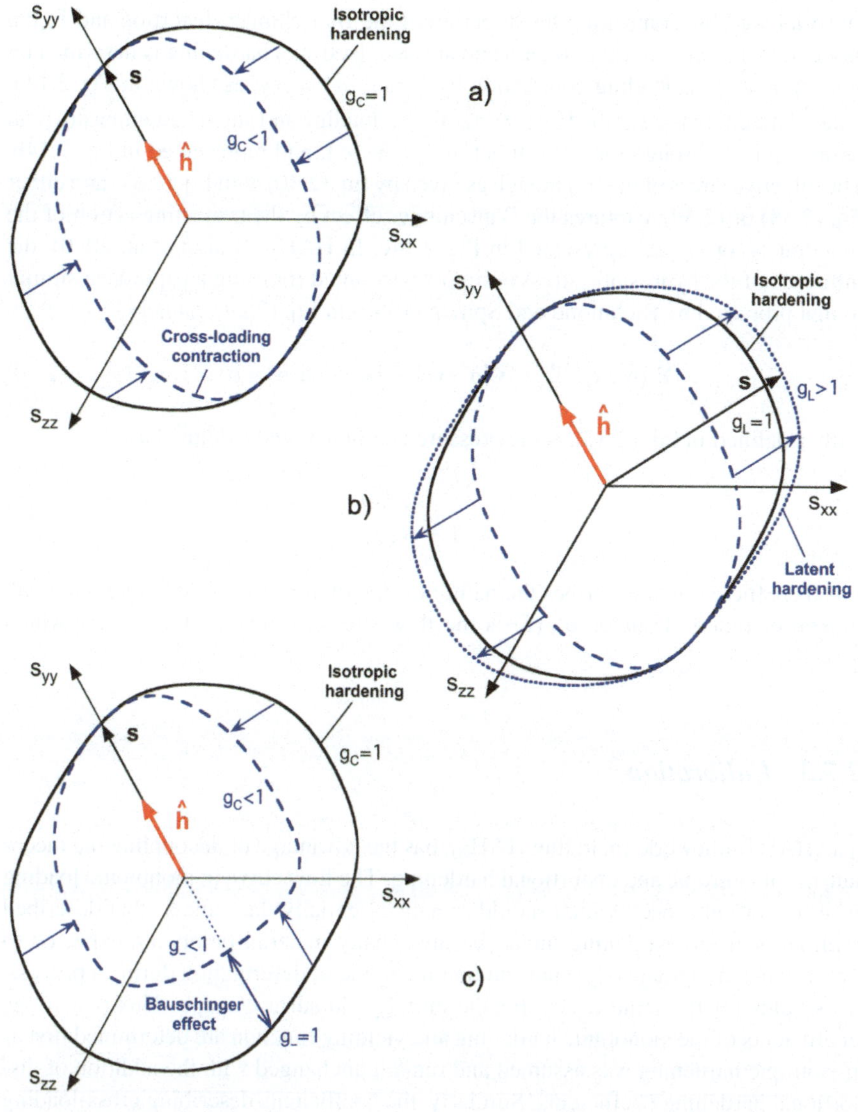

Fig. 2.14 Schematic yield condition in deviatoric plane, assuming isotropic hardening with; **a** cross-loading contraction; **b** latent hardening; **c** distortional hardening HAH. Adapted from Barlat et al. (2020)

tensorial variable controlling the direction of the Bauschinger distortion and $\hat{\boldsymbol{\varepsilon}}_h$ the associated direction normal to the yield surface. Isotropic hardening is described by $\bar{\phi} = \sigma_R$ and cross-loading contraction by $\bar{\xi} = f\left(\bar{\phi}\right) = \sigma_R$, as shown in Fig. 2.14a. Latent hardening is described by $\bar{\xi} = \sigma_R$. Cross-loading and latent hardening are both lumped into $\bar{\xi}$ through the yield function $\bar{\xi} = \psi\left(\bar{\phi}\right)$, and represented in Fig. 2.14b. The effective stress of the full model, as given by Eq. (2.40) with Eq. (2.37) and either Eq. (2.38) or (2.39), captures the Bauschinger effect by the truncating action of the function ϕ_h or ϕ_ε, as represented in Fig. 2.14c. In HAH$_{20}$ (Barlat et al. 2020), the influence of the hydrostatic stress is simply accounted for using an approach similar to that proposed by Richmond and Spitzig (1980) in Eq. (2.29), namely

$$\Sigma\left(\boldsymbol{\sigma}', x_k, \bar{\varepsilon}\right) = \bar{\sigma}\left(\boldsymbol{\sigma}', x_k\right) + \alpha c\left(\bar{\varepsilon}\right) \operatorname{tr} \boldsymbol{\sigma} = c\left(\bar{\varepsilon}\right) \tag{2.40}$$

with $\bar{\sigma}$ defined in Eq. (2.37), α the pressure coefficient and c defined as

$$c\left(\bar{\varepsilon}\right) = \frac{\sigma_R\left(\bar{\varepsilon}\right)}{1 - \alpha\zeta\sigma_R\left(\bar{\varepsilon}\right)} \tag{2.41}$$

ζ is a coefficient that describes the nature of the reference stress state, i.e. uniaxial, biaxial or others. Usually, $\sigma_R\left(\bar{\varepsilon}\right)$ is the flow stress in uniaxial tension for which $\zeta = 1$.

2.7.3 *Calibration*

The HAH framework, including HAH$_{20}$, has the advantage of decoupling the mechanisms of isotropic and distortional hardening. The flow curve in monotonic loading is an important aspect, which should remain essentially the same as that described with an isotropic hardening model because many material points are expected to deform in a monotonic or pseudo-monotonic mode of deformation during a process. This holds for the same reason for the initial yield surface shape. This is why the coefficients of the monotonic hardening and yielding function are determined first as if isotropic hardening was assumed and remain unchanged with the addition of distortional hardening coefficients. Similarly, the coefficients describing cross-loading contraction are decoupled from those ascribed to reverse loading. Thus the calibration procedure allows the separation of four different groups of coefficients that can be optimized separately.

2.8 Finite Element Implementation

2.8.1 Stress Integration Algorithm

The stress tensor should be determined at each time step through a numerical (piece-wise) integration in the finite element analysis due to the non-linearity of material plasticity. The stress increment should be such that the yield condition is satisfied at each individual simulation time step. In addition, the integrated stress must satisfy the global equilibrium in an implicit analysis. This is achieved by numerically solving a series of partial differential equations, which are established according to Euler's method. As a root-finding technique, the Newton-Raphson method has been widely used to solve the system of equations. This process is known as stress integration (update) algorithm. Generally, it is used to reproduce the response of a theoretical model within the finite element simulation program through an appropriate channel, namely, the user-defined material subroutine (UMAT). The stress integration algorithm mainly returns (1) the integrated stress ($\boldsymbol{\sigma}_{n+1}$), (2) the effective (accumulated) plastic strain ($\bar{\varepsilon}_{n+1}$), (3) the updated hardening state variables ($\mathbf{r}_{n+1}$) and, (4) the elasto-plastic tangent modulus ($\mathbf{C}^{ep}_{n+1}$) from the variables determined at the previous time step. The overall procedure of an implicit finite element analysis is schematically represented in Fig. 2.15.

One of the most popular stress integration methods is the closest point projection method (CPPM), which is fully-implicit (Simo and Hughes 2006). It determines the stress update direction with securing the minimum plastic work path (Chung and Richmond 1993). The CPPM leads to high accuracy for a wide range of time increments (Cardoso and Yoon 2009). However, it requires the calculation of the second-order derivative of the yield function, which increases the computational cost and implementation complexity for an arbitrary plasticity theory. The cutting plane method (CTM), which explicitly provides the stress increment, requires only the first-order derivative of the yield function (Ortiz and Simo 1986). The principles of both stress update methods are graphically compared in Fig. 2.16. For the purpose of alleviating the difficulty of calculating the derivative for plasticity models, a numerical differentiation technique, namely, the finite difference method, is introduced. The details are explained in the Appendix and successfully applied to anisotropic plasticity models (Choi and Yoon 2019; Yoon et al. 2020). A Newton-Raphson technique where the derivatives are numerically determined is called the Secant method. Figures 2.16a and c compares the Newton-Raphson and Secant methods during the stress updating process. Note that the current Newton-Raphson method (local) is independent of the Newton-Raphson method (global) for solving the global equilibrium within a static deformation problem, namely, an implicit analysis.

For technical reasons, the deformation state of a material point for the next time increment is first set by means of the elastically determined trial stress (elastic predictor $\Delta\boldsymbol{\sigma}^{ep}$). When the trial stress exceeds the flow stress defined in terms of the effective strain, the material point is subjected to plastic deformation. In order to determine the stress increment based on a specific constitutive model, the algorithm

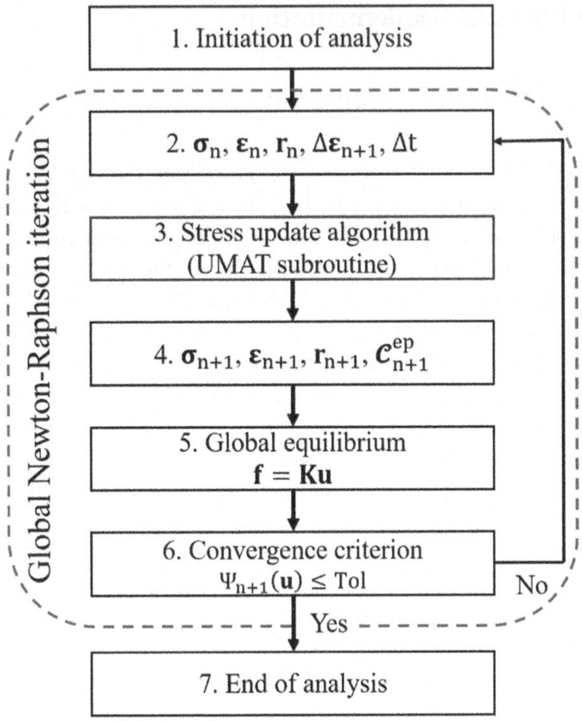

Fig. 2.15 Flowchart of implicit finite element analysis

should project the trial stress onto a strain-hardened yield surface by acquiring the plastic corrector ($\Delta\boldsymbol{\sigma}^{pc}$). This is called the elastic predictor-plastic corrector method. The trial stress ($\boldsymbol{\sigma}^{tr}$) is obtained by splitting the stress increment into an elastic predictor and a plastic corrector. The known variables are the fourth-order elastic stiffness tensor ($\mathbf{C}^e$) and the strain increment for the current time step ($\Delta\boldsymbol{\varepsilon}_{n+1}$). Therefore, the stress integration is achieved by calculating the unknown plastic corrector using the following system of equations

$$\boldsymbol{\sigma}_{n+1} = \boldsymbol{\sigma}_n + \Delta\boldsymbol{\sigma} = \boldsymbol{\sigma}_n + \Delta\boldsymbol{\sigma}^{ep} + \Delta\boldsymbol{\sigma}^{pc} \tag{2.42a}$$

$$\boldsymbol{\sigma}^{tr} = \boldsymbol{\sigma}_n + \Delta\boldsymbol{\sigma}^{ep} = \boldsymbol{\sigma}_n + \mathbf{C}^e : \Delta\boldsymbol{\varepsilon}_{n+1} \tag{2.42b}$$

Another primary variable necessary to determine is the increment of effective plastic strain ($\Delta\bar{\varepsilon}_{n+1}$), which represents the amount of plastic deformation. More generally, when the hardening state variables for anisotropic hardening models are formulated in terms of the effective strain, $\Delta\bar{\varepsilon}_{n+1}$, they are explicitly integrated regardless of the numerical scheme of the stress update algorithm

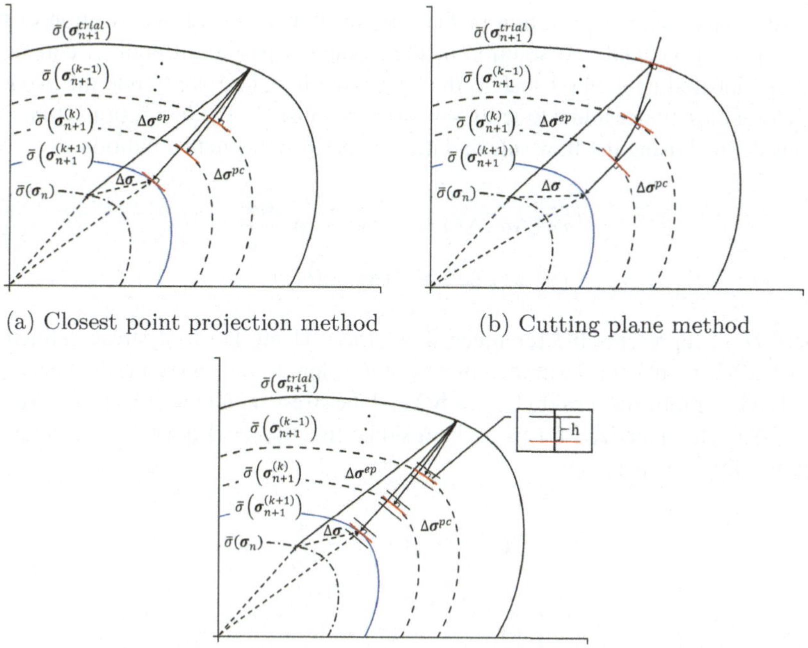

(a) Closest point projection method

(b) Cutting plane method

(c) Closest point projection method (Secant method)

Fig. 2.16 Graphical representation of stress update directions for the **a** Closest point projection, **b** Cutting plane, and **c** Closest point projection based on Secant method. Adapted from Simo and Hughes (2006)

$$\mathbf{r}_{n+1} = \mathbf{r}_n + \Delta\mathbf{r} = \mathbf{r}_n + \frac{\Delta\mathbf{r}}{\Delta\bar{\varepsilon}}\Delta\bar{\varepsilon} \tag{2.43}$$

The associated flow rule allows the replacement of the plastic multiplier with the effective strain increment by taking advantage of the plastic work equivalence ($\Delta W^p = \bar{\sigma}\Delta\bar{\varepsilon} = \boldsymbol{\sigma} : \Delta\boldsymbol{\varepsilon}^p$) where $\bar{\sigma}(\boldsymbol{\sigma})$ is the effective stress. Using Euler's theorem on homogeneous functions of first degree ($\bar{\sigma} = \boldsymbol{\sigma} : \partial\bar{\sigma}/\partial\boldsymbol{\sigma}$), it is simple to show that

$$\Delta\bar{\varepsilon} = \frac{\boldsymbol{\sigma} : \Delta\boldsymbol{\varepsilon}^p}{\bar{\sigma}} = \Delta\lambda\frac{\boldsymbol{\sigma} : \frac{\partial\bar{\sigma}}{\partial\boldsymbol{\sigma}}}{\bar{\sigma}} = \Delta\lambda \tag{2.44}$$

Note that this relationship is only valid for the associated flow rule where the plastic potential and the yield function are identical.

The two primary variables ($\Delta\boldsymbol{\sigma}^{pc}$ and $\Delta\bar{\varepsilon}$) are solved with the update algorithm. For simplicity, the plastic corrector ($\Delta\boldsymbol{\sigma}^{pc}$) is hereafter denoted as the stress increment ($\Delta\boldsymbol{\sigma}$) within the stress update algorithm formulation. In this subsection, the closest point projection method (CPPM), which is formulated on the basis of the Euler backward method (fully-implicit scheme), is briefly introduced. Euler's method consists of three steps, i.e., (1) definition of the residual functions; (2) linearization of the residual functions and; (3) solution of the linearized partial differential equations using a root-finding method such as the Secant method. In order to find the two primary variables, two residual functions must be established. The first residual function (ϕ_1) is defined using the flow rule and the second (ϕ_2) the yield condition

$$\begin{aligned} \phi_1(\Delta\boldsymbol{\sigma}, \Delta\bar{\varepsilon}) &= -\Delta\boldsymbol{\varepsilon}^p + \Delta\bar{\varepsilon}\frac{\partial\bar{\sigma}}{\partial\boldsymbol{\sigma}} \\ \phi_2(\Delta\boldsymbol{\sigma}, \Delta\bar{\varepsilon}) &= \bar{\sigma}(\boldsymbol{\sigma}) - H(\bar{\varepsilon}) \end{aligned} \tag{2.45}$$

where $H(\bar{\varepsilon})$ represents the reference flow stress. Using the trial stress defined in Eq. (2.42b), the plastic strain increment is equivalent to the tensor product between the elastic compliance tensor ($\mathbf{C}^{-e} = \mathbf{S}^e$) and the stress increment ($\Delta\boldsymbol{\sigma}$), i.e., $\Delta\boldsymbol{\varepsilon}^p = -\mathbf{S}^e : \Delta\boldsymbol{\sigma}$. The linearized forms of the residual functions in Eq. (2.45) are obtained with the Taylor expansion

$$\begin{aligned} \phi_1^{k+1} &= \phi_1^k + \mathcal{H}^{-1} : \delta(\Delta\boldsymbol{\sigma})_{k+1} + \frac{\partial\bar{\sigma}}{\partial\boldsymbol{\sigma}}\delta(\Delta\bar{\varepsilon})_{k+1} \approx 0 \\ \phi_2^{k+1} &= \phi_2^k + \frac{\partial\bar{\sigma}}{\partial\boldsymbol{\sigma}} : \delta(\Delta\boldsymbol{\sigma})_{k+1} - \frac{\partial H}{\partial\bar{\varepsilon}}\delta(\Delta\bar{\varepsilon})_{k+1} \approx 0 \end{aligned} \tag{2.46}$$

In the above equations, k is the index for the iteration of the Secant method. Note that the initial estimate of the stress tensor is the trial stress, i.e., $\boldsymbol{\sigma}_{k=0} = \boldsymbol{\sigma}^{tr}$. $\mathcal{H}$ represents the Hessian matrix and contains the second-order derivative of the yield function.

$$\mathcal{H}^{-1} = \frac{\partial\phi_1^k}{\partial\boldsymbol{\sigma}} = \mathbf{S}^e + \frac{\partial^2\bar{\sigma}}{\partial\boldsymbol{\sigma}\partial\boldsymbol{\sigma}} \tag{2.47}$$

Since the algorithm recursively updates the primary variables, the solutions are expressed in the form of the step size of the variables, i.e., $\delta(\Delta\boldsymbol{\sigma})_{k+1}$ and $\delta(\Delta\bar{\varepsilon})_{k+1}$. After rearrangement of Eq. (2.46), the partial differential equation problem is completely linearized and takes the form

$$\begin{Bmatrix} \phi_1^k \\ \phi_2^k \end{Bmatrix} = \begin{bmatrix} -\mathcal{H}^{-1} & -\frac{\partial\bar{\sigma}}{\partial\boldsymbol{\sigma}} \\ -\frac{\partial\bar{\sigma}}{\partial\boldsymbol{\sigma}} & \frac{\partial H}{\partial\bar{\varepsilon}} \end{bmatrix} \begin{Bmatrix} \delta(\Delta\boldsymbol{\sigma})_{k+1} \\ \delta(\Delta\bar{\varepsilon})_{k+1} \end{Bmatrix} = \mathcal{J} \begin{Bmatrix} \delta(\Delta\boldsymbol{\sigma})_{k+1} \\ \delta(\Delta\bar{\varepsilon})_{k+1} \end{Bmatrix} \tag{2.48}$$

Here, the step sizes of the primary variables (the solutions) are determined through the inversion of the Jacobian matrix ($\mathcal{J}$) by means of the lower-upper (LU) decomposition. The step sizes are used to integrate the associated primary variables. Particularly, the step size of the effective strain ($\delta(\Delta\bar{\varepsilon})_{k+1}$) is also used to update all the state variables ($\mathbf{r}_{k+1}$).

$$\boldsymbol{\sigma}_{k+1} = \boldsymbol{\sigma}_k + \delta(\Delta\boldsymbol{\sigma})_{k+1} \tag{2.49a}$$

$$\bar{\varepsilon}_{k+1} = \bar{\varepsilon}_k + \delta(\Delta\bar{\varepsilon})_{k+1} \tag{2.49b}$$

$$\mathbf{r}_{k+1} = \mathbf{r}_k + \frac{d\mathbf{r}}{d\bar{\varepsilon}}\delta(\Delta\bar{\varepsilon})_{k+1} \tag{2.49c}$$

The convergence criteria of the algorithm are validated with the updated variables, i.e.,

$$\left\|\phi_1^{k+1}\right\|_2 \leq \text{Tol}_1 \text{ and } \left|\phi_2^{k+1}\right| \leq \text{Tol}_2 \tag{2.50}$$

The L^2-norm of the first multidimensional residual value (ϕ_1^{k+1}) is used to check the convergence condition. In the present work, both tolerance values (Tol_1 and Tol_2) are set to 10^{-10}. When the convergence criteria are satisfied, the primary variables are integrated as follows.

$$\boldsymbol{\sigma}_{n+1} = \boldsymbol{\sigma}_{n+1} = \boldsymbol{\sigma}_{k=0} + \Delta\boldsymbol{\sigma}_{k+1} \tag{2.51a}$$

$$\bar{\varepsilon}_{n+1} = \bar{\varepsilon}_{k+1} = \bar{\varepsilon}_{k=0} + \Delta\bar{\varepsilon}_{k+1} \tag{2.51b}$$

$$\mathbf{r}_{n+1} = \mathbf{r}_{k+1} = \mathbf{r}_{k=0} + \Delta\mathbf{r}_{k+1} \tag{2.51c}$$

The current CPPM formulation is available for anisotropic yield functions and hardening models when the state variables are functions of the effective strain. The stress update algorithms for HAH_{20} with pressure-independent and dependent plasticity approaches are detailed in Yoon et al. (2020, 2022), respectively.

2.8.2 *Elasto-Plastic Tangent Modulus*

The tangent modulus represents the constitutive relationship between the stress and strain increments. In elastic deformation, for instance, the tangent modulus corresponds to the elastic stiffness tensor ($\mathbf{C}^e$), which is assumed to be constant. Meanwhile, the tangent modulus in elasto-plasticity is determined at every FE-simulation time step due to the non-linearity of the constitutive relationships. Two representative elasto-plastic tangent modulus tensors have been widely used, namely, (1) Continuum tangent modulus by Ortiz and Simo (1986) and; (2) Consistent tangent modulus by Simo and Hughes (2006). According to the quantitative study of Starman et al. (2014), the consistent tangent modulus leads to a better convergence in implicit FEA. Since the consistent tangent modulus demands the second-order derivative of the yield function, it is more suitable for CPPM than CTM. The continuum tangent modulus is sometimes utilized in direct method for stress integration (Dunne and Petrinic 2005). Nevertheless, for an iterative (indirect) stress integration process based Euler's method such as CTM and CPPM, the tangent modulus does not influence the accuracy of the FE-simulation results but only the convergence rate (Anandarajah 2010). The derivation of the consistent tangent modulus is detailed in this subsection.

The consistency condition is obtained from the material time derivative of the yield condition

$$\dot{F} = \frac{\partial \bar{\sigma}}{\partial \boldsymbol{\sigma}} : \dot{\boldsymbol{\sigma}} - \frac{\partial H}{\partial \bar{\varepsilon}} \dot{\bar{\varepsilon}} = 0 \tag{2.52}$$

The stress rate can be approximated from the generalized Hooke's law

$$\dot{\boldsymbol{\sigma}} = \mathbf{C}^e : \left\{ \dot{\boldsymbol{\varepsilon}} - \dot{\boldsymbol{\varepsilon}}^p \right\} \tag{2.53}$$

The strain rate is determined from the material time derivative of the associated flow rule

$$\dot{\boldsymbol{\varepsilon}}^p = \Delta \bar{\varepsilon} \frac{\partial^2 \bar{\sigma}}{\partial \boldsymbol{\sigma} \partial \boldsymbol{\sigma}} : \dot{\boldsymbol{\sigma}} + \frac{\partial \bar{\sigma}}{\partial \boldsymbol{\sigma}} \dot{\bar{\varepsilon}} \tag{2.54}$$

The stress rate can be rewritten by inserting Eq. (2.54) into Eq. (2.53).

$$\dot{\boldsymbol{\sigma}} = \mathcal{H} : \dot{\boldsymbol{\varepsilon}} - \mathcal{H} : \frac{\partial \bar{\sigma}}{\partial \boldsymbol{\sigma}} \dot{\bar{\varepsilon}} \tag{2.55}$$

The effective strain rate can be obtained from Eqs. (2.52) and (2.55).

$$\dot{\bar{\varepsilon}} = \frac{\frac{\partial \bar{\sigma}}{\partial \boldsymbol{\sigma}} : \mathcal{H} : \dot{\boldsymbol{\varepsilon}}}{\frac{\partial \bar{\sigma}}{\partial \boldsymbol{\sigma}} : \mathcal{H} : \frac{\partial \bar{\sigma}}{\partial \boldsymbol{\sigma}} + \frac{\partial H}{\partial \bar{\varepsilon}}} \tag{2.56}$$

Finally, the non-linear constitutive relationship for the elasto-plastic deformation is expressed by means of the consistent tangent modulus ($\mathbf{C}^{ep}$)

$$\dot{\boldsymbol{\sigma}} = \mathbf{C}^{ep} : \dot{\boldsymbol{\varepsilon}} = \left\{ \mathcal{H} - \frac{\mathcal{H} : \frac{\partial \bar{\sigma}}{\partial \boldsymbol{\sigma}} \otimes \mathcal{H} : \frac{\partial \bar{\sigma}}{\partial \boldsymbol{\sigma}}}{\frac{\partial \bar{\sigma}}{\partial \boldsymbol{\sigma}} : \mathcal{H} : \frac{\partial \bar{\sigma}}{\partial \boldsymbol{\sigma}} + \frac{\partial H}{\partial \bar{\varepsilon}}} \right\} : \dot{\boldsymbol{\varepsilon}} \tag{2.57}$$

2.8.3 FE-Application: Non-proportional Loading

In this subsection, the FE-implemented HAH_{20} model (FE-HAH_{20}) is validated with the help of the results of a variety of strain path change simulations in comparison with those of the associated stand-alone code (SA-HAH_{20}), and from experimental measurements. In addition, the FE-implemented pressure—dependent distortional plasticity hardening model (FE-HAH_{20P}) is also validated. The theoretical models are implemented into the FE-framework through the user-defined material (UMAT) subroutine provided by Abaqus 6.20 (Hibbitt et al. 2011). The FE-validation for the non-proportional loading predictions are conducted using a single element for:

1. reverse loading;
2. cross-loading with contraction and;
3. cross-loading with macroscopic latent hardening.

Besides, the performance of HAH_{20} is comparatively studied with that of HAH_{14} for the cross-loading tests. Unlike the isotropic hardening model, the FE-simulation accuracy strongly depends on the evolution of the hardening state variables. In most of the anisotropic hardening models, the hardening state variables are formulated with respect to the effective plastic strain. Consequently, the step size of the effective strain ($\delta(\Delta\bar{\varepsilon})_{k+1}$) must be carefully determined in order to achieve a close representation of the model (Yoon et al. 2020).

Reverse Loading

In reverse loading, the Bauschinger effect has been observed regardless of the microstructure of the metal as described in many experimental articles. The Bauschinger effect, which results from the reorganization of the internal stresses after unloading of a plastically deformed blank, plays a crucial role in the predictions of springback. In this subsection, the prediction of FE-HAH_{20} for tests with reversals is compared with the stand-alone code SA-HAH_{20} and the associated experimental data. The model coefficients for the current deformation mode are calibrated based on uniaxial tension-compression data of the investigated sheet metals.

The first example in Fig. 2.17 is a uniaxial tension-compression-tension (TCT) test with various strain amplitudes ($|\boldsymbol{\varepsilon}| = 0.03$ and 0.06) performed on a DP780 steel sheet sample. The plasticity model coefficients, calculated after the mechanical characterization of DP780, are listed in Table 2.2. The stress-strain curve predicted with the current hardening model are in good agreement with the experimental data for the two different strain ranges. The FE-implemented HAH model appears to behave

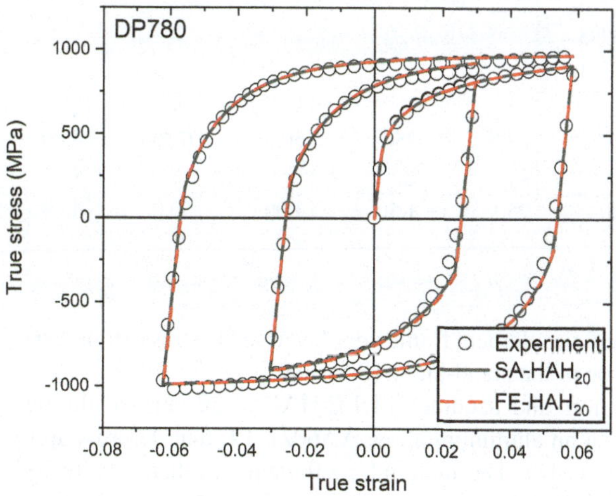

Fig. 2.17 Uniaxial tension-compression-tension predictions with FE-HAH_{20} and SA-HAH_{20} using DP780 for various strain ranges: $|\boldsymbol{\varepsilon}| = 0.03$ and 0.06

Table 2.2 Model coefficients for DP780 (Yoon et al. 2020)

Hockett-Sherby: $H(\bar{\varepsilon}) = \sigma_s - (\sigma_s - \sigma_y) e^{-(N\bar{\varepsilon})^P}$ (Unit: MPa)							
σ_s	σ_y	N	P				
1125.76	317.91	34.571	0.4526				
Yld2000-2d ($a = 6$)							
α_1	α_2	α_3	α_4	α_5	α_6	α_7	α_8
0.9577	0.9877	1.0291	0.9931	1.0113	0.9617	1.0266	0.9797
HAH$_{20}$ ($q = p = 3$)							
k	k'	ξ_R	k_1	k_2	k_3	ξ_B	ξ'_B
60.0	30.0	8.00	60.0	50.0	0.40	4.00	1.50
k_4	k_5	k_S	ξ_S	C	k_C	k'_C	ξ_C
0.80	4.00	60.0	1.50	0.65	60.0	28.0	6.00
L	k_L						
1.00	500						

Table 2.3 Model coefficients for AA6022-T4 (Feng et al. 2020)

Hockett-Sherby: $H(\bar{\varepsilon}) = \sigma_s - (\sigma_s - \sigma_y) e^{-(N\bar{\varepsilon})^P}$ (Unit: MPa)							
σ_s	σ_y	N	P				
429.92	174.31	8.0496	0.7396				
Yld2004-18p ($a = 8$)							
c'_{12}	c'_{13}	c'_{21}	c'_{23}	c'_{31}	c'_{32}	c'_{44}	c'_{55}
1.0000	1.0000	0.4396	0.4322	0.7911	1.1357	1.0724	1.0599
c'_{66}	c'_{12}	c'_{13}	c'_{21}	c'_{23}	c'_{31}	c'_{32}	c'_{44}
0.3736	1.4533	1.1995	0.9938	1.1294	1.0336	1.0901	0.8794
c'_{55}	c'_{66}						
0.8921	1.3795						
HAH$_{20}$ ($q = p = 3$)							
k	k'	ξ_R	k_1	k_2	k_3	ξ_B	ξ'_B
250	15.0	8.00	250	50.0	0.30	4.00	1.50
k_4	k_5	k_S	ξ_S	C	k_C	L	k_L
0.90	15.0	250	1.50	1.00	50.0	1.00	500

properly in the finite element framework because the stress-strain curve exactly coincides with those of the stand-alone (SA) code.

The robustness and accuracy of FE-HAH$_{20}$ are verified through the multiple cycle loading of an aluminum alloy, AA6022-T4, that deforms up to a large strain range ($\bar{\varepsilon}_{max} \approx 0.347$). The material coefficients are listed in Table 2.3. As shown in Fig. 2.18, the CPPM solution obtained with the Newton-Raphson method (NR-CPPM) does not converge after an accumulated plastic strain of about 0.148 in the multiple reversal test. This is directly related to the fluctuation of the hardening

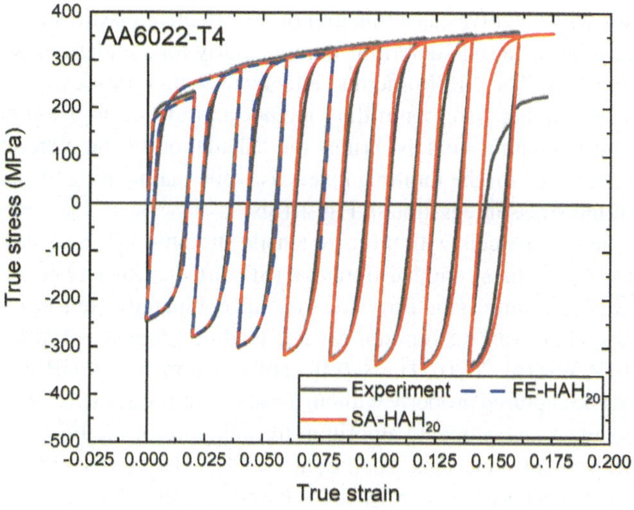

(a) Newton-Raphson

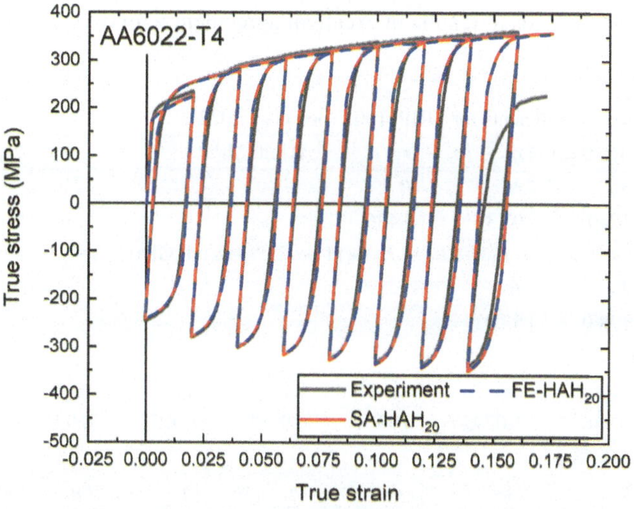

(b) Line-search

Fig. 2.18 Multiple cycle loading predictions with FE-HAH$_{20}$ and SA-HAH$_{20}$ compared with experimental measurements. FE stress update integrated with **a** Newton-Raphson and **b** Line-search based CPPM

state variables which evolves as a function of the effective plastic strain. The effective strain must be accurately determined, particularly for simulations with complex deformation history. The CPPM method integrated with a line-search method (LS-CPPM), which is a step size controlling method introduced in Yoon et al. (2020) for anisotropic hardening models, brings the calculation of the experimental flow curve to completion with the multiple reversals while maintaining the accuracy and robustness of the stress integration in Fig. 2.18b.

The so-called giga-steels with ultimate tensile strengths (UTS) exceeding 1 GPa have shown a significant strength differential (SD) effect. This physical feature indicates that a higher compression flow stress occurs compared to tension, not only for monotonic tests but even for forward-reverse loading sequences (Park et al. 2013; Jung et al. 2019; Wi et al. 2020). The SD effect of a 1.6 mm thick TRIP1180 steel sheet sample cannot be captured properly without the above introduced pressure-dependent plasticity model. A stress update algorithm formulation for the FE-implementation of this model was introduced in Yoon et al. (2022). The model coefficients of the TRIP1180 steel sheet are listed in Table 2.4 based on experimental data from a previous work (Wi et al. 2020). The FE-implementation of HAH_{20P} is verified through a forward-reverse loading within strain ranges of $|\varepsilon| = 0.03$ and 0.05 as depicted in Fig. 2.19. HAH_{20} leads to results in excellent agreement with two independent mea-

Table 2.4 Model coefficients for TRIP1180 (Yoon et al. 2022)

Hockett-Sherby: $H(\bar{\varepsilon}) = \sigma_s - (\sigma_s - \sigma_y)\, e^{-(N\bar{\varepsilon})^P}$ (Unit: MPa)							
σ_s	σ_y	N	P				
7284	1040	0.135	0.436				
Chord modulus: $E_{chord} = E_0 - (E_0 - E_s) * (1 - E^{-B\bar{\varepsilon}})$ (Unit: GPa)							
E_0	E_s	B					
197.4	178.0	57.02					
Yld2000-2d ($a = 6$)							
α_1	α_2	α_3	α_4	α_5	α_6	α_7	α_8
1.0331	0.9464	1.0295	0.9961	0.9981	1.0051	0.9913	0.9714
HAH_{20} ($q = p = 3$)							
k	k'	ξ_R	k_1	k_2	k_3	ξ_B	ξ'_B
100	15.0	8.00	100	150	0.50	4.00	1.50
k_4	k_5	k_S	ξ_S	C	k_C	k'_C	ξ_C
0.95	10.0	100	3.00	0.79	65.0	65.0	6.00
L	k_L	ξ_L	ξ'_L				
1.22	18	0.50	0.50				
Pressure-effect (Unit: TPa^{-1})							
α							
30							

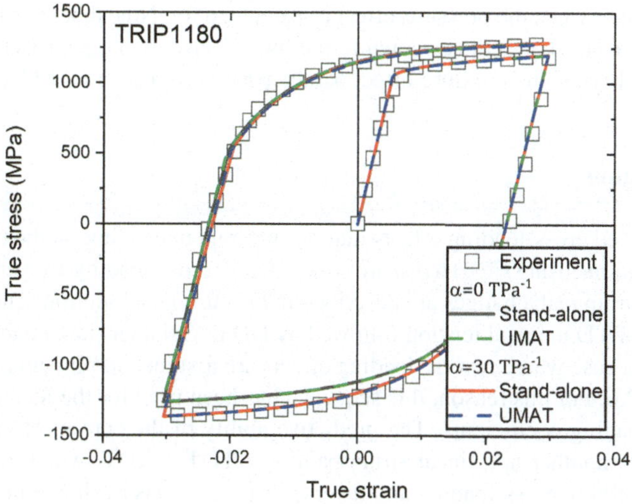

(a) $|\varepsilon| = 0.03$

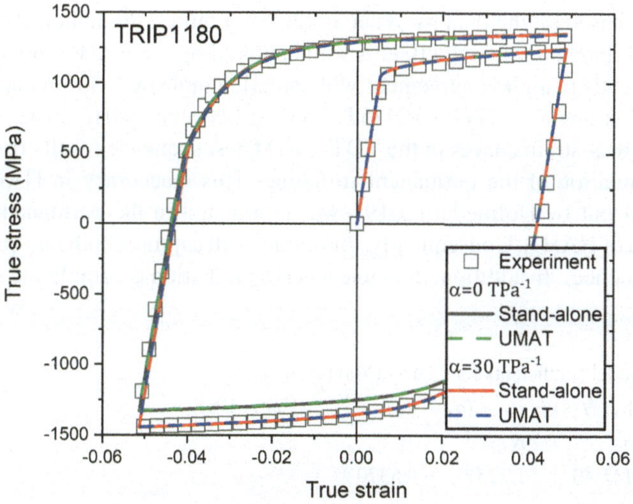

(b) $|\varepsilon| = 0.05$

Fig. 2.19 Comparison of FE-HAH$_{20P}$ and SA-HAH$_{20P}$ predictions with experimental data for TRIP1180 in forward-reverse loading with strain amplitudes ($|\varepsilon|$) of **a** 0.03 and **b** 0.05

surements only when the pressure-effect is accounted for. Note that the stress update algorithm for HAH_{20P} reduces down to the more conventional approach described in Sect. 2.8.1 when the pressure-effect parameter is zero, i.e., $\alpha = 0$ TPa^{-1}

Cross-Loading

Since observed cross-loading effects may be different depending on the microstructure, simulations using FE-HAH_{20} are conducted in cross-loading for various materials. Two strain path changes are described in the current subsection. The sequence consisting of TD uniaxial tension followed by DD uniaxial tension, denoted TDT + RD45T, is a case where reverse loading effects are assumed to play no role according to Fig. 2.6. For this reason, this sequence has been used for the identification of the cross-loading coefficients. The predictive ability of the current HAH model is verified with another non-linear strain path, i.e., TDT + RDT, which includes the influences of both cross-loading and reverse loading effects as shown in Fig. 2.6. In addition, the cross-loading predictions of HAH_{20} and HAH_{14} are compared.

The HAH_{20} coefficients of a 0.7 mm thick EDDQ steel sheet sample are listed in Table 2.5 based on the data by Wi et al. (2020). Figure 2.20 demonstrates that the macroscopic latent hardening effect is well captured by the HAH_{20} model.

HAH_{14} leads to a good agreement with experimental stress-strain curves only in pure-cross loading, i.e., TDT + RD45T. The gap between HAH_{14} experimental and predicted stress-strain curves in the TDT + RDT test sequence results from an issue in the formulation of the permanent softening. This inaccuracy in HAH_{14}, which was pointed out by Holmedal (2019), was eliminated in the permanent softening formulation of HAH_{20}. Consequently, this model well captures both non-proportional loading sequences. In addition, the same level of prediction is completely reproduced

Table 2.5 Model coefficients for EDDQ (Yoon et al., 2022)

Hockett-Sherby: $H(\bar{\varepsilon}) = \sigma_s - (\sigma_s - \sigma_y)\, e^{-(N\bar{\varepsilon})^P}$ (Unit: MPa)							
σ_s	σ_y	N	P				
645.68	141.40	1.2257	0.5370				
Yld2000-2d ($a = 6$)							
α_1	α_2	α_3	α_4	α_5	α_6	α_7	α_8
1.1946	0.9629	0.6930	0.8882	0.8273	0.9303	1.0463	0.9373
HAH_{20} ($q = p = 3$)							
k	k'	ξ_R	k_1	k_2	k_3	ξ_B	ξ'_B
350	15.0	8.00	350	350	0.60	4.00	1.50
k_4	k_5	k_S	ξ_S	C	k_C	L	k_L
0.90	30.0	350	3.00	1.00	60.0	1.40	450
ξ_L	ξ'_L						
0.50	0.50						

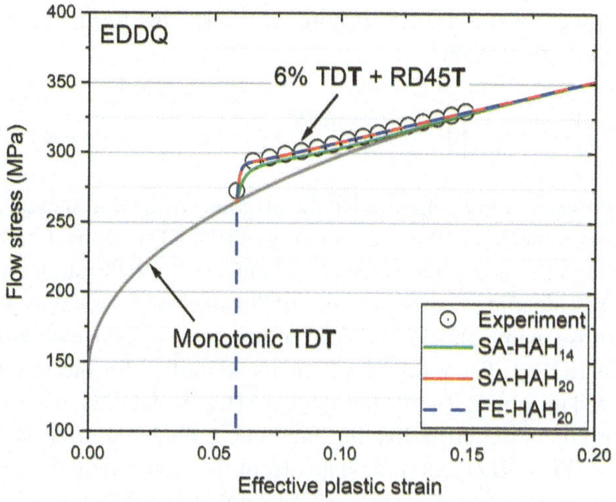

(a) 6% TDT + RD45T

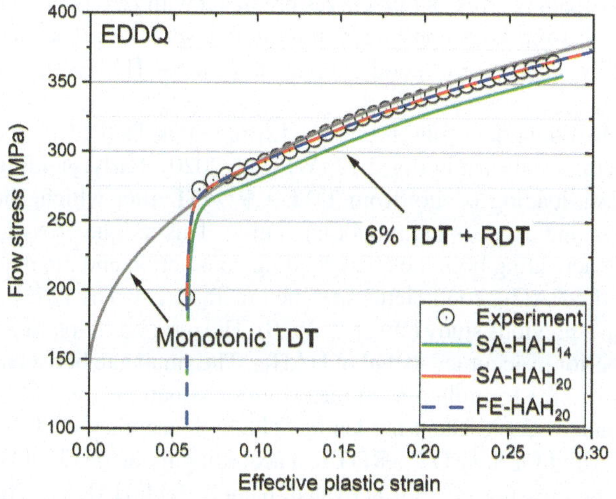

(b) 6% TDT + RDT

Fig. 2.20 Macroscopic latent hardening predictions with FE-HAH$_{20}$, compared with SA-HAH$_{20}$, SA-HAH$_{14}$, and experimental measurements for EDDQ steel deformed under non-linear strain paths: **a** TDT-RD45T and; **b** TDT-RDT

in the finite element method. Furthermore, an alternative formulation of the latent hardening evolution equation

$$\frac{dg_L}{d\bar{\varepsilon}} = k_L \left\{ (L - g_L) \left[1 - \left| \cos\chi' \right|^{\xi_L} \right] + (1 - g_L) \left[1 + \left| \cos\chi' \right|^{\xi'_L} \right] \right\} \quad (2.58)$$

is used in the present article compared to the original work (Barlat et al. 2020), which leads to a better prediction of the overshooting of EDDQ steel sheet in cross-loading.

Cross-loading tests were conducted on 1.4 mm thick DP780 steel sheet (Lee et al. 2020a). The cross-loading contraction of DP780 steel is observed as an early yielding after the subsequent reloading instead of the flow stress overshooting as shown in Fig. 2.21 for EDDQ. The model coefficients, identified for the HAH_{20} model for DP780, are listed in Table 2.2 while those for HAH_{14} are detailed in Lee et al. (2020a). HAH_{14} well predicts the cross-loading flow curve obtained after **TDT** + **RD45T**. However, the **TDT** + **RDT** curve deviates from the experimental data. Indeed, the amount of permanent softening is proportional only to the effective plastic strain but does not account for the deformation history (the loading direction change). In contrast, the cross-loading stress-strain curves predicted with HAH_{20} are in good agreement with those from both non-linear strain path tests. The associated FE-HAH_{20} simulations flow curve are in good agreement with SA-HAH_{20} and the measured data.

The cross-loading effects observed for a 1.6 mm thick TRIP1180 steel sheet sample, a giga-steel, are shown in Fig. 2.22 (Wi et al. 2020). Early-yielding is observed right after cross-loading occurs from **TDT** + **RD45T** after which, the flow stress increases up to and exceeds the monotonic curve. This unique cross-loading effect is captured by activating both the cross-loading contraction and the latent hardening effects using the HAH_{20} coefficients specified in Table 2.4. The HAH_{14} coefficients are those from a previous study (Wi et al. 2020). The pure-cross loading prediction of HAH_{20} is superior in accuracy to that of HAH_{14}. The small gap between both models are mainly due to the formulations of the evolution equations for cross-loading contraction and macroscopic latent hardening. The high accuracy of the cross-loading effects is also valid for the **TDT** + **RDT** test according to the result of HAH_{20} in particular when the pressure-effect is activated, namely, with HAH_{20P}. The predictions of FE-HAH_{20P} for the strain path change tests applied to TRIP1180 steel are well reproduced in the finite element simulations.

Miscellaneous Applications

It was shown above that FE-HAH_{20} simulation results are in good agreement with the stand-alone code SA-HAH_{20} predictions and experimental data in the various strain path change cases investigated. However, the main purpose of the FE-implementation is to use an advanced theoretical model for practical problems by taking advantage of the finite element framework to solve boundary-value problems with complex geometries. For this purpose, Lee et al. (2012) obtained an accurate prediction of

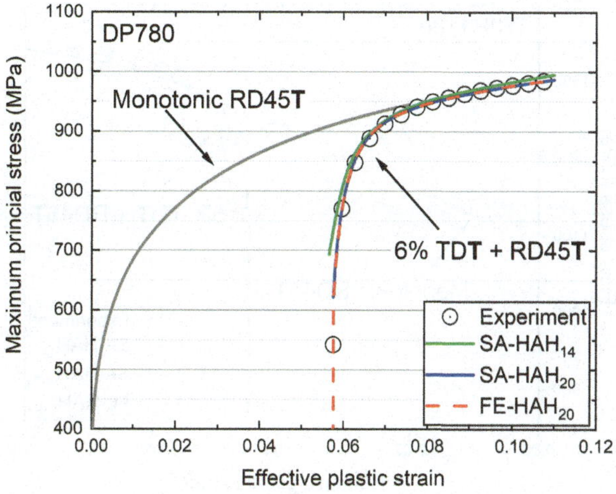

(a) 6% TD**T** + RD45**T**

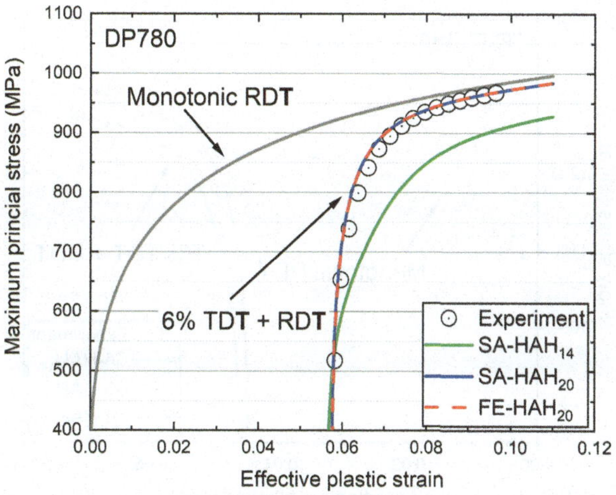

(b) 6% TD**T** + RD**T**

Fig. 2.21 Cross-loading contraction predicted with FE-HAH$_{20}$ and compared with SA-HAH$_{20}$, SA-HAH$_{14}$ and experimental measurement for DP780 steel sheet under non-linear strain paths: **a** TDT-RD45**T** and **b** TD**T**-RDT

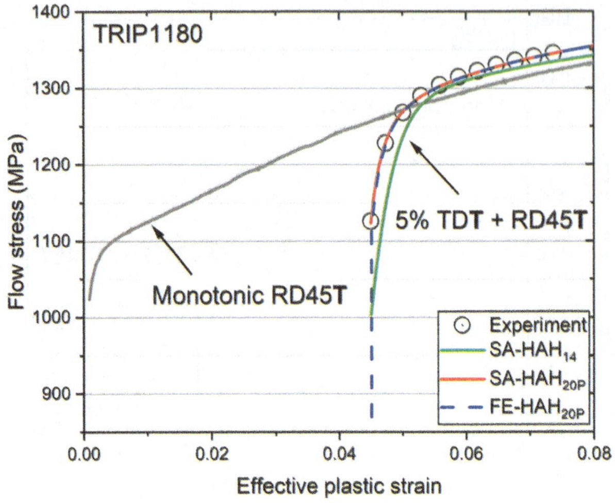

(a) 6% TDT + RD45T

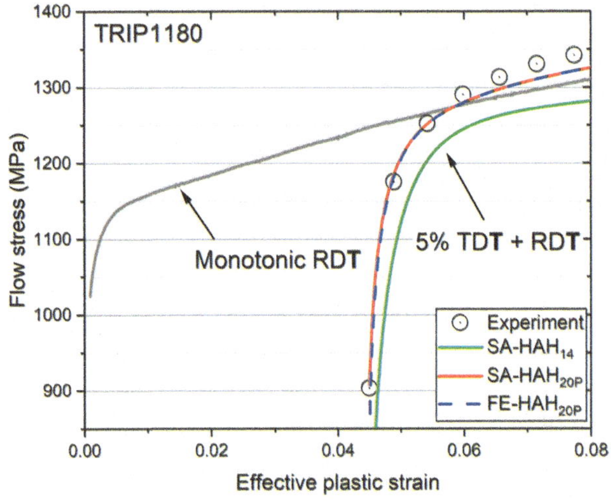

(b) 6% TDT + RDT

Fig. 2.22 Cross-loading effects of TRIP1180 captured by FE-HAH$_{20P}$ in comparison with SA-HAH$_{20P}$, SA-HAH$_{14}$, and experimental measurement for non-proportional loading paths: **a** TDT-RD45T and **b** TDT-RDT

springback after U-draw bending of a pre-strained DP780 steel sheet using HAH_{11}. Choi et al. (2016) applied HAH_{11} to predict the springback behavior after the forming of a S-rail. Choi and Yoon (2019) identified the HAH_{11} model via an inverse method based on 3-point bending simulations and experiments to predict the U-draw bending simulation of an ultra-thin stainless steel. Recently, Yoon et al. (2022) captured the flow curve of a TRIP1180 steel sheet that exhibits asymmetry due to the SD effect in U-draw bending by means of the pressure-dependent model HAH_{20}. Liao et al. (2017) researched the twist springback characteristics after forming simulations of simple shapes, namely, P-channel and C-channel using HAH_{14} model. In addition, HAH_{20P} was successfully applied to the forming simulation of a B-pillar (B-PLR, EV-POSCO concept) shown in Fig. 2.1 and the springback behavior of real part forming was discussed in depth to demonstrate the potential advantages of advanced hardening models for practical applications.

Acknowledgements The authors are very grateful to POSCO for financial and technical support.

Appendix

State Variable Evolution Equations

The yield condition is given above by combining Eq. (2.37) with (2.38) or (2.39). The development of the state variable evolution equations is explained in the original article (Barlat et al. 2020), and are only listed below.

1. *Transformed deviators*: A number of transformed stress deviators are introduced to describe cross-loading effects

$$\mathbf{s}_C = (\boldsymbol{\sigma}' : \hat{\mathbf{h}})\hat{\mathbf{h}} \tag{2.59}$$

$$\mathbf{s}_O = \boldsymbol{\sigma}' - \mathbf{s}_C \tag{2.60}$$

$$\mathbf{s}_X = \xi_C \left(\frac{1 - g_C}{g_L}\right) \mathbf{s}_O \tag{2.61}$$

$$\mathbf{s}_L = \frac{1}{\xi_L (g_L - 1) + 1} \mathbf{s}_C + \frac{1}{g_L} \mathbf{s}_O \tag{2.62}$$

2. *Cross-loading-modified effective stress*: The captioned effective stress is defined as

$$\bar{\xi}(\boldsymbol{\sigma}') = \left[\bar{\phi}(\mathbf{s}_L)^p + \bar{\phi}(\mathbf{s}_X)^p\right]^{1/p} \tag{2.63}$$

3. *Strain path change amplitude*: These parameters correspond to the severity of strain path change, i.e., 1 for monotonic, 0 for pure cross-loading and –1 for reverse loading

$$\cos\chi = \hat{\mathbf{h}} : \hat{\sigma}' \quad \cos\chi' = \hat{\mathbf{h}}' : \hat{\sigma}' \tag{2.64}$$

4. *Microstructure deviators*: These are the tensors, both set equal to the stress deviator at the first step, which account for a smooth evolution of the microstructure

$$\frac{d\hat{\mathbf{h}}'}{d\bar{\varepsilon}} = \lambda' k' \left(\hat{\sigma}' - \cos\chi' \hat{\mathbf{h}}'\right) \tag{2.65}$$

$$\frac{d\hat{\mathbf{h}}}{d\bar{\varepsilon}} = \lambda k \left(\cos\chi'^{2\xi_R} + \cos\chi^{2\xi_R}\right)\left(\hat{\sigma}' - \cos\chi\hat{\mathbf{h}}\right) \tag{2.66}$$

5. *Reverse loading*: Reverse loading is described in the yield condition by the two state variables, f_- and f_+, which are function of two other variables g_- and g_+

$$f_\omega = \sqrt{\frac{3}{8}}\left\{\frac{1}{g_\omega^q} - 1\right\}^{\frac{1}{q}} \tag{2.67}$$

where ω stands for the symbols $-$ or $+$. The evolution equations are expressed with the variables g_- and g_+

$$\begin{aligned}\frac{dg_-}{d\bar{\varepsilon}} &= \frac{1+\lambda}{2}\left\{k_1 \frac{1-g_-}{g_-^{\xi'_B}}\left[1-\left(\cos^2\chi\right)^{\xi_B}\right]\right.\\ &\left.+k_2\left(k_3\frac{\sigma_y}{\sigma(\bar{\varepsilon})} - g_-\right)\left(\cos^2\chi\right)^{\xi_B}\right\} + \frac{1-\lambda}{2}\left\{k_1\frac{1-g_-}{g_-^{\xi'_B}}\right\}\end{aligned} \tag{2.68}$$

$$\begin{aligned}\frac{dg_+}{d\bar{\varepsilon}} &= \frac{1-\lambda}{2}\left\{k_1 \frac{1-g_+}{g_+^{\xi'_B}}\left[1-\left(\cos^2\chi\right)^{\xi_B}\right]\right.\\ &\left.+k_2\left(k_3\frac{\sigma_y}{\sigma(\bar{\varepsilon})} - g_+\right)\left(\cos^2\chi\right)^{\xi_B}\right\} + \frac{1+\lambda}{2}\left\{k_1\frac{1-g_+}{g_+^{\xi'_B}}\right\}\end{aligned} \tag{2.69}$$

6. *Permanent softening*: Permanent softening should be maximum for strain reversals

$$g'_P = g_P^* - \left(g_P^* - g_3^*\right)\left|\hat{\sigma}' : \hat{\mathbf{h}}^*\right|^{2\xi_S} \tag{2.70}$$

$$\frac{dg_3}{d\bar{\varepsilon}} = k_5 g_S \left(k_4 - g_3\right) \tag{2.71}$$

$$\frac{dg_S}{d\bar{\varepsilon}} = -k_S\left[1 - \left|\hat{\sigma}' : \hat{\mathbf{h}}^*\right|\right] g_S \tag{2.72}$$

7. *Cross-loading contraction*: The contraction and recovery of the yield surface are controlled by the next equations, respectively

$$\frac{dg_C}{d\bar{\varepsilon}} = k_C\left(C - g_C\right) \tag{2.73}$$

$$\frac{dg_C}{d\bar{\varepsilon}} = k_C' \frac{1 - g_C}{g_C^{\xi_C}} \tag{2.74}$$

8. *Latent hardening*: Finally, latent hardening is defined by

$$\frac{dg_L}{d\bar{\varepsilon}} = k_L\left[\frac{\sigma_r(\bar{\varepsilon}) - \sigma_r(0)}{\sigma_r(\bar{\varepsilon})}\left(\sqrt{L + (1 - L)\cos^2\chi'^{\xi_L'}} - 1\right) + 1 - g_L\right] \tag{2.75}$$

Finite Difference Method

The first and second-order numerical derivatives of yield function are acquired based on the finite difference method (FDM) in a scaled stress space (Aretz 2007). The main advantages of the numerical derivatives for FE-implementation are that (1) the challenges for FE-implementation of advanced plasticity models are alleviated, (2) the singularities that appear in analytical derivatives do not require a special treatment, and (3) FDM is applicable to shell and solid elements corresponding to plane and 3D stress states, respectively. The scaled (normalized) stress space expands the utility of the FDM step size ($\delta\hat{\sigma}$) for various stress scales. As remarked in Choi and Yoon (2019), the midpoint rule (central difference method) accelerates the algorithm convergence and higher order of truncated error.

For the sake of the brevity, in this context, the stress tensor is expressed in the form of Voigt's notation (vectorized form).

$$[\sigma] = \begin{bmatrix} \sigma_{11} & \sigma_{12} & \sigma_{13} \\ \sigma_{12} & \sigma_{22} & \sigma_{23} \\ \sigma_{13} & \sigma_{23} & \sigma_{33} \end{bmatrix} = \begin{Bmatrix} \sigma_{11} \\ \sigma_{22} \\ \sigma_{33} \\ \sigma_{23} \\ \sigma_{13} \\ \sigma_{12} \end{Bmatrix} = \begin{Bmatrix} \sigma_1 \\ \sigma_2 \\ \sigma_3 \\ \sigma_4 \\ \sigma_5 \\ \sigma_6 \end{Bmatrix} \tag{2.76}$$

A stress tensor is scaled by the normalization factor (ξ).

$$\hat{\sigma} = \xi\sigma = \frac{\sigma}{\sqrt{\sigma : \sigma}} \tag{2.77}$$

The first-order derivative is obtained based on a midpoint rule

$$\frac{\partial\bar{\sigma}(\hat{\boldsymbol{\sigma}})}{\partial\hat{\sigma}_i} \cong \frac{\bar{\sigma}\left(\hat{\sigma}_i + \delta\hat{\sigma}, \hat{\sigma}_j\right) - \bar{\sigma}\left(\hat{\sigma}_i - \delta\hat{\sigma}, \hat{\sigma}_j\right)}{2(\delta\hat{\sigma})} \tag{2.78}$$

where $i \neq j$ and $\{i, j\} \in \{1, 2, 3, 4, 5, 6\}$. It is worth noting that the step size for the finite difference method $(\delta\hat{\sigma})$ is assumed to be constant for all components. $\delta\hat{\sigma} = 10^{-6}$ is recommended for the first-order derivative. The normalization factor does not influence the values of the first derivative because the yield function derivative is a homogeneous function of first degree.

$$\frac{\partial\bar{\sigma}(\hat{\boldsymbol{\sigma}})}{\partial\hat{\boldsymbol{\sigma}}} = \frac{\partial\xi\bar{\sigma}(\boldsymbol{\sigma})}{\partial\xi\boldsymbol{\sigma}} = \frac{\partial\bar{\sigma}(\boldsymbol{\sigma})}{\partial\boldsymbol{\sigma}} \tag{2.79}$$

The second-order derivative is defined differently based on the components, i.e., (1) diagonal $(i = j)$ and (2) off-diagonal $(i \neq j)$ and symmetry according to Schwarz' theorem.

$$\begin{aligned} &\frac{\partial^2\bar{\sigma}(\hat{\boldsymbol{\sigma}})}{\partial\hat{\sigma}_i\partial\hat{\sigma}_j} = \frac{\bar{\sigma}_+ - 2\bar{\sigma}(\hat{\boldsymbol{\sigma}}) + \bar{\sigma}_-}{(\delta\hat{\sigma})^2} && \textbf{IF} \quad i = j \\ &\frac{\partial^2\bar{\sigma}(\hat{\boldsymbol{\sigma}})}{\partial\hat{\sigma}_i\partial\hat{\sigma}_j} = \frac{\bar{\sigma}_{++} - \bar{\sigma}_{+-} - \bar{\sigma}_{-+} + \bar{\sigma}_{--}}{(\delta\hat{\sigma})^2} && \textbf{ELSE} \end{aligned} \tag{2.80}$$

References

Anandarajah, A. (2010). *Computational Methods in Elasticity and Plasticity: Solids and Porous Media*. New York: Springer. https://doi.org/10.1007/978-1-4419-6379-6.

Aretz, H. (2007). An advanced numerical differentiation scheme for plastic strain-rate computation. In *AIP Conference Proceedings* (Vol. 907, 151–156).

Armstrong, P. J., & Frederick, C. O. (1966). *A Mathematical Representation of the Multiaxial Bauschinger Effect* (Vol. 731). Berkley, UK: Central Electricity Generating Board & Berkeley Nuclear Laboratories.

Banabic, D., Barlat, F., Cazacu, O., & Kuwabara, T. (2020). Advances in anisotropy of plastic behaviour and formability of sheet metals. *International Journal of Material Forming, 13*(5), 749–787. https://doi.org/10.1007/s12289-020-01580-x

Barlat, F., & Lian, K. (1989). Plastic behavior and stretchability of sheet metals. Part I: A yield function for orthotropic sheets under plane stress conditions. *International Journal of Plasticity, 5*(1), 51–66.

Barlat, F., Lege, D. J., & Brem, J. C. (1991). A six-component yield function for anisotropic materials. *International Journal of Plasticity, 7*(7), 693–712. https://doi.org/10.1016/0749-6419(91)90052-Z

Barlat, F., Brem, J. C., Yoon, J. W., Chung, K., Dick, R. E., Lege, D. J., Pourboghrat, F., Choi, S.-H., & Chu, E. (2003a). Plane stress yield function for aluminum alloy sheets - Part 1: Theory. *International Journal of Plasticity, 19*(9), 1297–1319. https://doi.org/10.1016/S0749-6419(02)00019-0

Barlat, F., Duarte, J. M. F., J.J. Gracio, Lopes, A. B., & Rauch, E. F. (2003b). Plastic flow for non-monotonic loading conditions of an aluminum alloy sheet sample. *International Journal of Plasticity, 19*(8), 1215–1244. https://doi.org/10.1016/S0749-6419(02)00020-7

Barlat, F., Aretz, H., Yoon, J. W., Karabin, M. E., Brem, J. C., & Dick, R. E. (2005). Linear transfomation-based anisotropic yield functions. *International Journal of Plasticity, 21*(5), 1009–1039. https://doi.org/10.1016/j.ijplas.2004.06.004

Barlat, F., Gracio, J. J., Lee, M.-G., Rauch, E. F., & Vincze, G. (2011). An alternative to kinematic hardening in classical plasticity. *International Journal of Plasticity, 27*(9), 1309–1327. https://doi.org/10.1016/j.ijplas.2011.03.003

Barlat, F., Vincze, G., Grácio, J. J., Lee, M.-G., Rauch, E. F., & Tomé, C. N. (2014). Enhancements of homogenous anisotropic hardening model and application to mild and dual-phase steels. *International Journal of Plasticity, 58*, 201–218. https://doi.org/10.1016/j.ijplas.2013.11.002

Barlat, F., Yoon, S.-Y., Lee, S.-Y., Wi, M.-S., & Kim, J.-H. (2020). Distortional plasticity framework with application to advanced high strength steel. *International Journal of Solids and Structures, 202*, 947–962. https://doi.org/10.1016/j.ijsolstr.2020.05.014

Bishop, J. F. W., & Hill, R. (1951). XLVI. A theory of the plastic distortion of a polycrystalline aggregate under combined stresses. *The London, Edinburgh, and Dublin Philosophical Magazine and Journal of Science, 42*(327), 414–427. 1951. https://doi.org/10.1080/14786445108561065

Bulatov, V. V., Richmond, O., & Glazov, M. V. (1999). Atomistic dislocation mechanism of pressure-dependent plastic flow in aluminum. *Acta Materialia, 47*(12), 3507–3514. https://doi.org/10.1016/S1359-6454(99)00154-8

Cardoso, R. P. R., & Yoon, J. W. (2009). Stress integration method for a nonlinear kinematic/isotropic hardening model and its characterization based on polycrystal plasticity. *International Journal of Plasticity, 25*(9), 1684–1710. https://doi.org/10.1016/j.ijplas.2008.09.007

Cazacu, O., & Barlat, F. (2001). Generalization of Drucker's yield criterion to orthotropy. *Mathematics and Mechanics of Solids, 6*(6), 613–630. https://doi.org/10.1177/108128650100600603

Cazacu, O., & Barlat, F. (2003). Application of the theory of representation to describe yielding of anisotropic aluminum alloys. *International Journal of Engineering Science, 41*(12), 1367–1385. https://doi.org/10.1016/S0020-7225(03)00037-5

Cazacu, O., & Barlat, F. (2004). A criterion for description of anisotropy and yield differential effects in pressure-insensitive metals. *International Journal of Plasticity, 20*(11), 2027–2045. https://doi.org/10.1016/j.ijplas.2003.11.021

Chaboche, J. L. (1977). Sur l'utilisation des variables d'état interne pour la description de la viscoplasticité cyclique avec endommagement. In *Symposium Franco-Polonais de Rhéologie et Mécanique: Problèmes Non Linéaires de Méecanique* (pp. 137–159). Cracovie.

Chaboche, J. L. (2008). A review of some plasticity and viscoplasticity constitutive theories. *International Journal of Plasticity, 24*(10), 1642–1693. https://doi.org/10.1016/j.ijplas.2008.03.009

Choi, H., & Yoon, J. W. (2019). Stress integration-based on finite difference method and its application for anisotropic plasticity and distortional hardening under associated and non-associated flow rules. *Computer Methods in Applied Mechanics and Engineering, 345*, 123–160. https://doi.org/10.1016/j.cma.2018.10.031

Choi, J., Lee, J., Bae, G., Barlat, F., & Lee, M.-G. (2016). Evaluation of springback for DP980 S rail using anisotropic hardening models. *JOM, 68*(7), 1850–1857. https://doi.org/10.1007/s11837-016-1924-z

Choi, J. S., Lee, J. W., Kim, J.-H., Barlat, F., Lee, M. G., & Kim, D. (2015). Measurement and modeling of simple shear deformation under load reversal: Application to advanced high strength steels. *International Journal of Mechanical Sciences, 98*, 144–156. https://doi.org/10.1016/j.ijmecsci.2015.04.014

Chung, K., & Richmond, O. (1993). A deformation theory of plasticity based on minimum work paths. *International Journal of Plasticity, 9*(8), 907–920. https://doi.org/10.1016/0749-6419(93)90057-W

Dafalias, Y. F., & Popov, E. P. (1976). Plastic internal variables formalism of cyclic plasticity. *Transactions ASME: Journal of Applied Mechanics, 43*(4), 645–651. https://doi.org/10.1115/1.3423948

Dunne, F., & Petrinic, N. (2005). *Introduction to Computational Plasticity*. Oxford: Oxford University Press.

Feigenbaum, H. P., & Dafalias, Y. F. (2007). Directional distortional hardening in metal plasticity within thermodynamics. *International Journal of Solids and Structures, 44*(22–23), 7526–7542. https://doi.org/10.1016/j.ijsolstr.2007.04.025

Feng, Z., Yoon, S.-Y., Choi, J.-H., Barrett, T. J., Zecevic, M., Barlat, F., & Knezevic, M. (2020). A comparative study between elasto-plastic self-consistent crystal plasticity and anisotropic yield function with distortional hardening formulations for sheet metal forming. *Mechanics of Materials, 148*, 103422. https://doi.org/10.1016/j.mechmat.2020.103422

François, M. (2001). A plasticity model with yield surface distortion for non proportional loading. *International Journal of Plasticity, 17*(5), 703–717. https://doi.org/10.1016/S0749-6419(00)00025-5

Ha, J., Lee, M.-G., & Barlat, F. (2013). Strain hardening response and modeling of EDDQ and DP780 steel sheet under non-linear strain path. *Mechanics of Materials, 64*, 11–26. https://doi.org/10.1016/j.mechmat.2013.04.004

Hasegawa, T., & Yakou, T. (1980). Effects of stress reversal and thermal recovery on stress vs strain behavior in aluminum. *Scripta Metallurgica, 14*(10), 1083–1087. https://doi.org/10.1016/0036-9748(80)90210-0

Hasegawa, T., Yakou, T., & Karashima, S. (1975). Deformation behaviour and dislocation structures upon stress reversal in polycrystalline aluminium. *Materials Science and Engineering, 20*, 267–276. https://doi.org/10.1016/0025-5416(75)90159-7

He, W. J., Zhang, S. H., & Song, H. W. (2013). An extended homogenous yield function based anisotropic hardening model for description of anisotropic hardening behavior of materials. *International Journal of Mechanical Sciences, 77*, 343–355. https://doi.org/10.1016/j.ijmecsci.2013.05.018

Hershey, A. V. (1954). The plasticity of an isotropic aggregate of anisotropic face-centered cubic crystals. *JOurnal of Applied Mechanics, 21*(3), 241–249. https://doi.org/10.1115/1.4010900.

Hibbitt, H., Karlsson, B., & Sorensen, P. (2011). Abaqus user subroutine reference manual version 6.10. *Dassault Systemes Simulia Corp.: USA*.

Hill, R. (1948). A theory of the yielding and plastic flow of anisotropic metals. *Proceedings of the Royal Society of London. Series A. Mathematical and Physical Sciences, 193*(1033), 281–297.

Hockett, J. E., & Sherby, O. D. (1975). Large strain deformation of polycrystalline metals at low homologous temperatures. *Journal of the Mechanics and Physics of Solids, 23*(2), 87–98. https://doi.org/10.1016/0022-5096(75)90018-6

Hollomon, J. H. (1945). Tensile deformation. *Transactions of the Metallurgical Society of AIME, 162*, 268–290.

Holmedal, B. (2019). Bauschinger effect modelled by yield surface distortions. *International Journal of Plasticity, 123*, 86–100. https://doi.org/10.1016/j.ijplas.2019.07.009

Jeong, Y., Barlat, F., Tomé, C. N., & Wen, W. (2017). A comparative study between micro- and macro-mechanical constitutive models developed for complex loading scenarios. *International Journal of Plasticity, 93*, 212–228. https://doi.org/10.1016/j.ijplas.2016.07.015

Jung, J. (1981). A note on the influence of hydrostatic pressure on dislocations. *Philosophical Magazine A: Physics of Condensed Matter, Structure, Defects and Mechanical Properties, 43*(4), 1057–1061. https://doi.org/10.1080/01418618108239511

Jung, J., Hur, Y. C., Jun, S., Lee, H.-S., Kim, B.-M., & Kim, J. H. (2019). Constitutive modeling of asymmetric hardening behavior of transformation-induced plasticity steels. *International Journal of Automotive Technology, 20*, 19–30. https://doi.org/10.1007/s12239-019-0124-6

Kitayama, K., Tomé, C. N., Rauch, E. F., Gracio, J. J., & Barlat, F. (2013). A crystallographic dislocation model for describing hardening of polycrystals during strain path changes. Application to low carbon steels. *International Journal of Plasticity, 46*, 54–69 (2013). https://doi.org/10.1016/j.ijplas.2012.09.004

Krieg, R. D. (1975). A practical two surface plasticity theory. *Transactions ASME: Journal of Applied Mechanics, 42*(3), 641–646. https://doi.org/10.1115/1.3423656

Kurtyka, T., & Życzkowski, M. (1996). Evolution equations for distortional plastic hardening. *International Journal of Plasticity, 12*(2), 191–213. https://doi.org/10.1016/S0749-6419(96)00003-4

Kuwabara, T. (2013). Advanced material testing in support of accurate sheet metal forming simulations. In *The 11th International Conference on Numerical Methods in Industrial Forming Processes: Numiform 2013* (Vol. 1532, pp. 69–80). https://doi.org/10.1063/1.4806810

Lee, J.-Y., Lee, J.-W., Lee, M.-G., & Barlat, F. (2012). An application of homogeneous anisotropic hardening to springback prediction in pre-strained u-draw/bending. *International Journal of Solids and Structures, 49*(25), 3562–3572. https://doi.org/10.1016/j.ijsolstr.2012.03.042

Lee, S.-Y., Kim, J.-M., Kim, J.-H., & Barlat, F. (2020). Validation of homogeneous anisotropic hardening model using non-linear strain path experiments. *International Journal of Mechanical Sciences, 183*, 105769. https://doi.org/10.1016/j.ijmecsci.2020.105769

Lee, S.-Y., Yoon, S.-Y., Kim, J.-H., & Barlat, F. (2020). Calibration of distortional plasticity framework and application to U-draw bending simulations. *ISIJ International, 60*(12), 2927–2941. https://doi.org/10.2355/isijinternational.ISIJINT-2020-391

Liao, J., Xue, X., Lee, M.-G., Barlat, F., Vincze, G., & Pereira, A. B. (2017). Constitutive modeling for path-dependent behavior and its influence on twist springback. *International Journal of Plasticity, 93*, 64–88. https://doi.org/10.1016/j.ijplas.2017.02.009

Logan, R. W., & Hosford, W. F. (1980). Upper-bound anisotropic yield locus calculations assuming 111-pencil glide. *International Journal of Mechanical Sciences, 22*(7), 419–430. https://doi.org/10.1016/0020-7403(80)90011-9

Lopes, A. B., Barlat, F., Gracio, J. J., Duarte, J. F. F., & Rauch, E. F. (2003) Effect of texture and microstructure on strain hardening anisotropy for aluminum deformed in uniaxial tension and simple shear. *International Journal of Plasticity, 19*(1), 1–22.

Ludwik, P. (1909) Fließvorgänge bei einfachen Beanspruchungen. In *Elemente der Technologischen Mechanik* (pp. 11–35). Springer.

Mróz, Z. (1967). On the description of anisotropic workhardening. *Journal of the Mechanics and Physics of Solids, 15*(3), 163–175. https://doi.org/10.1016/0022-5096(67)90030-0

Ortiz, M., & Popov, E. P. (1983). Distortional hardening rules for metal plasticity. *Journal of Engineering Mechanics, 109*(4), 1042–1057. https://doi.org/10.1061/(ASCE)0733-9399(1983)109:4(1042)

Ortiz, M., & Simo, J. C. (1986). An analysis of a new class of integration algorithms for elastoplastic constitutive relations. *International Journal for Numerical Methods in Engineering, 23*(3), 353–366. https://doi.org/10.1002/nme.1620230303

Park, S. C., Park, T., Koh, Y., Seok, D. Y., Kuwabara, T., Noma, N., & Chung, K. (2013). Spring-back prediction of MS1470 steel sheets based on a non-linear kinematic hardening model. *Transactions of Materials Processing, 22*(6), 303–309. https://doi.org/10.5228/KSTP.2013.22.6.303

Peeters, B., Bacroix, B., Teodosiu, C., Van Houtte, P., & Aernoudt, E. (2001a) Work-hardening/softening behaviour of b.c.c. polycrystals during changing strain paths: II. TEM observations of dislocation sheets in an if steel during two-stage strain paths and their representation in terms of dislocation densities. *Acta Materialia, 49*(9), 1621–1632. https://doi.org/10.1016/S1359-6454(01)00067-2

Peeters, B., Seefeldt, M., Teodosiu, C., Kalidindi, S. R., Van Houtte, P., & Aernoudt, E. (2001b) Work-hardening/softening behaviour of b.c.c. polycrystals during changing strain paths: I. An integrated model based on substructure and texture evolution, and its prediction of the stress-strain behaviour of an if steel during two-stage strain paths. *Acta Materialia, 49*(9), 1607–1619. https://doi.org/10.1016/S1359-6454(01)00066-0

Peeters, B., Kalidindi, S. R., Teodosiu, C., Houtte, P. V., & Aernoudt, E. (2002). A theoretical investigation of the influence of dislocation sheets on evolution of yield surfaces in single-phase b.c.c. polycrystals. *Journal of the Mechanics and Physics of Solids, 50*(4), 783–807. https://doi.org/10.1016/S0022-5096(01)00094-1

Prager, W. (1949). Recent developments in the mathematical theory of plasticity. *Journal of Applied Physics, 20*(3), 235–241. https://doi.org/10.1063/1.1698348

Qin, J., Holmedal, B., Zhang, K., & Hopperstad, O. S. (2017). Modeling strain-path changes in aluminum and steel. *International Journal of Solids and Structures, 117*, 123–136. https://doi.org/10.1016/j.ijsolstr.2017.03.032

Qin, J., Holmedal, B., & Hopperstad, O. S. (2018). A combined isotropic, kinematic and distortional hardening model for aluminum and steels under complex strain-path changes. *International Journal of Plasticity, 101*, 156–169. https://doi.org/10.1016/j.ijplas.2017.10.013

Qin, J., Holmedal, B., & Hopperstad, O. S. (2019). Experimental characterization and modeling of aluminum alloy AA3103 for complex single and double strain-path changes. *International Journal of Plasticity, 112*, 158–171. https://doi.org/10.1016/j.ijplas.2018.08.011

Qin, J. S., Holmedal, B., & Hopperstad, O. (2017). Modelling of strain-path transients in commercially pure aluminium. *Materials Science Forum, 877*, 662–667. https://doi.org/10.4028/www.scientific.net/MSF.877.662

Rauch, E. F., Gracio, J. J., Barlat, F., & Vincze, G. (2011). Modelling the plastic behaviour of metals under complex loading conditions. *Modelling and Simulation in Materials Science and Engineering, 19*(3), 035009. https://doi.org/10.1088/0965-0393/19/3/035009

Richmond, O., & Spitzig, W. A. (1980). Pressure dependence and dilatancy of plastic flow. In *Proceedings ASME IUTAM Conference* (pp. 377–386).

Simo, J. C., & Hughes, T. J. R. (2006). *Computational Inelasticity*. Interdisciplinary Applied Mathematics (Vol. 7). Springer Science & Business Media.

Spitzig, W. A., & Richmond, O. (1984). The effect of pressure on the flow stress of metals. *Acta Metallurgica, 32*(3), 457–463. https://doi.org/10.1016/0001-6160(84)90119-6

Spitzig, W. A., Sober, R. J., & Richmond, O. (1975). Pressure dependence of yielding and associated volume expansion in tempered martensite. *Acta Metallurgica, 23*(7), 885–893. https://doi.org/10.1016/0001-6160(75)90205-9

Spitzig, W. A., Sober, R. J., & Richmond, O. (1976). The effect of hydrostatic pressure on the deformation behavior of maraging and HY-80 steels and its implications for plasticity theory. *Metallurgical Transactions A, 7*(10), 1703–1710. https://doi.org/10.1007/BF02817888

Starman, B., Halilovič, M., Vrh, M., & Štok, B. (2014). Consistent tangent operator for cutting-plane algorithm of elasto-plasticity. *Computer Methods in Applied Mechanics and Engineering, 272*, 214–232. https://doi.org/10.1016/j.cma.2013.12.012

Sun, L., & Wagoner, R. H. (2013). Proportional and non-proportional hardening behavior of dual-phase steels. *International Journal of Plasticity, 45*, 174–187. https://doi.org/10.1016/j.ijplas.2013.01.018

Swift, H. W. (1952). Plastic instability under plane stress. *Journal of the Mechanics and Physics of Solids, 1*(1), 1–18.

Teodosiu, C., & Hu, Z. (1998). Microstructure in the continuum modelling of plastic anisotropy. In J. V. Carstensen, T. Leffers, T. Lorentzen, O. B. Petersen, B. F. S. Sørensen, & G. Winkler (Eds.), *Nineteenth risø International Symposium on Materials Science 1998* (pp. 149–168).

Tozawa, Y. (1978). Plastic deformation behavior under conditions of combined stress. In D. Koistinen, & N. M. Wang, (Eds.), *Mechanics of Sheet Metal Forming* (pp. 81–110). Boston: Springer. https://doi.org/10.1007/978-1-4613-2880-3_4

Vincze, G., Barlat, F., Rauch, E. F., Tomé, C. N., Butuc, M. C., & Grácio, J. J. (2013). Experiments and modeling of low carbon steel sheet subjected to double strain path changes. *Metallurgical and Materials Transactions A, 44*(10), 4475–4479.

Voce, E. (1948). The relationship between stress and strain for homogeneous deformation. *The Journal of the Institute of Metals, 74*, 537–562.

Voyiadjis, G. Z., & Foroozesh, M. (1990). Anisotropic distortional yield model. *Transactions ASME: Journal of Applied Mechanics, 57*(3), 537–547. https://doi.org/10.1115/1.2897056

Wi, M.-S. (2021). Characterization and simulation of the plastic behavior of steels subject to complex loading histories. Ph.D. Thesis, Pohang University of Science and Technology.
Wi, M. S., Lee, S. Y., Kim, J. H., Kim, J. M., & Barlat, F. (2020). Experimental and theoretical plasticity analyses of steel materials deformed under a nonlinear strain path. *International Journal of Mechanical Sciences, 182*, 105770. https://doi.org/10.1016/j.ijmecsci.2020.105770
Yoon, S.-Y., Lee, S.-Y., & Barlat, F. (2020). Numerical integration algorithm of updated homogeneous anisotropic hardening model through finite element framework. *Computer Methods in Applied Mechanics and Engineering, 372*, 113449. https://doi.org/10.1016/j.cma.2020.113449
Yoon, S.-Y., Barlat, F., Lee, S.-Y., Kim, J.-H., Wi, M.-S., & Kim, D.-J. (2022). *Journal of Materials Processing Technology, 302*, 117494. https://doi.org/10.1016/j.jmatprotec.2022.117494
Yoshida, F., & Uemori, T. (2002). A model of large-strain cyclic plasticity describing the Bauschinger effect and workhardening stagnation. *International Journal of Plasticity, 18*(5–6), 661–686. https://doi.org/10.1016/S0749-6419(01)00050-X
Zaman, S. B., Barlat, F., & Kim, J.-H. (2018). Deformation-induced anisotropy of uniaxially prestrained steel sheets. *International Journal of Solids and Structures, 134*, 20–29.

Further readings

Barlat, F., Cazacu, O., Życzowski, M., Banabic, D., & Yoon, J.-W. (2004). Continuum scale simulation of engineering materials: Fundamentals-microstructures-process applications. In D. Raabe, F. Roters, F. Barlat, & L.-Q. Chen (Eds.), *Yield Surface Plasticity and Anisotropy* (pp. 145–177). Wiley.
Barlat, F., Kuwabara, T., & Korkolis, Y. P. (2018). Anisotropic plasticity and application to plane stress. In H. Altenbach, & A. Öchsner (Eds.), *Encyclopedia of Continuum Mechanics* (pp. 1–22). Berlin, Heidelberg: Springer Berlin Heidelberg. ISBN 978-3-662-53605-6. https://doi.org/10.1007/978-3-662-53605-6_225-1
Besson, J., Cailletaud, G., Chaboche, J.-L., & Forest, S. (2009). *Non-linear Mechanics of Materials*. Solid Mechanics and Its Applications (Vol. 167). Springer Science & Business Media.
Cazacu, O., & Revil-Baudard, B. (2020) *Plasticity of Metallic Materials: Modeling and Applications to Forming*. Elsevier.
Itskov, M. (2007). *Tensor Algebra and Tensor Analysis for Engineers*. Springer.
Lemaitre, J., & Chaboche, J.-L. (1994). *Mechanics of Solid Materials*. Cambridge University Press.
Malvern, L. E. (1969). *Introduction to the Mechanics of a Continuous Medium*. Englewood Cliffs, NJ: Prentice-Hall.
Nye, J. F. (1985). *Physical Properties of Crystals: Their Representation by Tensors and Matrices*. Oxford University Press.
Skrzypek, J. J., & Ganczarski, A. W. (2015). *Mechanics of Anisotropic Materials*. Springer.
Ziegler, H. (1959). A modification of prager's hardening rule. *Quarterly of Applied Mathematics, 17*(1), 55–65.

Chapter 3
Anisotropy of Yield/Failure Criteria—Comparison of Explicit and Implicit Formulations

Artur Ganczarski

Abstract Six lectures on anisotropic plasticity comprise the following subjects: description of anisotropy influence on limit criteria (yield/failure) for modern homogeneous metallic alloys, anisotropy of limit criteria, critical comparison of explicit versus implicit approaches, discussion on of physical interpretation and convexity of implicit approach.

3.1 Lecture—Preliminaries

3.1.1 Even Order Tensors—Invariants and Matrix Representations

Yield/failure criteria presented in this work use either invariants of second order tensors or common invariants of two or more tensors of even orders. Therefore, it is essential to recall some basic rules of the tensor algebra.

3.1.1.1 Second-Order Tensors

A second-order tensor $\boldsymbol{T}$, or t_{ij} with indices $i, j = 1, 2, 3$ or $i, j = x, y, z$, is represented by a 3×3 square matrix

$$[t] = [t_{ij}] = \begin{bmatrix} t_{11} & t_{12} & t_{13} \\ t_{21} & t_{22} & t_{23} \\ t_{31} & t_{32} & t_{33} \end{bmatrix} = \begin{bmatrix} t_{xx} & t_{xy} & t_{xz} \\ t_{yx} & t_{yy} & t_{yz} \\ t_{zx} & t_{zy} & t_{zz} \end{bmatrix} \tag{3.1}$$

where x, y, z refer to cartesian co-ordinate system.

A. Ganczarski (✉)
Chair of Applied Mechanics and Bomechanics, Department of Mechanical Engineering, Cracow University of Technology, Al. Jana Pawła II, 31-864 Kraków, Poland
e-mail: artur.ganczarski@pk.edu.pl

H. Altenbach and A. Ganczarski (eds.), *Advanced Theories for Deformation, Damage and Failure in Materials*, CISM International Centre for Mechanical Sciences 605, https://doi.org/10.1007/978-3-031-04354-3_3

In case of symmetry $t_{ij} = t_{ji}$ the tensor $\boldsymbol{T}$ has only 6 independent components and it may be represented as columnar vector

$$\{\boldsymbol{T}\} = \{t_{11}, t_{22}, t_{33}, t_{23}, t_{13}, t_{12}\}^{\mathrm{T}} \tag{3.2}$$

Any second-order tensor $\boldsymbol{T}$ may be decomposed into an axiator (spherical tensor) $\boldsymbol{A}_T$ and a deviator $\boldsymbol{D}_T$ as follows

$$\boldsymbol{T} = \boldsymbol{A}_T + \boldsymbol{D}_T \tag{3.3}$$

where

$$[\boldsymbol{A}_T] = \begin{bmatrix} t_{\mathrm{m}} & 0 & 0 \\ 0 & t_{\mathrm{m}} & 0 \\ 0 & 0 & t_{\mathrm{m}} \end{bmatrix}, \qquad [\boldsymbol{D}_T] = [d_{ij}] = \begin{bmatrix} t_{xx} - t_{\mathrm{m}} & t_{xy} & t_{xz} \\ t_{yx} & t_{yy} - t_{\mathrm{m}} & t_{yz} \\ t_{zx} & t_{zy} & t_{zz} - t_{\mathrm{m}} \end{bmatrix} \tag{3.4}$$

where $t_{\mathrm{m}} = \frac{1}{3}\mathrm{tr}(\boldsymbol{T})$ denotes the mean value of the diagonal components. In case of index notation corresponding decomposition into the axiator and the deviator is as follows

$$t_{ij} = \frac{1}{3}t_{kk}\delta_{ij} + d_{ij} = t_{\mathrm{m}}\delta_{ij} + d_{ij} = \frac{1}{3}\mathrm{tr}\,(\boldsymbol{T})\,\delta_{ij} + d_{ij} \tag{3.5}$$

where $\delta_{ij} = \begin{cases} 1 & i = j \\ 0 & i \neq j \end{cases}$ denotes Kronecker's symbol.

Classical tensorial transformation rule from i, j to k, l directions is

$$t_{kl} = n_{ki}n_{lj}t_{ij} \tag{3.6}$$

where second-order tensor transformation rule is applied and n_{ki}, n_{lj} denote direction cosines of the transformation from the original frame $i, j = x, y, z$ in the new reference frame $k, l = \xi, \eta, \zeta$. It is possible to distinguish the specific transformation into eigendirections (principal directions) for which the corresponding tensor takes the diagonal representation

$$\begin{bmatrix} t_1 & 0 & 0 \\ 0 & t_2 & 0 \\ 0 & 0 & t_3 \end{bmatrix} \tag{3.7}$$

Three principal components are determined as real roots of the cubic equation, being solution of eigenproblem for the tensor $\boldsymbol{T}$

$$\boldsymbol{T} = \lambda\mathbf{1} \tag{3.8}$$

where $\lambda_i = t_1, t_2, t_3$ stand for eigenvalues. These principal components are real roots of the characteristic equation $\lambda_i = t_i$

$$\det(\boldsymbol{T} - \lambda \mathbf{1}) = 0 \tag{3.9}$$

which can be rewritten in the equivalent fashion

$$\lambda^3 - I_{1T}\lambda^2 + I_{2T}\lambda - I_{3T} = 0 \tag{3.10}$$

Three coefficients of the characteristic equation (3.10) I_{1T}, I_{2T}, I_{3T} are called the principal invariants and may be defined in terms of tensor components

$$\begin{aligned} I_{1T} &= \mathrm{tr}(\boldsymbol{T}) = t_{ii} = t_{xx} + t_{yy} + t_{zz} \\ I_{2T} &= t_{xx}t_{yy} + t_{yy}t_{zz} + t_{zz}t_{xx} - (t_{xy}^2 + t_{yz}^2 + t_{zx}^2) \\ I_{3T} &= \det\boldsymbol{T} = t_{xx}t_{yy}t_{zz} + 2t_{xy}t_{yz}t_{zx} - (t_{xx}t_{yz}^2 + t_{yy}t_{xz}^2 + t_{zz}t_{xy}^2) \end{aligned} \tag{3.11}$$

Apart from the principal invariants, the basic invariants also called the generic invariants are of particular importance, namely

$$\begin{aligned} J_{1T} &= t_{ii} = \mathrm{tr}\,(\boldsymbol{T}), \\ J_{2T} &= \frac{1}{2}t_{ij}t_{ji} = \frac{1}{2}\mathrm{tr}\left(\boldsymbol{T}^2\right), \\ J_{3T} &= \frac{1}{3}t_{ij}t_{jk}t_{ki} = \frac{1}{3}\mathrm{tr}\left(\boldsymbol{T}^3\right) \end{aligned} \tag{3.12}$$

It is seen, that the basic invariants can be interpreted as traces of subsequent powers of the tensor $\boldsymbol{T}$, $\boldsymbol{T}^2 = \boldsymbol{T} \cdot \boldsymbol{T}$, $\boldsymbol{T}^3 = \boldsymbol{T} \cdot \boldsymbol{T} \cdot \boldsymbol{T}$, if appropriate coefficients 1, 1/2, 1/3 are used. Note that the basic invariants differ from the principal invariants which are coefficients of the characteristic equation (3.10).

The basic invariants J_{1T}, J_{2T}, J_{3T} are expressed in terms of the principal invariants I_{1T}, I_{2T}, I_{3T} as follows

$$J_{1T} = I_{1T}, \quad J_{2T} = \frac{1}{2}I_{1T}^2 - I_{2T}, \quad J_{3T} = \frac{1}{3}I_{1T}^3 - I_{1T}^2 I_{2T} + I_{3T} \tag{3.13}$$

consequently the reciprocal relations are given by equations

$$I_{1T} = J_{1T}, \quad I_{2T} = \frac{1}{2}J_{1T}^2 - J_{2T}, \quad I_{3T} = \frac{1}{6}J_{1T}^3 - J_{1T}^2 J_{2T} + J_{3T} \tag{3.14}$$

Decomposition of the tensor into the axiator (spherical tensor) and the deviator (3.4, 3.5) leads to the following system of the principal or the generic invariants of the deviator

$$\begin{aligned} J_{1D} &= d_{ii} = \text{tr}(\boldsymbol{D}_T) = 0, \\ J_{2D} &= \frac{1}{2} d_{ij} ds_{ji} = \frac{1}{2} \text{tr}\left(\boldsymbol{D}_T^2\right) \\ J_{3D} &= \frac{1}{3} d_{ij} d_{jk} d_{ki} = \frac{1}{3} \text{tr}\left(\boldsymbol{D}_T^3\right) \end{aligned} \tag{3.15}$$

where, in similar fashion as in Eq. (3.12), subsequent powers of the deviator $\boldsymbol{D}_T$, $\boldsymbol{D}_T^2 = \boldsymbol{D}_T \cdot \boldsymbol{D}_T$, $\boldsymbol{D}_T^3 = \boldsymbol{D}_T \cdot \boldsymbol{D}_T \cdot \boldsymbol{D}_T$ are used. Note that the first basic deviatoric stress invariant J_{1D} is equal to zero according to definition (3.5).

For purpose of further applications it is convenient to introduce additionally the cylindrical invariants of the stress tensor, called also the Haigh–Westergaard coordinates ξ, ρ and θ defined as follows

$$\xi = \frac{J_{1\sigma}}{\sqrt{3}}, \quad \rho = \sqrt{2J_{2s}}, \quad \cos(3\theta) = \frac{3\sqrt{3}}{2} \frac{J_{3s}}{(J_{2s})^{3/2}} \quad \text{for} \quad 0 \leq \theta \leq \frac{\pi}{3} \tag{3.16}$$

(cf. Haigh 1920; Westergaard 1920).

3.1.2 Fourth-Order Tensors

In many equations of plasticity common invariants of two or more tensors are used. The general theory of such invariants is rather complicated and some results were published by Goldenblat (1955); Spencer (1971). For the further application we quote here example of two symmetric second-order tensors and one symmetric fourth-order tensor, that is typical situation for anisotropy

$$t_{ij} B_{ijkl} t_{kl} \tag{3.17}$$

where the fourth-rank compliance tensor B_{ijkl} is uniquely defined by $3^4 = 81$ components, since each of indices i, j, k, l runs through 1, 2, 3. Conditions of symmetry of both second-order tensors $t_{ij} = t_{ji}$ and $t_{kl} = t_{lk}$ requires symmetry with respect to change of indices in pairs $i \leftrightarrow j$ and $k \leftrightarrow l$

$$B_{ijkl} = B_{jikl} = B_{ijlk} \tag{3.18}$$

Additionally, the positive definiteness of quadratic form requires the symmetry with respect to change of indices between pairs $ij \leftrightarrow kl$

$$B_{ijkl} = B_{klij} \tag{3.19}$$

In this way, from among 81 components of compliance tensor, only 21 are independent. In order to perform the common invariant (3.17) by use of vector-matrix Voigt's notation, both second-order tensors are rewritten as columnar vectors, if the following scheme of change between tensor $i, j = 1, 2, 3$ and vectors $k = 1, 2, \ldots, 6$ indices holds

$$\begin{array}{ccccccc} ij & 11 & 22 & 33 & 23, 32 & 31, 13 & 12, 21 \\ & \downarrow & \downarrow & \downarrow & \downarrow & \downarrow & \downarrow \\ k & 1 & 2 & 3 & 4 & 5 & 6 \end{array} \tag{3.20}$$

From the above scheme we get the representations of second-order tensors previously demonstrated by (3.2). Application of analogous scheme to the first and the second pairs of indices of the fourth-order tensor gives

$$\begin{aligned} B_{ijkl} &= B_{mn} \text{ if } m \text{ and } n \text{ go through } 1, 2, 3 \\ 2B_{ijkl} &= B_{mn} \text{ if } m \text{ or } n \text{ go through } 1, 2, 3 \\ 4B_{ijkl} &= B_{mn} \text{ if } m = n \text{ or } m \text{ or } n \text{ go through } 4, 5, 6 \\ 8B_{ijkl} &= B_{mn} \text{ if both } m \text{ and } n \text{ go through } 4, 5, 6 \end{aligned} \tag{3.21}$$

where appropriate factors 2, 4 or 8 are applied.

Next, application of aforementioned scheme to the common invariant (3.17) yields

$$\begin{array}{l} B_{1111}t_{11}^2 + B_{2222}t_{22}^2 + B_{3333}t_{33}^2 + 2B_{1122}t_{11}t_{22} + 2B_{2233}t_{11}t_{33} + 2B_{3311}t_{33}t_{11} \\ \quad\downarrow \qquad\quad \downarrow \qquad\quad \downarrow \qquad\quad \downarrow \qquad\quad \downarrow \qquad\quad \downarrow \\ B_{11}t_1^2 \;+\; B_{22}t_2^2 \;+\; B_{33}t_3^2 \;+\; B_{12}t_1t_2 \;+\; B_{23}t_2t_3 \;+\; B_{31}t_3t_1 \\[2ex] +4B_{1123}t_{11}t_{23} + 4B_{1131}t_{11}t_{31} + 4B_{1112}t_{11}t_{12} + 4B_{2223}t_{22}t_{23} + 4B_{2231}t_{22}t_{31} \\ \quad\downarrow \qquad\quad \downarrow \qquad\quad \downarrow \qquad\quad \downarrow \qquad\quad \downarrow \\ +B_{14}t_1t_4 \;+\; B_{15}t_1t_5 \;+\; B_{16}t_1t_6 \;+\; B_{24}t_2t_4 \;+\; B_{25}t_2t_5 \\[2ex] +4B_{2312}t_{23}t_{12} + 4B_{3323}t_{33}t_{23} + 4B_{3331}t_{33}t_{31} + 4B_{3312}t_{33}t_{12} + 4B_{2323}t_{23}^2 \\ \quad\downarrow \qquad\quad \downarrow \qquad\quad \downarrow \qquad\quad \downarrow \qquad\quad \downarrow \\ +B_{46}t_4t_6 \;+\; B_{34}t_3t_4 \;+\; B_{35}t_3t_5 \;+\; B_{36}t_3t_6 \;+\; B_{44}t_4^2 \\[2ex] +4B_{3131}t_{31}^2 + 4B_{1212}t_{12}^2 + 8B_{1223}t_{12}t_{23} + 8B_{2331}t_{23}t_{31} + 8B_{3112}t_{31}t_{12} \\ \quad\downarrow \qquad\quad \downarrow \qquad\quad \downarrow \qquad\quad \downarrow \qquad\quad \downarrow \\ +B_{55}t_5^2 \;+\; B_{66}t_6^2 \;+\; B_{64}t_6t_4 \;+\; B_{45}t_4t_5 \;+\; B_{56}t_5t_6 \end{array} \tag{3.22}$$

Finally, the common invariant is represented in vector-matrix notation as follows

$$\begin{Bmatrix} t_1 \\ t_2 \\ t_3 \\ t_4 \\ t_5 \\ t_6 \end{Bmatrix}^{\mathrm{T}} \begin{bmatrix} B_{11} & B_{12} & B_{13} & B_{14} & B_{15} & B_{16} \\ & B_{22} & B_{23} & B_{24} & B_{25} & B_{26} \\ & & B_{33} & B_{34} & B_{35} & B_{36} \\ & & & B_{44} & B_{45} & B_{46} \\ & & & & B_{55} & B_{56} \\ & & & & & B_{66} \end{bmatrix} \begin{Bmatrix} t_1 \\ t_2 \\ t_3 \\ t_4 \\ t_5 \\ t_6 \end{Bmatrix} \tag{3.23}$$

where, with respect of symmetry $B_{mn} = B_{nm}$, only the upper triangle is filled in.

3.1.3 Positive Definiteness of Quadratic Form $\{t\}^{\mathrm{T}}[B]\{t\}$ in Sylvester's Sense

Common invariant of two symmetric second-order tensors and the fourth-order tensor (3.17) turns out to be the quadratic form which in Voigt's notation is equivalent to (3.23). According to the Sylvester criterion, necessary and sufficient condition for positive definiteness of the quadratic form (3.23) is follows

$$\det[\boldsymbol{B}]_k > 0 \qquad k = 1, 2, \ldots, 6 \tag{3.24}$$

for arbitrary arguments $\{\boldsymbol{t}\}$, where $[\boldsymbol{B}]_k$ stands for minors (sub-matrices of dimension $k \times k$). Therefore, the Sylvester criterion (3.24) performs in unabbreviated format system of six inequalities, which guarantees positive definiteness

$$\begin{aligned} & B_{11} > 0 \\ & \det \begin{bmatrix} B_{11} & B_{12} \\ & B_{22} \end{bmatrix} > 0 \\ & \vdots \\ & \det \begin{bmatrix} B_{11} & B_{12} & B_{13} & B_{14} & B_{15} & B_{16} \\ & B_{22} & B_{23} & B_{24} & B_{25} & B_{26} \\ & & B_{33} & B_{34} & B_{35} & B_{36} \\ & & & B_{44} & B_{45} & B_{46} \\ & & & & B_{55} & B_{56} \\ & & & & & B_{66} \end{bmatrix} > 0 \end{aligned} \tag{3.25}$$

3.2 Lecture—General Concept of Limit Surfaces

3.2.1 Pressure Sensitive or Insensitive Yield Criteria

Invariant description of any limit surface has to be performed by the use of irreducible set of invariants being arguments of a scalar function defining limit surface. For anisotropic materials equation of limit surface is a scalar function of common stress and structural anisotropy tensor invariants (cf. Sayir 1970, etc.)

$$f[\boldsymbol{\sigma} : \overset{<2>}{\boldsymbol{A}}, \boldsymbol{\sigma} : \overset{<4>}{\boldsymbol{A}} : \boldsymbol{\sigma}, \boldsymbol{\sigma} : \overset{<6>}{\boldsymbol{A}} : \boldsymbol{\sigma} : \boldsymbol{\sigma}, \ldots] \tag{3.26}$$

where only even ranks of anisotropy tensors are taken into account.

In some cases of anisotropic alloys exhibiting tension/compression asymmetry it is convenient to consider a scalar function of selected (mixed) stress invariants and common invariants (cf. Khan and Liu 2012, etc.)

$$f\left[\mathrm{tr}(\boldsymbol{\sigma}),\ \frac{1}{2}\mathrm{tr}(\boldsymbol{s}\cdot\boldsymbol{s}),\ \frac{1}{3}\mathrm{tr}(\boldsymbol{s}\cdot\boldsymbol{s}\cdot\boldsymbol{s});\ldots,\ \boldsymbol{\sigma}:\overset{<4>}{\boldsymbol{A}}:\boldsymbol{\sigma},\ \ldots\right] \tag{3.27}$$

The following effects are of particular importance when describing features of the limit surfaces:

- hydrostatic pressure dependence
- tension/compression asymmetry

In order to properly capture all features considered the limit criteria have to include the second common invariant (either $\boldsymbol{\sigma}:\overset{<4>}{\boldsymbol{A}}:\boldsymbol{\sigma}$ or $\boldsymbol{s}:\overset{<4>}{\boldsymbol{A}}:\boldsymbol{s}$). This is a direct consequence of the necessity to include total or pure shear elastic energy. The presence of the first and the third common invariants $\boldsymbol{\sigma}:\overset{<2>}{\boldsymbol{A}}, \boldsymbol{\sigma}:\overset{<6>}{\boldsymbol{A}}:\boldsymbol{\sigma}:\boldsymbol{\sigma})$ is necessary to capture dependence on hydrostatic pressure and tension/compression asymmetry.

In general, materials can be classified into two groups: hydrostatic pressure dependent and hydrostatic pressure independent materials, alternatively called pressure sensitive and pressure insensitive materials. Traditionally, ductile materials can be considered as hydrostatic pressure independent. On the other hand, brittle materials should be treated as hydrostatic pressure dependent ones. Hydrostatic pressure dependence of anisotropic limit criteria can be captured in the two different manners:

- direct dependence on the hydrostatic pressure through both the first and the second common deviatoric invariants

$$f[\boldsymbol{\sigma}:\overset{<2>}{\boldsymbol{A}},\boldsymbol{s}:\overset{<4>}{\boldsymbol{A}}:\boldsymbol{s}] \tag{3.28}$$

- indirect dependence on the hydrostatic pressure through the second common stress invariant

$$f(\boldsymbol{\sigma}:\overset{<4>}{\boldsymbol{A}}:\boldsymbol{\sigma}) \qquad (\boldsymbol{\sigma}=\boldsymbol{s}+\sigma_{\mathrm{h}}\mathbf{1}) \tag{3.29}$$

There exists a broad class of engineering materials which do not exhibit any dependence on the hydrostatic pressure, neither direct nor indirect. This means that in case of the isotropic hydrostatic pressure independent materials the corresponding limit surfaces have to include the second deviatoric invariant exclusively. In case of anisotropy limit surfaces can include the second common deviatoric invariant and additionally the first common deviatoric invariant. However in all cases considered equation of limit surface has to include the second stress or the second common invariants which results from the quadratic form of energy representation. Exemplary equations of limit surfaces that found confirmation in engineering materials are presented in Table 3.1 according to aforementioned classification.

Tension/compression asymmetry, also called strength differential effect is included in a natural way in limit criteria for anisotropic materials. However limit surfaces based on the second common invariants exclusively are not capable of capture shape change due to distortion. By contrast the limit criteria based on the second and the

Table 3.1 Hydrostatic pressure dependence of initial yield/failure criteria

Dependence on hydrostatic pressure σ_h		
Direct	Indirect	Lack of dependence
Tsai–Wu	von Mises	Hill
$\boldsymbol{\sigma} : \overset{<2>}{\boldsymbol{A}} + \boldsymbol{s} : \overset{<4>}{\boldsymbol{A}^{\mathrm{H}}} : \boldsymbol{s} = 1$	$\boldsymbol{\sigma} : \overset{<4>}{\boldsymbol{A}} : \boldsymbol{\sigma} = 1$	$\boldsymbol{s} : \overset{<4>}{\boldsymbol{A}^{\mathrm{H}}} : \boldsymbol{s} = 1$

Table 3.2 Effect of first and third common invariants on tension/compression asymmetry and distortion of limit surfaces

Lack of asymmetry and distortion	Asymmetry without distortion	Asymmetry and distortion
Hill	Tsai–Wu	Kowalsky et al.
$\boldsymbol{s} : \overset{<4>}{\boldsymbol{A}^{\mathrm{H}}} : \boldsymbol{s} = 1$	$\boldsymbol{\sigma} : \overset{<2>}{\boldsymbol{A}} + \boldsymbol{s} : \overset{<4>}{\boldsymbol{A}^{\mathrm{H}}} : \boldsymbol{s} = 1$	$\boldsymbol{s} : \overset{<4>}{\boldsymbol{A}} : \boldsymbol{s} + \boldsymbol{s} : \overset{<6>}{\boldsymbol{A}} : \boldsymbol{s} : \boldsymbol{s} = 1$

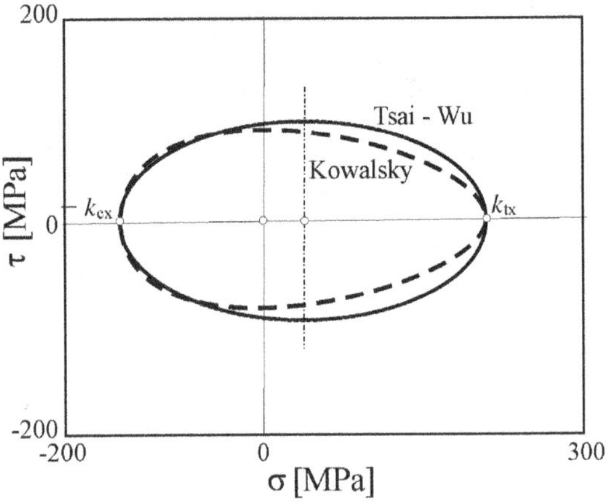

Fig. 3.1 Comparision of tension/compression asymmetry and distortion of limit curves in case of Tsai–Wu citerion versus Kowalsky et al. criterion

third common invariants are capable of capture both tension/compression asymmetry and distortion. Table 3.2 shows selected anisotropic criteria that correspond to the lack of tension/compression asymmetry and distortion, tension/compression asymmetry with distortion ruled out and tension/compression asymmetry with distortion accounted for. To illustrate classification described in Table 3.2 a comparison between asymmetry without distortion and asymmetry with distortion accounted for is presented in Fig. 3.1. The Tsai–Wu criterion is compared with the Kowalsky et al. criterion. The Tsai–Wu criterion accounts for tension/compression asymmetry without distortion (only translation through the first common invariant accounted for). By contrast when the Kowalsky et al. the six order criterion is used the tension/compression asymmetry and shape distortion are coupled in an anisotropic fashion through the third common invariant.

In general, a material anisotropy can be captured by use of the two approaches. In the first mathematically consistent approach called the explicit anisotropy approach the system of stress invariants $J_{1\sigma}$, J_{2s}, J_{3s} is substituted by the corresponding system of common invariants $\boldsymbol{\sigma} : \overset{<2>}{\boldsymbol{A}}$, $\boldsymbol{s} : \overset{<4>}{\boldsymbol{A}} : \boldsymbol{s}$, $\boldsymbol{s} : \overset{<6>}{\boldsymbol{A}} : \boldsymbol{s} : \boldsymbol{s}$ according to the Goldenblat, Kopnov and Sayir concept when formulating anisotropic limit criteria. In the other currently dynamically developed approach called the implicit anisotropy approach by Barlat and Khan either the second J_{2s} and the third J_{3s} stress invariants are substituted by the corresponding transformed deviatoric invariants J^0_{2s}, J^0_{3s} or the stress deviator is transformed by use of the two independent 4-rank transformation tensors $\boldsymbol{\Sigma}' = \overset{<4>}{\boldsymbol{C}'} : \boldsymbol{s}$ and $\boldsymbol{\Sigma}'' = \overset{<4>}{\boldsymbol{C}''} : \boldsymbol{s}$ and next they are inserted to one of well know isotropic criteria, either Drucker's one or Hosford's one respectively. These linear transformations correspond to mapping of the deviatoric Cauchy stress tensor $\boldsymbol{\sigma}$ to other two deviatoric stresses $\boldsymbol{\Sigma}'$, $\boldsymbol{\Sigma}''$ referring to the material anisotropy (orthotropy) frame.

The implicit approach is able to capture the full material orthotropy with distortion effect included by use of two 4-rank orthotropic transformation tensors $\overset{<4>}{\boldsymbol{C}'}$, $\overset{<4>}{\boldsymbol{C}''}$ containing $2 \times 9 = 18$ independent material constants by contrast to the explicit common invariant based approach which requires in case of material orthotropy 4-rank tensor $\overset{<4>}{\boldsymbol{A}}$ and 6-rank tensor $\overset{<6>}{\boldsymbol{A}}$ containing $9 + 56 = 65$ material constants. Although the explicit approach is more mathematically rigorous than the implicit one but simultaneously it is much more cumbersome and open to misunderstandings. Both approaches, the explicit and the implicit, are alternative ones and obviously they lead to different approximations.

Comparison of the explicit and the implicit approaches to capture anisotropy is schematically presented for selected criteria in Table 3.3. A major difficulty for the limit yield/failure description is caused by the coupling between anisotropy and strong tension/compression asymmetry as discussed by Khan et al. (2012). Such significant coupling can lead to a complete distortion of the limit surface (possible lack of any axis of symmetry) as it is presented in Fig. 3.2 based on Luo et al. (2007) experimental findings for AZ31B Mg alloy fitted by Plunkett et al. (2008).

Table 3.3 Explicit or implicit anisotropy of limit surfaces

Explicit	Implicit
Hill $\boldsymbol{s} : \overset{<4>}{\boldsymbol{A}^{\mathrm{H}}} : \boldsymbol{s} - 1$	Cazacu and Barlat $(J^0_2)^{3/2} - cJ^0_3 = k^3$ Plunkett et al. $\sum_{i=1}^{3}(\lvert\Sigma'_i\rvert - \widehat{k}'\Sigma'_i)^a + \sum_{i=1}^{3}(\lvert\Sigma''_i\rvert - \widehat{k}''\Sigma''_i)^a = 2k^a$

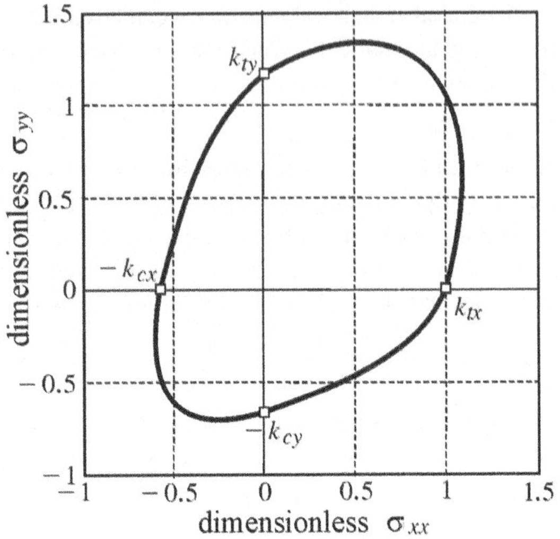

Fig. 3.2 Fitting of Luo et al. experimental data for AZ31B Mg alloy to the implicit anisotropic yield criterion by Cazacu and Barlat

3.2.2 *Survey of Symmetry Groups*

Identification of material symmetry in plastic range of deformation (anisotropy, orthotropy, transverse isotropy, etc.) is a starting point to appropriate description of the limit yield/failure criteria.

Analogy between chosen symmetry groups of constitutive matrices of elasticity and initiation of plasticity is presented in Table 3.4. In the fundamental book by Love (1944) the analogy between crystal symmetry classes and groups from one hand and appropriate forms of elastic strain energy function $\frac{1}{2}\{\varepsilon\}^{\mathrm{T}}\,[\overset{<4>}{\boldsymbol{E}}]\,\{\varepsilon\}$ from the other, is demonstrated. In the present lecture an extension of the aforementioned analogy also for symmetry of constitutive matrix of plastic yield initiation $[\overset{<4>}{\boldsymbol{A}}]$ appearing in the von Mises criterion $\{\sigma\}^{\mathrm{T}}\,[\overset{<4>}{\boldsymbol{A}}]\,\{\sigma\} = 1$ is proposed. The corresponding constitutive elasticity matrices have schematically been presented applying Nye (1957) graphics (symbol ● refers to independent element, symbol ○ refers to dependent element, whereas symbols ●—● or ○—○ represent pairs of identical matrix elements).

In case of full anisotropy the complete analogy between the Hooke matrix and the von Mises plasticity matrix holds (21 independent matrix elements in both classes). However, when narrower symmetry groups are considered: orthotropic, transversely isotropic of tetragonal or hexagonal classes, it is necessary to notice that elastic matrices are usually defined in stress tensor coordinates, whereas plastic constitutive matrices are often defined in the narrower stress deviator coordinates.

Reduction of the tensorial space to the deviatoric one is always equivalent to imposing additional constraints, hence number of independent elements of plasticity matrix is always lower than corresponding number of independent elements of elasticity matrix. Namely, it is clear that the 6–element orthotropic deviatoric

Table 3.4 Analogy between chosen symmetry groups of Hooke's matrix and plastic yield initiation von Mises' matrix

Chosen constitutive matrix symmetry			
Elastic Hooke's matrix [*E*]	Plastic yield initiation matrix [*A*]	Elastic Hooke's matrix [*E*]	Plastic yield initiation matrix [*A*]
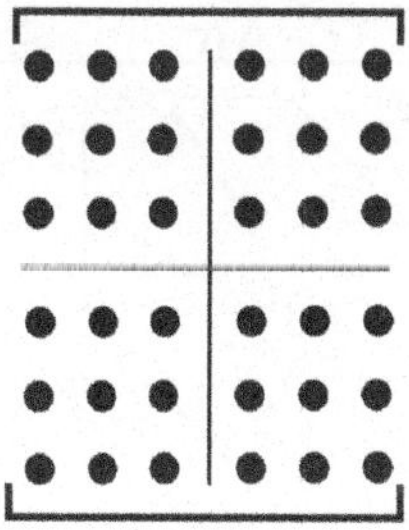 Hooke's (21 constants)	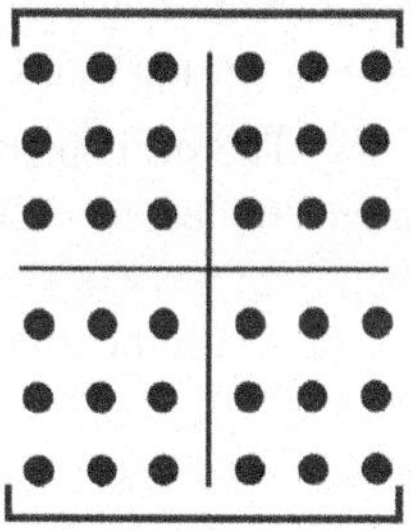 von Mises (21 constants)	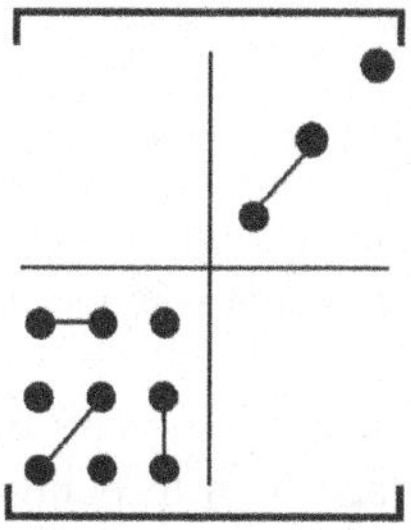 transversely isotropic tetragonal Hooke's (6 constants)	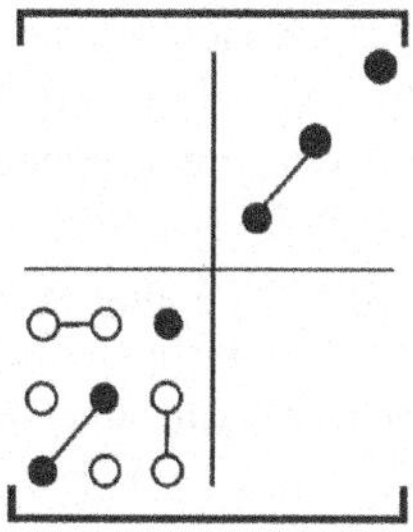 deviatoric transversely isotropic teragonal Hill's (4 constants)
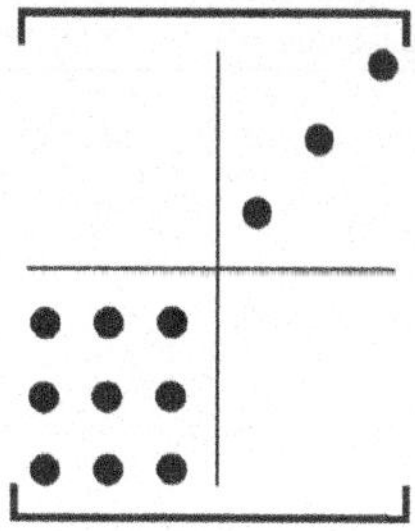 orthotropic Hooke's (9 constants)	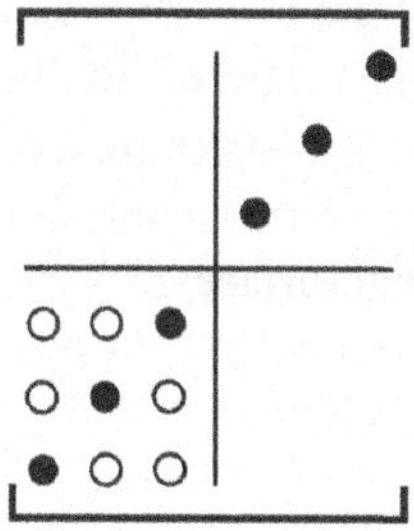 deviatorc Hill's (6 constants)	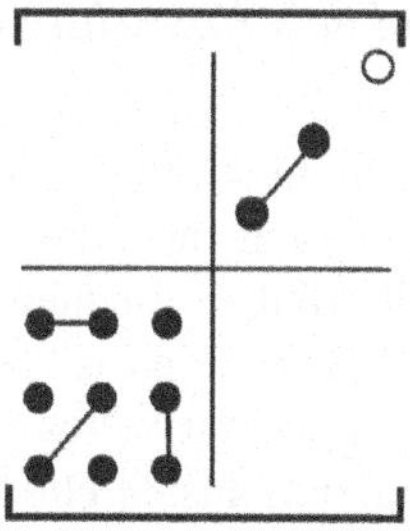 transversely isotropic hexagonal Hooke's (5 constants)	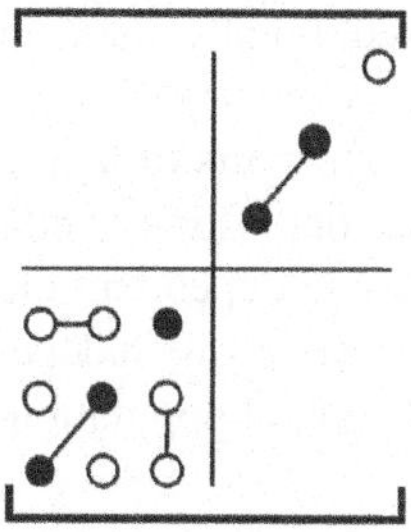 pseudodeviatoric transversely isotropic hexagonal Hu-Marin's (3 constants)

Hill's matrix corresponds to the 9–element orthotropic Hooke's matrix. Similarly, the 4–element transversely isotropic tetragonal class Hill's matrix corresponds to the 6–element Hooke's matrix, when the independence of Hill's matrix of hydrostatic stress is imposed. Finally, the 3–element transversely isotropic hexagonal class Hu–Marin's matrix corresponds to the 5–element transversely isotropic hexagonal class Hooke matrix. The pairs of identical matrix elements are arranged in the same way in both matrices of elasticity and plasticity.

Although, some dependent elements in the plasticity matrix, as represented by symbol ○, correspond to independent elements of elasticity matrix, sketched by symbol ●, but general population of both matrices remains unchanged.

3.2.3 Drucker's Postulate of Stability

Formulation of yield/failure criteria requires application of Drucker's postulate that guarantees, as consequence, the convexity of the yield/failure surface and the associated flow rule. Both consequences play fundamental role in a proof of existence and the uniqueness of solution.

Consider first arbitrary stress cycle 0AB0 that consists of: elastic loading 0A from the initial state inside current yield surface (σ_{ij}^*) to point belonging to this surface (σ_{ij}), subsequent elementary loading AB corresponding to stress increment $\mathrm{d}\sigma_{ij}$ during which the yield surface is rebuilt $f_i \rightarrow f_{i+1}$, and final unloading B0 to the initial stress level (σ_{ij}^*) as shown in Fig. 3.3. Note however that this process describes the closed cycle only in stress space $\sigma_{ij}^0 = \sigma_{ij}^*$ but the final state (point 0) corresponds to changed strain state $\varepsilon_{ij}^0 = \varepsilon_{ij}^{\mathrm{e}} + \mathrm{d}\varepsilon_{ij}^{\mathrm{p}}$. The strain increment $\mathrm{d}\varepsilon_{ij}^{\mathrm{p}}$ stands for permanent and irreversible plastic strain change connected with rebuilding of the subsequent yield surface.

According to the Drucker postulate work per unit volume done by stress quasi-cycle on total deformation 0AB0 is nonnegative

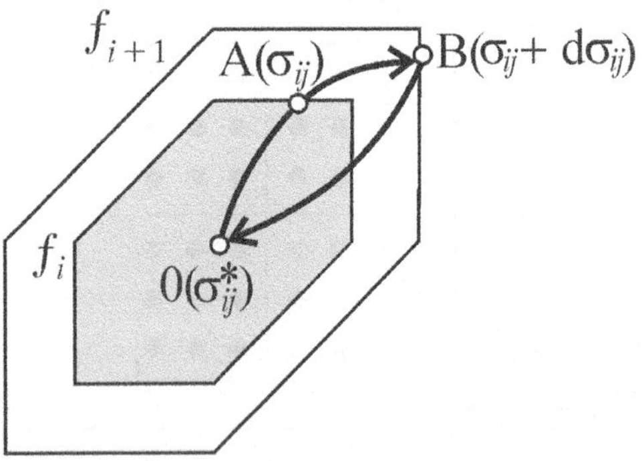

Fig. 3.3 Ilustration of the closed stress quasi-cycle

$$\oint \left(\sigma_{ij} - \sigma_{ij}^*\right) \mathrm{d}\varepsilon_{ij} \geq 0 \tag{3.30}$$

The additional load carried by the material over a complete stress quasi-cycle is called the external agency. In other words, when the work done by an external agency over the stress quasi-cycle would be negative a subsequent equilibrium state would have been reached in a spontaneous way associated with an energy dissipation. When according to small strain theory the additive decomposition of the total strain increment into the reversible and irreversible terms is done $\mathrm{d}\varepsilon_{ij} = \mathrm{d}\varepsilon_{ij}^{\mathrm{e}} + \mathrm{d}\varepsilon_{ij}^{\mathrm{p}}$ we arrive at

$$\oint \left(\sigma_{ij} - \sigma_{ij}^*\right) \mathrm{d}\varepsilon_{ij}^{\mathrm{e}} + \oint \left(\sigma_{ij} - \sigma_{ij}^*\right) \mathrm{d}\varepsilon_{ij}^{\mathrm{p}} \tag{3.31}$$

The first of above integrals is equal to zero since, according to the Hooke law $\varepsilon_{ij}^{\mathrm{e}} = C_{ijkl}\sigma_{kl}$ holds. Hence, keeping in mind that plastic strain is different from zero $\varepsilon_{ij}^{\mathrm{p}} \neq 0$ only along path BC the Drucker postulate (3.30) is finally expressed by inequality for the following simple integral (not circular integral)

$$\int_{\mathrm{BC}} \left(\sigma_{ij} - \sigma_{ij}^*\right) \mathrm{d}\varepsilon_{ij}^{\mathrm{p}} \geq 0 \tag{3.32}$$

This means that the work done by the external agency on plastic strain is nonnegative and corresponds to rebuilding of yield surface $f_i \longrightarrow f_{i+1}$. Applying expansion of integral (3.32) in the Taylor series around the initial point $\sigma_{ij} = \sigma_{ij}^*$ and saving only two first non-zero terms we arrive at

$$\left(\sigma_{ij} - \sigma_{ij}^*\right) \mathrm{d}\varepsilon_{ij}^{\mathrm{p}} + \frac{1}{2}\mathrm{d}\sigma_{ij}\mathrm{d}\varepsilon_{ij}^{\mathrm{p}} \geq 0 \tag{3.33}$$

which must hold for arbitrary initial stress state σ_{ij}^*, inside or on the current yield surface. Therefore the inequality (3.33) that expresses the condition of stability of elastic-plastic material in Drucker's sense can be finally furnished in the form of two following inequalities

$$(\sigma_{ij} - \sigma_{ij}^*)\mathrm{d}\varepsilon_{ij}^{\mathrm{p}} \geq 0 \quad \text{and} \quad \mathrm{d}\sigma_{ij}\mathrm{d}\varepsilon_{ij}^{\mathrm{p}} \geq 0 \tag{3.34}$$

The above inequalities must simultaneously hold which in mathematical sense corresponds to non-negative value of the first and the second energy differential in the neighborhood of the initial point $\sigma_{ij} = \sigma_{ij}^*$.

Conditions (3.34) can be interpreted in a geometric way regarding convexity of a yield surface and normality of vector of plastic strain increment.

First of the inequalities (3.34) can interpreted as non-negative value of the scalar product of two vectors $(\boldsymbol{\sigma} - \boldsymbol{\sigma}^*)$ and $\mathrm{d}\boldsymbol{\varepsilon}^{\mathrm{p}}$. Hence, the angle ψ between these two vectors in the stress space σ_{ij} has to be either acute or right angle $\psi \leq \pi/2$ (Fig. 3.4a). This condition holds for each vector $\boldsymbol{\sigma}^*$ which is located on or within the yield surface.

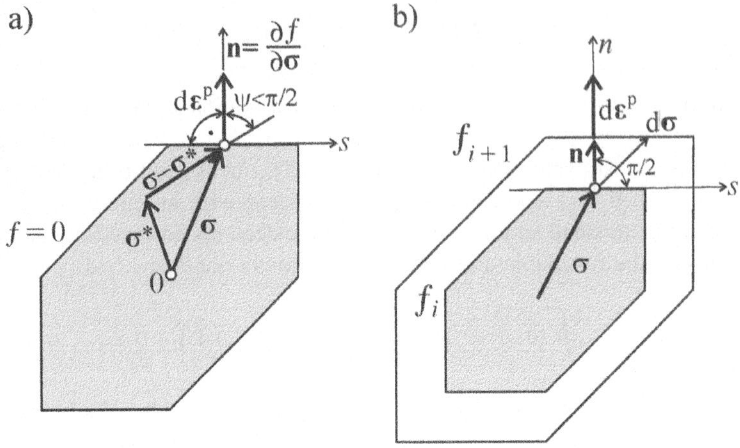

Fig. 3.4 Interpretation of Drucker's postulate consequences: **a** convexity, **b** normality

This implies that the yield surface must be convex surface in the stress space f. It is called in literature the convexity postulate of yield surface f.

The second of inequalities (3.34) can be interpreted as the scalar product of two vectors $\mathrm{d}\boldsymbol{\sigma}$ and $\mathrm{d}\boldsymbol{\varepsilon}^{\mathrm{p}}$ which must also be non-negative for arbitrarily chosen stress increment $\mathrm{d}\boldsymbol{\sigma}$ (Fig. 3.4b). This requirement must hold for arbitrary vectors $\mathrm{d}\boldsymbol{\sigma}$ connected with transition of f_i surface into f_{i+1} surface (Fig. 3.4b) which belong to the half-space tangent to ith surface, hence the only one possible vector $\mathrm{d}\boldsymbol{\varepsilon}^{\mathrm{p}}$ which always ensures condition (3.34) must be normal to this surface f, $\boldsymbol{n} = \boldsymbol{n}_f$

$$\mathrm{d}\boldsymbol{\varepsilon}^{\mathrm{p}} = \lambda \boldsymbol{n} = \lambda \frac{\partial f}{\partial \boldsymbol{\sigma}} = \lambda \mathbf{grad} f \tag{3.35}$$

Symbol λ is the scalar multiplier the magnitude of which ensures that the new point at the stress trajectory belongs to the new yield surface f. The condition (3.35) determines direction of the plastic strain increment $\mathrm{d}\boldsymbol{\varepsilon}^{\mathrm{p}}$ consistent with the gradient $\boldsymbol{n} = \partial f / \partial \boldsymbol{\sigma} = \lambda \mathbf{grad} f$ which is normal to the yield surface. This requirement is equivalent to the so called flow rule associated with the yield surface.

Drucker's postulate of stability (3.34) assumes that both convexity and normality rules must hold. In case when the normality does not hold it is possible to choose $\boldsymbol{\sigma}^* \neq \boldsymbol{\sigma}$ that the angle between vectors $(\boldsymbol{\sigma} - \boldsymbol{\sigma}^*)$ and $\mathrm{d}\boldsymbol{\varepsilon}^{\mathrm{p}}$ is greater than $\pi/2$ such that the scalar product of these two vectors is negative. This means that in this case Drucker's postulate is not satisfied. By contrast when the convexity of a yield surface is violated it is possible to choose $\boldsymbol{\sigma}^*$ such that the scalar product is negative $(\boldsymbol{\sigma} - \boldsymbol{\sigma}^*)\mathrm{d}\boldsymbol{\varepsilon}^{\mathrm{p}} < 0$. Both above negative examples are an indirect proof that violation even one of normality or convexity conditions means violation of Drucker's postulate as a whole.

At the end of this lecture it is essential to briefly compare the Sylvester criterion of positive definiteness with Drucker's postulate of stability. In both cases, the convexity of the yield/failure surface is guaranteed, however, from the mathematical point of view, Drucker's postulate is weaker constraint than the Sylvester criterion. In fact, the first one is requirement of nonnegative elementary work of plastic strain which can be satisfied even for a case when only one scalar product is positive and dominant over others. By contrast, the other one is stronger constraint since it requires positive definiteness of the quadratic from (3.23).

3.3 Lecture – Initial Yield Criteria of Pressure Insensitive Materials

3.3.1 von Mises Anisotropic Criterion

In a general case of material anisotropy, extension of the isotropic yield initiation criteria to the anisotropic yield/failure behaviour, by the use of common invariants of the stress tensor and of the structural tensors of plastic anisotropy (cf. Hill 1948; Sayir 1970; Betten 1988; Życzkowski 2001), can be shown in a general fashion

$$f\left(A, A_{ij}\sigma_{ij}, A_{ijkl}\sigma_{ij}\sigma_{kl}, A_{ijklmn}\sigma_{ij}\sigma_{kl}\sigma_{mn}, \ldots\right) = 0 \tag{3.36}$$

where Einstein's summation convention holds. In such a case, initiation of plastic flow or failure is governed by the structural tensors of material anisotropy of even-ranks: $\overset{<0>}{\boldsymbol{A}} = A$, $\overset{<2>}{\boldsymbol{A}} = A_{ij}$, $\overset{<4>}{\boldsymbol{A}} = A_{ijkl}$, $\overset{<6>}{\boldsymbol{A}} = A_{ijklmn}, \ldots$, etc.

In a particular case when a general tonsorially-polynomial form of Eq. (3.36) is assumed (cf. Sayir 1970; Kowalsky et al. 1999; Życzkowski 2001; Ganczarski and Skrzypek 2014) the polynomial anisotropic yield criterion is furnished

$$(A_{ij}\sigma_{ij})^{\alpha} + (A_{ijkl}\sigma_{ij}\sigma_{kl})^{\beta} + (A_{ijklmn}\sigma_{ij}\sigma_{kl}\sigma_{mn})^{\gamma} + \ldots - 1 = 0 \tag{3.37}$$

where if the Voigt notation is used and the structural anisotropy tensors take corresponding matrix forms

$$[\overset{<2>}{\boldsymbol{A}}] = \begin{bmatrix} a_{11} & a_{12} & a_{13} \\ & a_{22} & a_{23} \\ & & a_{33} \end{bmatrix} \tag{3.38}$$

and

$$[\overset{<4>}{\boldsymbol{A}}] = \begin{bmatrix} A_{11} & A_{12} & A_{13} & A_{14} & A_{15} & A_{16} \\ & A_{22} & A_{23} & A_{24} & A_{25} & A_{26} \\ & & A_{33} & A_{34} & A_{35} & A_{36} \\ & & & A_{44} & A_{45} & A_{46} \\ & & & & A_{55} & A_{56} \\ & & & & & A_{66} \end{bmatrix} \tag{3.39}$$

The even-rank structural anisotropy tensors A_{ij}, A_{ijkl}, A_{ijklmn}, ..., in Eq. (3.37) are normalized by the common constant A and $\alpha, \beta, \gamma \ldots$ etc., are arbitrary exponents of a polynomial representation. In a narrower case if $\alpha = 1, \beta = 1/2, \gamma = 1/3$, and limiting an infinite form (3.37) to the equation that contains only three common invariants, we arrive at the narrower form known as the Goldenblat and Kopnov criterion (Goldenblat and Kopnov 1966)

$$A_{ij}\sigma_{ij} + (A_{ijkl}\sigma_{ij}\sigma_{kl})^{1/2} + (A_{ijklmn}\sigma_{ij}\sigma_{kl}\sigma_{mn})^{1/3} - 1 = 0 \tag{3.40}$$

which satisfies the dimensional homogeneity of three polynomial components.

Equation (3.40), when limited only to three common invariants of the stress tensor $\boldsymbol{\sigma}$ and structural anisotropy tensors of even orders: 2nd A_{ij}, 4th A_{ijkl} and 6th A_{ijklmn} is not the most general one, in the meaning of the representation theorems, which determine the most general irreducible representation of the scalar and tensor functions that satisfy the invariance with respect to change of coordinates and material symmetry properties (cf. e.g. Spencer 1971; Rymarz 1993; Rogers 1990). However, 2nd, 4th and 6th order structural anisotropy tensors, which are used in (3.40) or in case if $\alpha = 1, \beta = 1, \gamma = 1$ and the deviatoric stress representation is used by Kowalsky et al. (1999)

$$h^{(1)}_{ij} s_{ij} + h^{(2)}_{ijkl} s_{ij} s_{kl} + h^{(3)}_{ijklmn} s_{ij} s_{kl} s_{mn} - h^{(0)} = 0 \tag{3.41}$$

are found satisfactory for describing fundamental transformation modes of limit surfaces caused by plastic or failure processes, namely: isotropic change of size, kinematic translation and rotation, as well as surface distortion (cf. Betten 1988; Kowalsky et al. 1999).

In what follows, we shall reduce class of the limit surface from the general tensorially-polynomial representation to the forms independent of both the first $A_{ij}\sigma_{ij}$ and the third $A_{ijklmn}\sigma_{ij}\sigma_{kl}\sigma_{mn}$ common invariants, but preserving the most general representation for the second common invariant, according to von Mises (1913, 1928). In such a case the 4th rank tensor of material anisotropy A_{ijkl} is, in general, defined by 21 anisotropy modules (but 18 of them independent), since the anisotropy 6×6 matrix $[\boldsymbol{A}]_{ij}$ (3.39) can completely be populated. Further reduction of the number of modules to 15 will be achieved, when the insensitivity of general von Mises quadratic form with respect to the change of hydrostatic stress will be assumed. In such a way the general tensorial von Mises criterion will be reduced to the deviatoric von Mises form defined by 15 anisotropy modules. A choice of 15 anisotropy modules considered as independent is, in general, not unique (cf. Szczepiński 1993; Ganczarski and Skrzypek 2013). However, the 15–parameter deviatoric von Mises criterion is sensitive to the change of sign of shear stresses, which may be considered as questionable (cf. e.g. Malinin and Rżysko 1981). Simplest way to avoid a doubtful physical explanation for existence of terms linear for shear stresses τ_{ij}, a reduction of the 15–parameter von Mises equation to the 9–parameter orthotropic von Mises criterion can be done. This form does not satisfy the deviatoric property, but when the constraint of independence of the hydrostatic stress is consistently applied, it is easily reduced to the deviatoric form, known as orthotropic Hill's criterion, with only 6 independent moduli of orthotropy (cf. Hill 1948).

Limiting ourselves to plastic yield initiation in ductile materials, a consecutive reduction of the general tensorially-polynomial anisotropic criterion (3.40) to the form dependent only on the 4th rank common invariant $\sigma_{ij} A_{ijkl} \sigma_{kl}$ holds, as it was proposed in the von Mises criterion for anisotropic yield initiation (cf. von Mises 1913, 1928).

$$\sigma_{ij} A_{ijkl} \sigma_{kl} - 1 = 0 \tag{3.42}$$

When the more convenient Voigt's vector-matrix notation is used, the form equivalent to (3.42) is obtained

$$\{\boldsymbol{\sigma}\}^{\mathrm{T}} [\overset{<4>}{\boldsymbol{A}}] \{\boldsymbol{\sigma}\} - 1 = 0 \tag{3.43}$$

where only one fourth-rank tensor of plastic anisotropy $\boldsymbol{A}$ is saved.

The structural 4th rank tensor of plastic anisotropy in equation (3.42) must be symmetric: $A_{ijkl} = A_{klij} = A_{jikl} = A_{ijlk}$, if stress tensor symmetry is assumed. Hence, in case if none other symmetry properties are implied, the von Mises plastic anisotropy tensor is defined by 21 modules. Finally, the general anisotropic von Mises criterion can be furnished as

$$\begin{aligned}
&A_{xxxx}\sigma_x^2 + A_{yyyy}\sigma_y^2 + A_{zzzz}\sigma_z^2 + 2A_{xxyy}\sigma_x\sigma_y + 2A_{yyzz}\sigma_y\sigma_z \\
&+2A_{zzxx}\sigma_z\sigma_x + 4A_{xxyz}\sigma_x\tau_{yz} + 4A_{xxzx}\sigma_x\tau_{zx} + 4A_{xxxy}\sigma_x\tau_{xy} \\
&+4A_{yyyz}\sigma_y\tau_{yz} + 4A_{yyzx}\sigma_y\tau_{zx} + 4A_{yyxy}\sigma_y\tau_{xy} + 4A_{zzyz}\sigma_z\tau_{yz} \\
&+4A_{zzzx}\sigma_z\tau_{zx} + 4A_{zzxy}\sigma_z\tau_{xy} + 8A_{xyyz}\tau_{xy}\tau_{yz} + 8A_{yzzx}\tau_{yz}\tau_{zx} \\
&+8A_{zxxy}\tau_{zx}\tau_{xy} + 4A_{yzyz}\tau_{yz}^2 + 4A_{zxzx}\tau_{zx}^2 + 4A_{xyxy}\tau_{xy}^2 = 1
\end{aligned} \tag{3.44}$$

where A_{ijkl} denote 21 components of the von Mises plastic anisotropy tensor.

The von Mises 6 × 6 matrix of plastic anisotropy, being symmetric and fully populated matrix representation of the 4th rank anisotropy tensor A_{ijkl} shown in (3.42), is furnished as follows

$$[\overset{<4>}{\boldsymbol{A}}] = \begin{bmatrix} A_{11} & A_{12} & A_{13} & A_{14} & A_{15} & A_{16} \\ & A_{22} & A_{23} & A_{24} & A_{25} & A_{26} \\ & & A_{33} & A_{34} & A_{35} & A_{36} \\ & & & A_{44} & A_{45} & A_{46} \\ & & & & A_{55} & A_{56} \\ & & & & & A_{66} \end{bmatrix} \tag{3.45}$$

The matrix coordinates A_{mn} are consistently defined by the tensorial coordinates A_{ijkl} according to scheme (3.21) and the general anisotropic von Mises equation equivalent to (3.44)

$$
\begin{aligned}
&A_{11}\sigma_x^2 + A_{22}\sigma_y^2 + A_{33}\sigma_z^2 + 2(A_{12}\sigma_x\sigma_y + A_{23}\sigma_y\sigma_z + A_{31}\sigma_z\sigma_x + A_{14}\sigma_x\tau_{yz} + \\
&A_{15}\sigma_x\tau_{zx} + A_{16}\sigma_x\tau_{xy} + A_{24}\sigma_y\tau_{yz} + A_{25}\sigma_y\tau_{zx} + A_{26}\sigma_y\tau_{xy} + A_{34}\sigma_z\tau_{yz} + \\
&A_{35}\sigma_z\tau_{zx} + A_{36}\sigma_z\tau_{xy} + A_{45}\tau_{yz}\tau_{zx} + A_{46}\tau_{xy}\tau_{yz} + A_{56}\tau_{zx}\tau_{xy}) + \\
&A_{44}\tau_{yz}^2 + A_{55}\tau_{zx}^2 + A_{66}\tau_{xy}^2 = 1
\end{aligned}
\tag{3.46}
$$

Representation of the anisotropic von Mises condition (3.43) in deviatoric form is done by stress decomposition $\boldsymbol{\sigma} = \boldsymbol{s} + \sigma_{\mathrm{h}}\mathbf{1}$

$$
\{s\}^{\mathrm{T}} [\overset{<4>}{\boldsymbol{A}}] \{s\} + \underline{\left(2\,\{s\}^{\mathrm{T}} + \sigma_{\mathrm{h}}\,\{\mathbf{1}\}^{\mathrm{T}}\right)\left([\overset{<4>}{\boldsymbol{A}}]\{\mathbf{1}\}\,\sigma_{\mathrm{h}}\right)} - 1 = 0 \tag{3.47}
$$

and reduction of the underlined term. This means that the tensorial von Mises equation (3.47) reduces to the deviatoric form independent of the hydrostatic pressure as follows

$$
\{s\}^{\mathrm{T}}\,[{}_{\mathrm{dev}}\boldsymbol{A}]\,\{s\} - 1 = 0 \tag{3.48}
$$

only if the constraint

$$
[\overset{<4>}{\boldsymbol{A}}]\{\mathbf{1}\} = 0 \tag{3.49}
$$

is consistently applied. The constraint (3.49) guarantees the deviatoric von Mises equation (3.48) be represented in the reduced 6–dimensional stress space by a cylindrical surface defined by 15 independent anisotropy modules, when 6 constraints are satisfied

$$
\begin{aligned}
A_{11} + A_{12} + A_{13} = 0, \quad A_{12} + A_{22} + A_{23} = 0 \\
A_{13} + A_{23} + A_{33} = 0, \quad A_{14} + A_{24} + A_{34} = 0 \\
A_{15} + A_{25} + A_{35} = 0, \quad A_{16} + A_{26} + A_{36} = 0
\end{aligned}
\tag{3.50}
$$

However, the final matrix representation (3.45) with (3.50) employed depends on a choice of independent elements. Two of such representations are of special importance.

In the first case, the elements of matrix (3.45) considered as independent are: $A_{12}, A_{13}, A_{23}; A_{15}, A_{16}, A_{24}, A_{26}, A_{34}, A_{35}$ and $A_{44}, A_{55}, A_{66}; A_{45}, A_{46}, A_{56}$, such that the following first representation for the deviatoric von Mises matrix is furnished

$$
[{}_{\mathrm{dev}}\boldsymbol{A}] =
$$

$$
\begin{bmatrix}
-A_{12} - A_{13} & A_{12} & A_{13} & -A_{24} - A_{34} & A_{15} & A_{16} \\
 & -A_{12} - A_{23} & A_{23} & A_{24} & -A_{15} - A_{35} & A_{26} \\
 & & -A_{13} - A_{23} & A_{34} & A_{35} & -A_{16} - A_{26} \\
 & & & A_{44} & A_{45} & A_{46} \\
 & & & & A_{55} & A_{56} \\
 & & & & & A_{66}
\end{bmatrix}
\tag{3.51}
$$

if constraints (3.50) are applied as follows

$$\begin{aligned} A_{11} &= -A_{12} - A_{13}, \quad A_{14} = -A_{24} - A_{34} \\ A_{22} &= -A_{12} - A_{23}, \quad A_{25} = -A_{15} - A_{35} \\ A_{33} &= -A_{13} - A_{23}, \quad A_{36} = -A_{16} - A_{26} \end{aligned} \tag{3.52}$$

In the second case, the elements of matrix (3.45) chosen as independent are: A_{11}, A_{22}, A_{33}; A_{15}, A_{16}, A_{24}, A_{26}, A_{34}, A_{35} and A_{44}, A_{55}, A_{66}; A_{45}, A_{46}, A_{56}, hence we arrive at the second representation of the deviatoric von Mises matrix as follows

$$[{}_{\text{dev}}\boldsymbol{A}] = \begin{bmatrix} A_{11} & \frac{A_{33}-A_{11}-A_{22}}{2} & \frac{A_{22}-A_{11}-A_{33}}{2} & -A_{24}-A_{34} & A_{15} & A_{16} \\ & A_{22} & \frac{A_{11}-A_{22}-A_{33}}{2} & A_{24} & -A_{15}-A_{35} & A_{26} \\ & & A_{33} & A_{34} & A_{35} & -A_{16}-A_{26} \\ & & & A_{44} & A_{45} & A_{46} \\ & & & & A_{55} & A_{56} \\ & & & & & A_{66} \end{bmatrix} \tag{3.53}$$

if, instead of (3.52), other substitution is used

$$\begin{aligned} A_{12} &= \frac{1}{2}(A_{33} - A_{11} - A_{22}), \quad A_{14} = -A_{24} - A_{34} \\ A_{13} &= \frac{1}{2}(A_{22} - A_{11} - A_{33}), \quad A_{25} = -A_{15} - A_{35} \\ A_{23} &= \frac{1}{2}(A_{11} - A_{22} - A_{33}), \quad A_{36} = -A_{16} - A_{26} \end{aligned} \tag{3.54}$$

A choice of 15 elements in the von Mises matrix (3.45) considered as independent is not a unique procedure and can result in the different deviatoric von Mises equation forms. In particular, when a more convenient representation (3.51) is substituted for $[{}_{\text{dev}}\boldsymbol{A}]$ in (3.48) we arrive at the following von Mises equation expressed in the deviatoric stress space

$$\begin{aligned} &-A_{12}\left(s_x - s_y\right)^2 - A_{13}\left(s_x - s_z\right)^2 - A_{23}\left(s_y - s_z\right)^2 \\ &\underline{+2\left\{\tau_{yz}\left[A_{24}\left(s_y - s_x\right) + A_{34}\left(s_z - s_x\right)\right] + \tau_{zx}\left[A_{15}\left(s_x - s_y\right)\right.\right.} \\ &\underline{\left.+A_{35}\left(s_z - s_y\right)\right] + \tau_{xy}\left[A_{16}\left(s_x - s_z\right) + A_{26}\left(s_y - s_z\right)\right]} \\ &\underline{\left.+A_{45}\tau_{yz}\tau_{zx} + A_{46}\tau_{xy}\tau_{yz} + A_{56}\tau_{zx}\tau_{xy}\right\}} \\ &+A_{44}\tau_{yz}^2 + A_{55}\tau_{zx}^2 + A_{66}\tau_{xy}^2 = 1 \end{aligned} \tag{3.55}$$

It is visible that above equation owns the clear deviatoric structure hence, when the tensorial stress space is used instead of the deviatoric one, the analogous equivalent to (3.55) representation of the deviatoric von Mises equation is also true in terms of stress components (cf. Szczepiński 1993)

$$\begin{aligned}
&-A_{12}\left(\sigma_x-\sigma_y\right)^2-A_{13}\left(\sigma_x-\sigma_z\right)^2-A_{23}\left(\sigma_y-\sigma_z\right)^2 \\
&\underline{+2\left\{\tau_{yz}\left[A_{24}\left(\sigma_y-\sigma_x\right)+A_{34}\left(\sigma_z-\sigma_x\right)\right]+\tau_{zx}\left[A_{15}\left(\sigma_x-\sigma_y\right)\right.\right.} \\
&\underline{\left.+A_{35}\left(\sigma_z-\sigma_y\right)\right]+\tau_{xy}\left[A_{16}\left(\sigma_x-\sigma_z\right)+A_{26}\left(\sigma_y-\sigma_z\right)\right]} \\
&\underline{\left.+A_{26}\left(\sigma_y-\sigma_z\right)\right]+A_{45}\tau_{yz}\tau_{zx}+A_{46}\tau_{xy}\tau_{yz}+A_{56}\tau_{zx}\tau_{xy}\}} \\
&+A_{44}\tau_{yz}^2+A_{55}\tau_{zx}^2+A_{66}\tau_{xy}^2=1
\end{aligned} \tag{3.56}$$

Note, that Eqs. (3.55) or (3.56) are defined by 15 elements A_{ij}. However, the underlined terms are sensitive to change of sign of shear stresses, e.g. $\tau_{yz}(\sigma_y-\sigma_x)$ etc., which is physically questionable and, finally, such terms are consequently omitted is some cases (cf. e.g. Malinin and Rżysko 1981).

3.3.2 von Mises Orthotropic Criterion, the Hill Deviatoric Criterion

General form of the 21–parameter anisotropic von Mises criterion (3.46) involves none material symmetry property. In a particular case if plastic orthotropy is assumed the 9–parameter orthotropic von Mises matrix (3.47) takes the form

$$[{}_{\mathrm{ort}}\boldsymbol{A}]=\begin{bmatrix} A_{11} & A_{12} & A_{13} & 0 & 0 & 0 \\ & A_{22} & A_{23} & 0 & 0 & 0 \\ & & A_{33} & 0 & 0 & 0 \\ & & & A_{44} & 0 & 0 \\ & & & & A_{55} & 0 \\ & & & & & A_{66} \end{bmatrix} \tag{3.57}$$

In such a case the general anisotropic von Mises equation (3.46) is reduced to the narrower 9–parameter orthotropic von Mises criterion

$$\begin{aligned}
&A_{11}\sigma_x^2+A_{22}\sigma_y^2+A_{33}\sigma_z^2+2(A_{12}\sigma_x\sigma_y+A_{23}\sigma_y\sigma_z+A_{31}\sigma_z\sigma_x) \\
&\quad+A_{44}\tau_{yz}^2+A_{55}\tau_{zx}^2+A_{66}\tau_{xy}^2=1
\end{aligned} \tag{3.58}$$

When the Voigt notation is used, the 9–parameter orthotropic von Mises criterion takes the form

$$\{\boldsymbol{\sigma}\}^{\mathrm{T}}\,[{}_{\mathrm{ort}}\boldsymbol{A}]\,\{\boldsymbol{\sigma}\}-1=0 \tag{3.59}$$

clearly pointing out that Eq. (3.59) belongs to the class of hydrostatic pressure sensitive criteria.

In order to achieve pressure insensitive orthotropic criterion we apply a procedure previously described and based on stress tensor decomposition into deviatoric and volumetric parts in the orthotropic von Mises equation (3.59) we arrive at the equation analogous to (3.47). Next, assuming the hydrostatic pressure insensitivity

$$[_{\mathrm{ort}}\boldsymbol{A}]\,\{\mathbf{1}\} = 0 \tag{3.60}$$

the following three constraints have to be satisfied

$$A_{11} + A_{12} + A_{13} = 0, \quad A_{12} + A_{22} + A_{23} = 0, \quad A_{13} + A_{23} + A_{33} = 0 \tag{3.61}$$

In this way the orthotropic von Mises criterion (3.59) reduces to the pressure insensitive criterion called Hill's criterion (cf. Hill 1948, 1950) that contains 6 independent modules

$$\{s\}^{\mathrm{T}}\,[\boldsymbol{A}^{\mathrm{H}}]\,\{s\} - 1 = 0 \tag{3.62}$$

Hill's matrix $[\boldsymbol{A}^{\mathrm{H}}]$ appearing in Eq. (3.62) contains 6 independent modules. A choice of the three independent modules form six involved in Eqs. (3.61) is not unique.

In this way we arrive at the following Hill's matrices

$$[\boldsymbol{A}^{\mathrm{H}}] = \begin{bmatrix} -A_{12} - A_{13} & A_{12} & A_{13} & & & \\ & -A_{12} - A_{23} & A_{23} & & & \\ & & -A_{13} - A_{23} & & & \\ & & & A_{44} & & \\ & & & & A_{55} & \\ & & & & & A_{66} \end{bmatrix} \tag{3.63}$$

or

$$[\boldsymbol{A}^{\mathrm{H}}] = \begin{bmatrix} A_{11} & \dfrac{A_{33} - A_{11} - A_{22}}{2} & \dfrac{A_{22} - A_{11} - A_{33}}{2} & & & \\ & A_{22} & \dfrac{A_{11} - A_{22} - A_{33}}{2} & & & \\ & & A_{33} & & & \\ & & & A_{44} & & \\ & & & & A_{55} & \\ & & & & & A_{66} \end{bmatrix} \tag{3.64}$$

When the engineering notation is used, corresponding representations of the Hill's criterion are

$$\begin{gathered} -A_{23}\left(\sigma_y - \sigma_z\right)^2 - A_{13}\left(\sigma_z - \sigma_x\right)^2 - A_{12}\left(\sigma_x - \sigma_y\right)^2 \\ +A_{44}\tau_{yz}^2 + A_{55}\tau_{zx}^2 + A_{66}\tau_{xy}^2 = 1 \end{gathered} \tag{3.65}$$

or

$$
\begin{aligned}
& A_{11}\sigma_x^2 + A_{22}\sigma_y^2 + A_{33}\sigma_z^2 + (A_{33} - A_{11} - A_{22})\,\sigma_x\sigma_y \\
& + (A_{22} - A_{11} - A_{33})\,\sigma_x\sigma_z + (A_{11} - A_{22} - A_{33})\,\sigma_y\sigma_z \\
& + A_{44}\tau_{yz}^2 + A_{55}\tau_{zx}^2 + A_{66}\tau_{xy}^2 = 1
\end{aligned}
\tag{3.66}
$$

Both representations (3.65) or (3.66) describe the same Hill's limit surface, but applying two different choices of six independent elements of the Hill matrices (3.63) or (3.64). In order to calibrate Hill's criterion in the form (3.65) or (3.66) three tests of uniaxial tension $\sigma_x = k_x, \sigma_y = k_y, \sigma_z = k_z$ and three tests of pure shear $\tau_{xy} = k_{xy}, \tau_{yz} = k_{yz}, \tau_{zx} = k_{zx}$, in directions and planes of material orthotropy, must be performed. These tests allow to express 6 modules of material orthotropy in Eqs. (3.65) and (3.66) in terms of 3 independent plastic tension limits k_x, k_y, k_z (in directions of orthotropy), and 3 independent plastic shear limits k_{yz}, k_{zx}, k_{xy} (in planes of material orthotropy). Hence,

$$
\begin{aligned}
-A_{23} &= \frac{1}{2}\left(\frac{1}{k_y^2} + \frac{1}{k_z^2} - \frac{1}{k_x^2}\right), \quad A_{44} = \frac{1}{k_{yz}^2} \\
-A_{13} &= \frac{1}{2}\left(\frac{1}{k_z^2} + \frac{1}{k_x^2} - \frac{1}{k_y^2}\right), \quad A_{55} = \frac{1}{k_{zx}^2} \\
-A_{12} &= \frac{1}{2}\left(\frac{1}{k_x^2} + \frac{1}{k_y^2} - \frac{1}{k_z^2}\right), \quad A_{66} = \frac{1}{k_{xy}^2}
\end{aligned}
\tag{3.67}
$$

such that orthotropic Hill's criteria equivalent to (3.65) or (3.66) can be furnished in terms of plastic anisotropy limits as follows

$$
\begin{aligned}
& \frac{1}{2}\left(\frac{1}{k_y^2} + \frac{1}{k_z^2} - \frac{1}{k_x^2}\right)(\sigma_y - \sigma_z)^2 + \frac{1}{2}\left(\frac{1}{k_z^2} + \frac{1}{k_x^2} - \frac{1}{k_y^2}\right)(\sigma_z - \sigma_x)^2 + \\
& \frac{1}{2}\left(\frac{1}{k_x^2} + \frac{1}{k_y^2} - \frac{1}{k_z^2}\right)(\sigma_x - \sigma_y)^2 + \left(\frac{\tau_{yz}}{k_{yz}}\right)^2 + \left(\frac{\tau_{zx}}{k_{zx}}\right)^2 + \left(\frac{\tau_{xy}}{k_{xy}}\right)^2 = 1
\end{aligned}
\tag{3.68}
$$

or

$$
\begin{aligned}
& \left(\frac{\sigma_x}{k_x}\right)^2 + \left(\frac{\sigma_y}{k_y}\right)^2 + \left(\frac{\sigma_z}{k_z}\right)^2 \\
& -\left(\frac{1}{k_x^2} + \frac{1}{k_y^2} - \frac{1}{k_z^2}\right)\sigma_x\sigma_y - \left(\frac{1}{k_y^2} + \frac{1}{k_z^2} - \frac{1}{k_x^2}\right)\sigma_y\sigma_z - \left(\frac{1}{k_z^2} + \frac{1}{k_x^2} - \frac{1}{k_y^2}\right)\sigma_z\sigma_x \\
& + \left(\frac{\tau_{yz}}{k_{yz}}\right)^2 + \left(\frac{\tau_{zx}}{k_{zx}}\right)^2 + \left(\frac{\tau_{xy}}{k_{xy}}\right)^2 = 1
\end{aligned}
\tag{3.69}
$$

Note that under a particular plane stress condition, e.g. in the x, y plane, when $\sigma_z = \tau_{zx} = \tau_{yz} = 0$, both formulas (3.68) and (3.69) reduce to the 4–parameter orthotropic Hill's condition

$$\frac{\sigma_x^2}{k_x^2} + \frac{\sigma_y^2}{k_y^2} - \left(\frac{1}{k_x^2} + \frac{1}{k_y^2} - \frac{1}{k_z^2}\right)\sigma_x\sigma_y + \frac{\tau_{xy}^2}{k_{xy}^2} = 1 \tag{3.70}$$

where initiation of plastic flow in the x, y plane is controlled not only by the in-plane limits k_x, k_y and k_{xy}, but also by the out-of-plane limit k_z, which may finally lead to inadmissible loss of convexity by the yield surface.

The Hill criterion (3.65) is formulated in the space of principal material directions of orthotropy which in general do not coincide with directions of principal stresses. In the particular case when the coaxiality holds $\sigma_x = \sigma_1$, $\sigma_y = \sigma_2$, $\sigma_z = \sigma_3$, $\tau_{xy} = \tau_{yz} = \tau_{zx} = 0$ we arrive at simplified

$$-A_{23}(\sigma_2 - \sigma_3)^2 - A_{13}(\sigma_3 - \sigma_1)^2 - A_{12}(\sigma_1 - \sigma_2)^2 = 1 \tag{3.71}$$

or when calibration (3.67) is used the explicit form of (3.71) is finally furnished

$$\begin{aligned}&\frac{1}{2}\left(\frac{1}{k_2^2} + \frac{1}{k_3^2} - \frac{1}{k_1^2}\right)(\sigma_2 - \sigma_3)^2 + \frac{1}{2}\left(\frac{1}{k_3^2} + \frac{1}{k_1^2} - \frac{1}{k_2^2}\right)(\sigma_3 - \sigma_1)^2 + \\ &\frac{1}{2}\left(\frac{1}{k_1^2} + \frac{1}{k_2^2} - \frac{1}{k_3^2}\right)(\sigma_1 - \sigma_2)^2 = 1\end{aligned} \tag{3.72}$$

Hill's condition (3.72) represents cylindrical elliptic surface the axis of which coincides with the hydrostatic axis. Nevertheless in some cases the limit surface looses closed form for high othotropy degree which may occur when one of following expressions

$$\frac{1}{k_2^2} + \frac{1}{k_3^2} - \frac{1}{k_1^2}$$

elsewhere

$$\frac{1}{k_3^2} + \frac{1}{k_1^2} - \frac{1}{k_2^2}$$

or

$$\frac{1}{k_1^2} + \frac{1}{k_2^2} - \frac{1}{k_3^2}$$

changes the sign.

It is convenient to express Hill's limit surface by use of the Haigh–Westergaard coordinates (cf. Ganczarski and Lenczowski 1997)

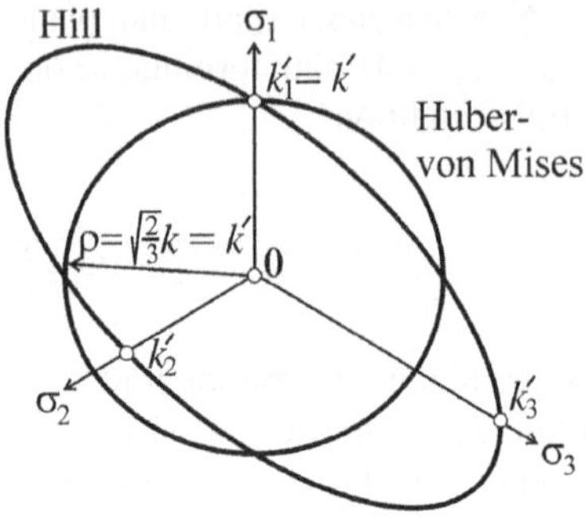

Fig. 3.5 Comparison of the Huber–von Mises and the Hill criteria ($k_1 = k$, $k_2 = 0.8k$, $k_3 = 1.5k$)

$$\begin{Bmatrix} \sigma_1 \\ \sigma_2 \\ \sigma_3 \end{Bmatrix} = \frac{\xi}{\sqrt{3}} \begin{Bmatrix} 1 \\ 1 \\ 1 \end{Bmatrix} + \sqrt{\frac{2}{3}}\rho(\theta) \begin{Bmatrix} \cos\theta \\ \cos\left(\theta - \dfrac{2\pi}{3}\right) \\ \cos\left(\theta + \dfrac{2\pi}{3}\right) \end{Bmatrix} \tag{3.73}$$

to finally obtain Hill's criterion in form $\rho(\theta)$

$$\rho(\theta) = \left[\frac{2}{\left(\frac{1}{k_2^2} + \frac{1}{k_3^2} - \frac{1}{k_1^2}\right)\sin^2\left(\theta + \frac{\pi}{3}\right) + \left(\frac{1}{k_3^2} + \frac{1}{k_1^2} - \frac{1}{k_2^2}\right)\sin^2\left(\theta - \frac{\pi}{3}\right) + \left(\frac{1}{k_1^2} + \frac{1}{k_2^2} - \frac{1}{k_3^2}\right)\sin^2\theta} \right]^{1/2} \tag{3.74}$$

Note that in case if $k_1 = k_2 = k_3 = k$ the Huber–von Mises circular cylinder is recovered Fig. 3.5

$$\rho = \sqrt{\frac{2}{3}}k = \text{const} \tag{3.75}$$

3.3.3 *Comparison of Hill's Criterion Versus Hu–Marin's Concept*

Classical orthotropic Hill's criterion (Hill 1948), despite obvious advantages and wide technical applications, is limited however by some constraints of applicability, which are discussed in the present section following Ganczarski and Skrzypek (2014).

First limitation of applicability range of the classical Hill criterion is established through the inequality bounding the magnitudes of the engineering orthotropy constants k_1, k_2 and k_3 in order to avoid ellipticity loss of the limit surface in the stress space when the coordinate axes are aligned with the material axes of orthotropy (see, e.g. Ottosen and Ristinmaa 2005; Ganczarski and Skrzypek 2013, 2014). Such limit bounds put upon the orthotropy limits usually hold in case if the degree of material orthotropy is moderate. For example if the material ensures the transverse isotropy symmetry, transverse isotropy symmetry it is shown that the orthotropy degree bounded by the inequality $\frac{k_{\max}}{k_{\min}} < 2$ guarantees ellipticity of the limit surface to be saved. However, if the orthotropy bound is violated, the Hill criterion becomes useless when a possible degeneration of the elliptic cylindrical surface into two concave hyperbolic cylinders occurs, what is inadmissible in the light of Drucker's postulate.

To illustrate this restriction, we consider two types of true materials for which the classical Hill criterion occurs to be: either useful, if material orthotropy degree is not very high such that the ellipticity property of the limit surface is preserved, or useless if the orthotropy degree is as high as the described limit surface no longer holds the ellipticity requirement. Other words, a physically inadmissible degeneration of the single convex and simply connected elliptical limit surface into two concave hyperbolic surfaces occurs.

The following inequality bounds the range of applicability for Hill's criterion (cf. e.g. Ottosen and Ristinmaa (2005))

$$\frac{2}{k_1^2k_2^2} + \frac{2}{k_2^2k_3^2} + \frac{2}{k_3^2k_1^2} > \frac{1}{k_1^4} + \frac{1}{k_2^4} + \frac{1}{k_3^4} \tag{3.76}$$

For simplicity, a coincidence of the principal stress axes with the material orthotropy axes is assumed in (3.76). In the narrower case of transverse isotropy $k_1 = k_2$, condition (3.76) reduces to the simple form

$$\frac{1}{k_3^2}\left(\frac{4}{k_1^2} - \frac{1}{k_3^2}\right) > 0 \tag{3.77}$$

Substitution of the dimensionless parameter $R = 2(\frac{k_3}{k_1})^2 - 1$, after Hosford and Backhofen (1964), leads to the simplified restriction

$$R > -0.5 \tag{3.78}$$

If the above inequalities (3.76)–(3.78) do not hold, elliptic cross sections of the limit surface degenerate to two hyperbolic branches and the lack of convexity occurs. To illustrate this limitation, the yield curves in two planes: the transverse isotropy (σ_1, σ_2) and the orthotropy plane (σ_1, σ_3) for various R–values, are sketched in Fig. 3.6a, b respectively. It is observed that when R starting from $R = 3$ approaches

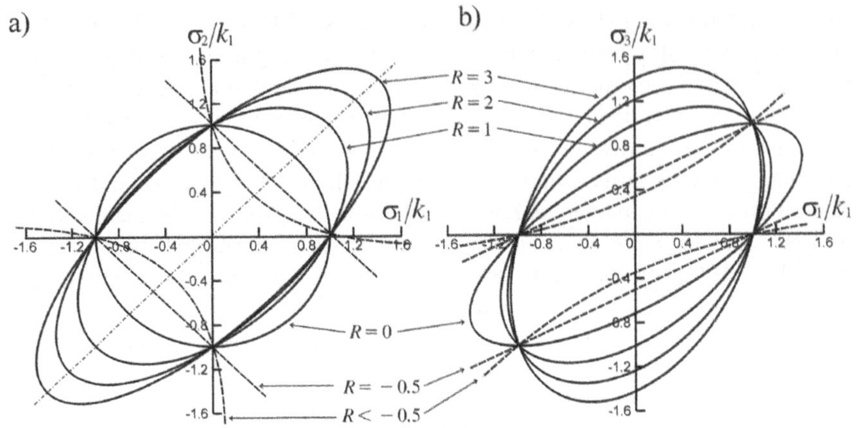

Fig. 3.6 Degeneration of the Hill's limit surface with respect to parameter R: **a** transverse isotropy plane, **b** orthotropy plane (after Ganczarski and Skrzypek 2014)

the limit $R = -0.5$, the curves change from closed ellipses to two parallel lines, whereas for $R < -0.5$ concave hyperbolas appear.

In a case when Hill's condition is not convex the other concept is proposed. This new approach suggests formulation of limit criterion based on the 9–parameter von Mises condition, but enhanced by the Hu–Marin type biaxial orthotropic loading conditions (cf. Hu and Marin 1956; Skrzypek and Ganczarski 2013).

In general case of strong orthotropy, when the ellipticity condition (3.76) does not hold, the deviatoric Hill criterion (3.68) or (3.69) becomes useless. Hence, in order to describe physically admissible close and convex limit surface, the more general 9–parameter orthotropic von Mises equation (3.57) has to be recalled. In a narrower case of the principal stress axes coinciding with the material orthotropy axes the Eq. (3.58) reads as

$$A_{11}\sigma_1^2 + A_{22}\sigma_2^2 + A_{33}\sigma_3^2 + 2(A_{12}\sigma_1\sigma_2 + A_{23}\sigma_2\sigma_3 + A_{31}\sigma_3\sigma_1) = 1 \tag{3.79}$$

The condition (3.79) is defined by six material parameters only, because $\tau_{23} \equiv \tau_{31} \equiv \tau_{12} \equiv 0$, hence its calibration requires six conditions:

- three tests of uniaxial tension along the orthotropy axes

$$\begin{array}{l} \sigma_1 = k_1\ \sigma_2 = 0\ \sigma_3 = 0 \longrightarrow A_{11} = 1/k_1^2 \\ \sigma_2 = k_2\ \sigma_1 = 0\ \sigma_3 = 0 \longrightarrow A_{22} = 1/k_2^2 \\ \sigma_3 = k_3\ \sigma_1 = 0\ \sigma_2 = 0 \longrightarrow A_{33} = 1/k_3^2 \end{array} \tag{3.80}$$

- and three orthotropic biaxial tension loading conditions (k_i, k_j)

$$\begin{array}{l} \sigma_1 = k_1\ \sigma_2 = k_2\ \sigma_3 = 0 \longrightarrow A_{12} = -1/2k_1k_2 \\ \sigma_1 = k_1\ \sigma_3 = k_3\ \sigma_2 = 0 \longrightarrow A_{13} = -1/2k_1k_3 \\ \sigma_2 = k_2\ \sigma_3 = k_3\ \sigma_1 = 0 \longrightarrow A_{23} = -1/2k_2k_3 \end{array} \tag{3.81}$$

Calibration of the orthotropic von Mises criterion (3.79), performed with conditions (3.80) and (3.81) used, leads to the three-axial extension of the Hu–Marin type criterion (cf. Ganczarski and Skrzypek 2011; Skrzypek and Ganczarski 2013)

$$\left(\frac{\sigma_1}{k_1}\right)^2 - \frac{\sigma_1\sigma_2}{k_1k_2} + \left(\frac{\sigma_2}{k_2}\right)^2 - \frac{\sigma_2\sigma_3}{k_2k_3} + \left(\frac{\sigma_3}{k_3}\right)^2 - \frac{\sigma_1\sigma_3}{k_1k_3} = 1 \tag{3.82}$$

The enhanced Mises–Hu–Marin type criterion (3.82) is free from Hill's deficiency even in case of arbitrarily strong orthotropy degree, since it never violates the Drucker stability postulate, which is not guaranteed by Hill's type equations. The Hu–Marin type Eq. (3.82) can easily be presented in the "pseudo-deviatoric" format

$$\left(\frac{\sigma_1}{k_1} - \frac{\sigma_2}{k_2}\right)^2 + \left(\frac{\sigma_2}{k_2} - \frac{\sigma_3}{k_3}\right)^2 + \left(\frac{\sigma_3}{k_3} - \frac{\sigma_1}{k_1}\right)^2 = 2 \tag{3.83}$$

Three orthotropy limit yield points k_1, k_2 and k_3 establish the proportional stress/strength axis of cylindrical Hu–Marin's surface. Note that this proportional stress/strength axis, which determines a position of the limit surface axis in the principal stress space, is different from the hydrostatic axis, but the condition of equal ratios $\sigma_i/k_i = \alpha$ holds at all points belonging to this axis. The extended von Mises–Hu–Marin type criteria (3.82)–(3.83) are always "unconditionally stable" criteria, it remains convex even for very strong orthotropy, by contrast to the classical Hill condition in which the possible loss of convexity can be met in the case of highly orthotropic materials. However, the fully deviatoric format of the Hill criteria (3.62)–(3.68) is lost in the Hu–Marin type format (3.83) where the hydrostatic pressure insensitivity is relaxed.

In the particular case of plane stress state $\sigma_3 = 0$ the three–parameter enhanced von Mises–Hu–Marin equation (3.82) is reduced to the two–parameter one, as proposed by Hu–Marin (Hu and Marin 1956)

$$\left(\frac{\sigma_1}{k_1}\right)^2 - \frac{\sigma_1\sigma_2}{k_1k_2} + \left(\frac{\sigma_2}{k_2}\right)^2 = 1 \tag{3.84}$$

Comparison of the 2–parameter Hu–Marin plane stress equation (3.84) with the simplified 4–parameter plane stress Hill's equation (3.70) written for principal stress axes, leads to the 3–parameter form

$$\left(\frac{\sigma_1}{k_1}\right)^2 - \left(\frac{1}{k_1^2} + \frac{1}{k_2^2} - \frac{1}{k_3^2}\right)\sigma_1\sigma_2 + \left(\frac{\sigma_2}{k_2}\right)^2 = 1 \tag{3.85}$$

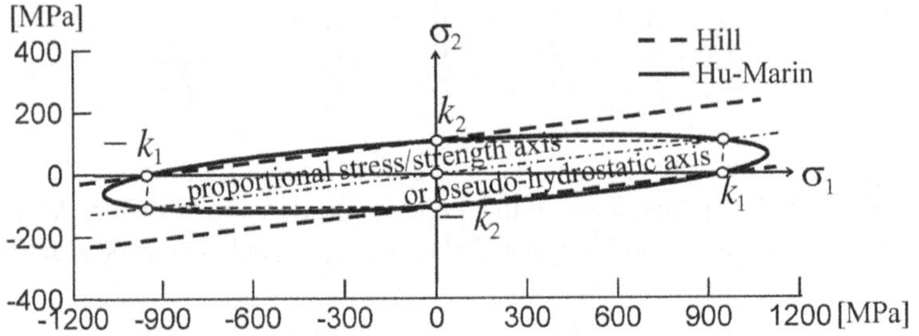

Fig. 3.7 Comparison of the Hill and the Hu–Marin plastic yield criteria for Ł62 brass (after Ganczarski and Skrzypek 2014)

which becomes identical to the von Mises–Hu–Marin equation (3.84) only if following constraint holds

$$\frac{1}{k_3^2} = \frac{1}{k_1^2} + \frac{1}{k_2^2} - \frac{1}{k_1 k_2} \tag{3.86}$$

which is usually not true.

In order to illustrate a suitability of the von Mises–Hu–Marin orthotropic equation (3.82), when compared to certain limitations of the Hill deviatoric equation (3.70) Ł62 brass is studied. The results are presented in Fig. 3.7 on the planes σ_1, σ_2. In case when the inequality (3.76) is not satisfied, following the Hill concept two concave hyperbolic cylinders are formed by opening of the elliptic cylinder towards the proportional stress/strength axis. On the other hand, the Hu–Marin type surface saves the ellipticity property regardless of the magnitude of orthotropy degree considered. In other words the Hu–Marin surface is "unconditionally stable" which remains convex for very strong orthotropy. It is possible due to three additional constraints (3.81) satisfied for the pairs of orthotropy yield limits (k_1, k_2), (k_2, k_3) and (k_3, k_1). But, it should be pointed out, that the Hu–Marin cylindrical surface does not satisfy the condition of deviatoricity, hence this condition in fact should be classified as a specific representative of the hydrostatic pressure sensitive class of materials where the independence of the hydrostatic stress constraint is relaxed.

The aforementioned possible loss of the convexity of classical Hill's criterion (Hill 1948) (3.68) in case of highly orthotropic materials is even more pronounced when the orthotropic generalization of the isotropic Hosford criterion (Hosford and Backhofen 1964) for higher (even) exponents is done

$$\begin{aligned} &-A_{23}|\sigma_y - \sigma_z|^m - A_{13}|\sigma_z - \sigma_x|^m - A_{12}|\sigma_x - \sigma_y|^m \\ &+A_{44}|\tau_{yz}|^m + A_{55}|\tau_{zx}|^m + A_{66}|\tau_{xy}|^m = 1 \end{aligned} \tag{3.87}$$

Six generalized orthotropy modules $A_{23}, \ldots, A_{66}$ can be expressed in terms of six yield point stresses $k_x, \ldots, k_{xy}$ in analogous fashion as previously discussed manner of calibration for Hill's criterion (3.67), namely

Table 3.5 Yield point stresses [MPa] for brass Ł22 after Malinin and Rżysko (1981)

m	k_x	k_y	k_z	k_{zy}	k_{zx}	k_{xy}
2	120	105	950	182	194	64.8
6 or 8	120	105	950	157	168	56.1

$$\begin{aligned}
-A_{23} &= \frac{1}{2}\left(\frac{1}{|k_y|^m} + \frac{1}{|k_z|^m} - \frac{1}{|k_x|^m}\right), \quad A_{44} = \frac{1}{|k_{yz}|^m} \\
-A_{13} &= \frac{1}{2}\left(\frac{1}{|k_z|^m} + \frac{1}{|k_x|^m} - \frac{1}{|k_y|^m}\right), \quad A_{55} = \frac{1}{|k_{zx}|^m} \\
-A_{12} &= \frac{1}{2}\left(\frac{1}{|k_x|^m} + \frac{1}{|k_y|^m} - \frac{1}{|k_z|^m}\right), \quad A_{66} = \frac{1}{|k_{xy}|^m}
\end{aligned} \tag{3.88}$$

The more general case when axes of material orthotropy are different from axes of principal stresses was considered by Ganczarski and Lenczowski (1997). It was shown that although the limit surface is closed and convex in space of principal material orthotropy frame it occurs that lack of convexity is met when transformation to the space of principal stress frame is done in terms of three angles defining the mutual configuration of these two frames. This type convexity loss was examined for the brass sheet Ł22 the six orthotropic yield points of which are given in Table 3.5 after Malinin and Rżysko (1981), who gave three axial yield point stresses, whereas three shear yield point stresses were estimated in Ganczarski and Lenczowski (1997) using simplified formulas $k_{ij} = \sqrt{\frac{k_i k_j}{3}}$ for $m = 2$ and $k_{ij} = \frac{\sqrt{k_i k_j}}{2}$ for $m = 6, 8$. For simplicity the evolution of the generalized orthotropic Hosford yield condition ($m = 8$) with respect to only one of the Euler angles ϑ was considered. It represents a prism of the semi-hexagonal cross-section with oval corners as presented in Fig. 3.8. The loss of convexity is observed for $18° \geq \vartheta \geq 26°$.

3.3.4 *Transverse Isotropy of Hill's Type Tetragonal Symmetry Versus Hu–Marin's Type Hexagonal Symmetry*

The second limitation of applicability range of classical Hill's criterion arises when the transverse isotropy property is considered. In this section it will be shown that, if reduction of Hill's criterion to the transverse isotropy symmetry is performed, the 4–parameter form that satisfies the tetragonal symmetry class is furnished (cf. e.g. Voyiadjis and Thiagarajan 1995; Sun and Vaidya 1996). This type of symmetry is of particular importance in case of unidirectional fiber reinforced composites. In such a case moduli: k_x, k_y, k_z and k_{xy} are considered as independent (z is the orthotropy axis), which makes impossible to reduce classical Hill's criterion to the isotropic von Mises condition in the plane of transverse isotropy.

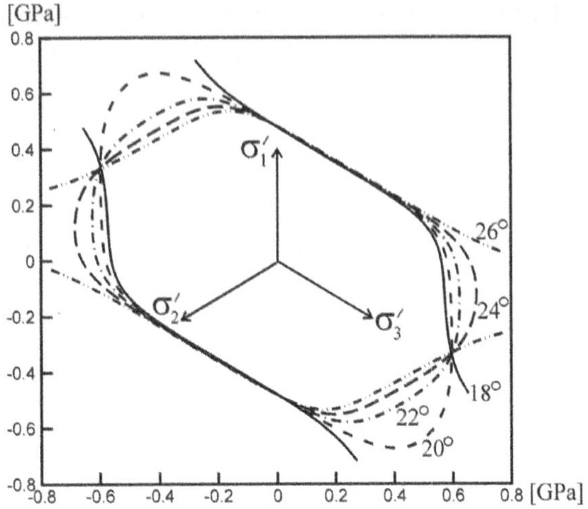

Fig. 3.8 Evolution of the generalized orthotropic Hosford yield condition versus the Euler angle ϑ for brass Ł22, after Ganczarski and Lenczowski (1997)

To avoid this irreducibility to isotropic von Mises', the new Hu–Marin's based transversely isotropic criterion exhibiting hexagonal symmetry is proposed instead of deviatoric transversely isotropic Hill's criterion exhibiting tetragonal symmetry. It enables to achieve reducibility to the isotropic von Mises condition in the transverse isotropy plane, preserving cylindricity regardless of the magnitude of orthotropy degree.

Finally, it will be demonstrated that, for some composite materials it is necessary to further modify the 3–parameter Hu–Marin type criterion to the new 4–parameter intermediate type criterion between classical Hill's and hexagonal Hu–Marin's concepts, taking advantage of the bulge test. This new hybrid-type criterion differs essentially from both the Hu–Marin hexagonal type criterion and the isotropic von Mises criterion in the isotropy plane. Bulge tests have been performed and described e.g. by Jackson et al. (1948) with equipment used by Lankford et al. (1947). This new criterion is capable of properly describing the SiC/Ti long fiber reinforced composite examined by Herakovich and Aboudi (1999).

Classical Hill's equation (3.65)–(3.66), which is expressed in terms of six independent plastic yield limits k_x, k_y, k_z, k_{yz}, k_{zx} and k_{xy}, (3.68) is often too general for engineering applications. Orthotropic structural materials usually exhibit the transversely isotropic symmetry, basically due to either fabrication process or microstructure texture, as often observed in many long parallel fiber reinforced composites.

Assume that the z–axis is the orthotropy axis, whereas x, y is the transverse isotropy plane. When applying Eq. (3.66) with calibrations (3.68) or (3.69) and additionally assuming $k_x = k_y \neq k_z$, $k_{zx} = k_{zy} \neq k_{xy}$, the number of independent limits in transversely isotropic Hill's equation reduces to four for instance: two axial yield limits k_x and k_z, and two shear yield limits k_{zx} and k_{xy}. In this way the following is furnished

$$-A_{13}=-A_{23}=\frac{1}{2k_z^2},\ A_{44}=A_{55}=\frac{1}{k_{zx}^2},\ -A_{12}=\frac{1}{k_x^2}-\frac{1}{2k_z^2},\ A_{66}=\frac{1}{2k_{xy}^2} \tag{3.89}$$

Substitution of (3.89) into (3.63) and (3.64) yields to transversely isotropic Hill's matrices

$$[{}_{\rm tris}\boldsymbol{A}^{\rm H}]=\begin{bmatrix} -A_{12}-A_{13} & A_{12} & A_{13} & & & \\ & -A_{12}-A_{23} & A_{13} & & & \\ & & -2A_{13} & & & \\ & & & A_{44} & & \\ & & & & A_{44} & \\ & & & & & A_{66} \end{bmatrix} \tag{3.90}$$

or

$$[{}_{\rm tris}\boldsymbol{A}^{\rm H}]=\begin{bmatrix} A_{11} & \frac{A_{33}-2A_{11}}{2} & -\frac{A_{33}}{2} & & & \\ & A_{11} & -\frac{A_{33}}{2} & & & \\ & & A_{33} & & & \\ & & & A_{44} & & \\ & & & & A_{44} & \\ & & & & & A_{66} \end{bmatrix} \tag{3.91}$$

The transversely isotropic 4–parameter Hill criteria corresponding to orthotropic Hill's criteria (3.68) and (3.69) take the following representations

$$\begin{aligned}&\frac{(\sigma_y-\sigma_z)^2+(\sigma_z-\sigma_x)^2}{2k_z^2}+\underline{\left(\frac{1}{k_x^2}-\frac{1}{2k_z^2}\right)}(\sigma_x-\sigma_y)^2\\&+\frac{\tau_{yz}^2+\tau_{zx}^2}{k_{zx}^2}+\frac{\tau_{xy}^2}{k_{xy}^2}=1\end{aligned} \tag{3.92}$$

or equivalently

$$\frac{\sigma_x^2+\sigma_y^2}{k_x^2}+\frac{\sigma_z^2}{k_z^2}-\underline{\left(\frac{2}{k_x^2}-\frac{1}{k_z^2}\right)}\sigma_x\sigma_y-\frac{\sigma_y\sigma_z+\sigma_z\sigma_x}{k_z^2}+\frac{\tau_{yz}^2+\tau_{zx}^2}{k_{zx}^2}+\frac{\tau_{xy}^2}{k_{xy}^2}=1 \tag{3.93}$$

Both forms involve four plastic limits k_x, k_z, k_{zx} and k_{xy} considered as independent parameters. Underlined factor in (3.93) includes not only k_x but also k_z. The explicitly deviatoric form (3.92) exhibits the similar feature. The plastic state in the transverse isotropy plane x, y is controlled not only by the tensile yield limit in this plane k_x, but also by the out-of-plane tensile yield limit k_z. Concluding, transversely isotropic Hill's criteria (3.92) or (3.93) have to be classified as the tetragonal symmetry format. The assumption of tetragonal symmetry of the criteria (3.92)–(3.93) was also considered by Voyiadjis and Thiagarajan (1995) in case of directionally reinforced metal matrix composites (Boron-Aluminium). Broader discussion that relates to distinction between the tetragonal versus hexagonal symmetry in the yield/failure criteria will be presented in the next lecture, where additional constraint for case if $A_{66} = -2(A_{13} + 2A_{12})$ is assumed, such that A_{66} has to be considered dependent plastic modulus. To this end, if aforementioned constraint postulated by Chen and Han (1995) is applied, the equality holds

$$A_{66} = \frac{4}{k_x^2} - \frac{1}{k_z^2} \tag{3.94}$$

instead of (3.89_4) and transversely isotropic 3–parameter Hill's criteria corresponding to (3.92) and (3.93) in following format

$$\begin{aligned} &\frac{\left(\sigma_y - \sigma_z\right)^2 + (\sigma_z - \sigma_x)^2}{2k_z^2} + \underline{\left(\frac{1}{k_x^2} - \frac{1}{2k_z^2}\right)\left(\sigma_x - \sigma_y\right)^2} \\ &+ \frac{\tau_{yz}^2 + \tau_{zx}^2}{k_{zx}^2} + \left(\frac{4}{k_x^2} - \frac{1}{k_z^2}\right)\tau_{xy}^2 = 1 \end{aligned} \tag{3.95}$$

or equivalently

$$\begin{aligned} &\frac{\sigma_x^2 + \sigma_y^2}{k_x^2} + \frac{\sigma_z^2}{k_z^2} - \underline{\left(\frac{2}{k_x^2} - \frac{1}{k_z^2}\right)}\sigma_x\sigma_y \\ &- \frac{\sigma_y\sigma_z + \sigma_z\sigma_x}{k_z^2} + \frac{\tau_{yz}^2 + \tau_{zx}^2}{k_{zx}^2} + \left(\frac{4}{k_x^2} - \frac{1}{k_z^2}\right)\tau_{xy}^2 = 1 \end{aligned} \tag{3.96}$$

can be written down.

In the particular case of plane stress state in the transverse isotropy plane (x, y) $\sigma_x, \sigma_y, \tau_{xy} \neq 0$ Eqs. (3.92) or (3.93) reduce to (3.70) with additional condition $k_x = k_y$

$$\frac{\sigma_x^2 + \sigma_y^2}{k_x^2} - \left(\frac{2}{k_x^2} - \frac{1}{k_z^2}\right)\sigma_x\sigma_y + \frac{\tau_{xy}^2}{k_{xy}^2} = 1 \tag{3.97}$$

The above form simply means that commonly used "transversely isotropic Hill's criterion" does not coincide in the "transverse isotropy plane" with the isotropic Huber–von Mises equation

$$\frac{\sigma_x^2 + \sigma_y^2}{k_x^2} - \frac{\sigma_x\sigma_y}{k_x^2} + 3\frac{\tau_{xy}^2}{k_{xy}^2} = 1 \tag{3.98}$$

In other words, when the new transversely isotropic yield criterion, that is free from inconsistencies between (3.97) and (3.98) is sought for, the material parameter preceding term $\sigma_x\sigma_y$ must be equal to $A_{33} - 2A_{11} = 1/k_x^2$ and not depend of k_z and, simultaneously, the material parameter $A_{66} = 3/k_x^2$ must depend on k_x only.

In order to derive the transversely isotropic yield criterion reducible to coincidence with the Huber–von Mises criterion in the isotropy plane, the new transversely isotropic hexagonal Hu–Marin equation will be postulated. To obtain this criterion, the general orthotropic von Mises equation (3.58), which is not deviatoric, can be calibrated in the way analogous to presented in (3.80) and (3.81). Namely, when the constraints of transverse isotropy are imposed, we invoke:

- the two tensile tests in the x- and the orthotropy z-axes and the shear test in the orthotropy zx-plane

$$\begin{aligned}
&\sigma_x = k_x,\quad \sigma_y=\sigma_z=\tau_{xy}=\tau_{yz}=\tau_{zx}=0 \longrightarrow A_{11} = A_{22} = \frac{1}{k_x^2}\\
&\sigma_z = k_z,\quad \sigma_x=\sigma_y=\tau_{xy}=\tau_{yz}=\tau_{zx}=0 \longrightarrow A_{33} = \frac{1}{k_z^2}\\
&\tau_{zx} = k_{zx},\ \sigma_x=\sigma_y=\sigma_z=\tau_{xy}=\tau_{yz}=0 \longrightarrow A_{44} = A_{55} = \frac{1}{k_{zx}^2}
\end{aligned} \tag{3.99}$$

- and the three bi-axial conditions for coincidence of appropriate pairs of yield limits

$$\begin{aligned}
&\sigma_x = k_x, \sigma_y = k_x,\quad \sigma_z=\tau_{xy}=\tau_{yz}=\tau_{zx}=0 \longrightarrow A_{12} = -\frac{1}{2k_x^2}\\
&\sigma_x = k_x, \sigma_z = k_z,\quad \sigma_y=\tau_{xy}=\tau_{yz}=\tau_{zx}=0 \longrightarrow A_{13} = -\frac{1}{2k_x k_z}\\
&\sigma_x = k_x, \tau_{xy} = \frac{k_x}{\sqrt{3}},\ \sigma_y=\sigma_z=\tau_{yz}=\tau_{zx}=0 \longrightarrow A_{66} = \frac{3}{k_x^2}
\end{aligned} \tag{3.100}$$

Introduction of (3.99) and (3.100) into orthotropic von Mises' criterion (3.58) leads to transversely isotropic 3–parameter hexagonal Hu–Marin's criterion as follows

$$\frac{\sigma_x^2+\sigma_y^2}{k_x^2} - \underline{\frac{\sigma_x\sigma_y}{k_x^2}} + \frac{\sigma_z^2}{k_z^2} - \frac{\sigma_y\sigma_z+\sigma_z\sigma_x}{k_z k_x} + \frac{\tau_{yz}^2+\tau_{zx}^2}{k_{zx}^2} + \underline{\underline{3\frac{\tau_{xy}^2}{k_x^2}}} = 1 \tag{3.101}$$

or

$$\left(\frac{\sigma_x-\sigma_y}{k_x}\right)^2 + \left(\frac{\sigma_y}{k_x}-\frac{\sigma_z}{k_z}\right)^2 + \left(\frac{\sigma_z}{k_z}-\frac{\sigma_x}{k_x}\right)^2 + 3\frac{\tau_{yz}^2+\tau_{zx}^2}{k_{zx}^2} + 6\frac{\tau_{xy}^2}{k_x^2} = 2 \tag{3.102}$$

Note, that the above conditions correspond to generalized Hu–Marin's equations (3.82) or (3.83) with $k_1 = k_2$, but enhanced by the additional shear terms and referring to optional directions x, y, z. Equations (3.101) or (3.102) reduce to the Huber–von Mises equation (3.98) in case of plane stress state in the transverse isotropy plane (x, y), which means that this new criterion can finally be recognized as transversely isotropic hexagonal symmetry von Mises–Hu–Marin's based criterion.

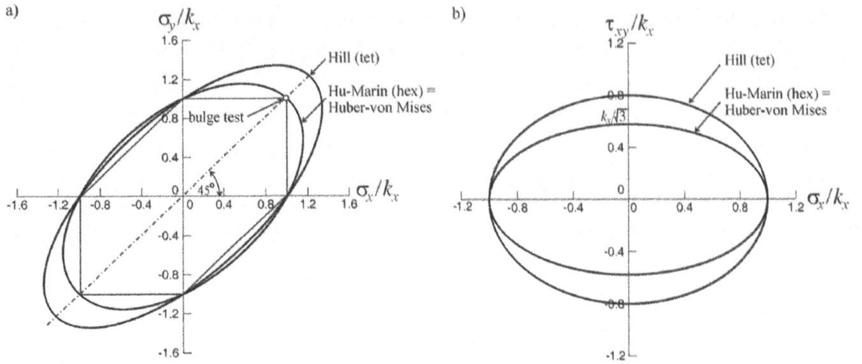

Fig. 3.9 Comparison of transversely isotropic criteria: Hill's tetragonal, Hu–Marin's hexagonal and Huber–von Mises' in case of 2D states of stress: **a** bi-axial normal stresses (σ_x, σ_y) and **b** combined normal with shear stresses (σ_x, τ_{xy}) (after Ganczarski and Skrzypek 2014)

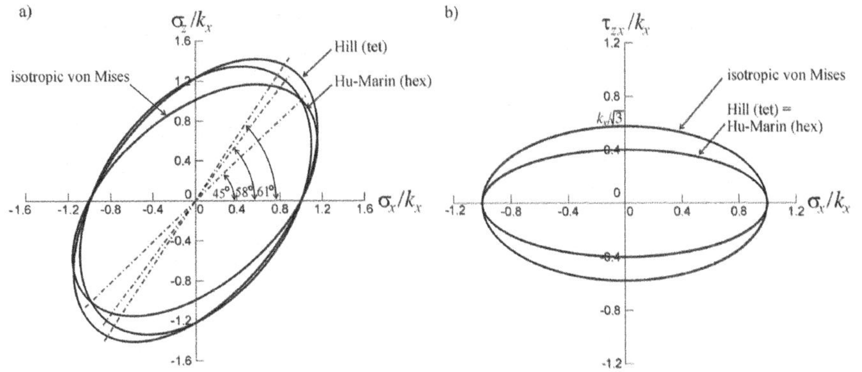

Fig. 3.10 Comparison of transversely isotropic criteria: Hill's tetragonal, Hu–Marin's hexagonal and Huber–von Mises' in case of 2D states of stress: **a** bi-axial normal stresses (σ_x, σ_z) and **b** combined normal with shear stresses (σ_x, τ_{zx}) (after Ganczarski and Skrzypek 2014)

Transversely isotropic conditions—tetragonal Hill's (3.82) or (3.83) and hexagonal Hu–Marin's (3.101) or (3.102), are examined for given orthotropy degrees $R = 2(\frac{k_z}{k_x})^2 - 1 = 2$, $k_{xy}/k_x = 0.8$, $k_{(xy)}/k_x = 0.9$ and $k_{zx}/k_x = 0.8$, for following stress states: bi-axial normal stresses (σ_x, σ_y) and combined normal with shear stresses (σ_x, τ_{xy}) in the transverse isotropy plane (see Fig. 3.9a–b), as well as bi-axial normal stresses (σ_x, σ_z) and combined normal with shear stresses (σ_x, τ_{zx}) in the orthotropy plane (see Fig. 3.10a, b). It is worth to mention that transversely isotropic Hill's condition of tetragonal symmetry (3.92) or (3.93) comprises four independent plastic yield limits: k_x, k_z, k_{zx} and k_{xy}, because shear yield limit in isotropy plane k_{xy} is considered as independent.

Contrary, transversely isotropic enhanced Hu–Marin's type condition, the symmetry class of which is hexagonal, is defined by three independent yield limits only: k_x,

k_z and k_{zx}, since in-plane shear yield limit k_{xy} must agree with the Huber–von Mises criterion in the isotropy plane $k_{xy} = \frac{k_x}{\sqrt{3}}$. Hence, representation of the transversely isotropic hexagonal symmetry Hu–Marin's type constitutive matrix of plasticity is as follows:

$$[{}_{\text{tris}}\boldsymbol{A}^{\text{HM}}] = \begin{bmatrix} \frac{1}{k_x^2} & -\frac{1}{2k_x^2} & -\frac{1}{2k_xk_z} & & & \\ & \frac{1}{k_x^2} & -\frac{1}{2k_xk_z} & & & \\ & & \frac{1}{k_z^2} & & & \\ & & & \frac{1}{k_{zx}^2} & & \\ & & & & \frac{1}{k_{zx}^2} & \\ & & & & & \frac{3}{k_x^2} \end{bmatrix} \tag{3.103}$$

The general case of transversely isotropic 4–parameter tetragonal symmetry Hu–Marin's type yield criterion that preserves convexity but lost property of reducibility to the isotropic von Mises condition in the plane of transverse isotropy is considered by Voyiadjis and Thiagarajan (1995). The corresponding matrix of plasticity used by authors results from the general orthotropic matrix when four independent plastic onset limits are $k_1, k_2 = k_3, k_4 = k_5$ and k_6 (if original notation is saved 1–fiber direction)

$$[{}_{\text{tris}}\boldsymbol{A}^{\text{VT}}] = \begin{bmatrix} \frac{4}{9}k_1^2 & -\frac{2}{9}k_1k_2 & -\frac{2}{9}k_1k_2 & & & \\ & \frac{4}{9}k_2^2 & -\frac{2}{9}k_1^2 & & & \\ & & \frac{4}{9}k_2^2 & & & \\ & & & \frac{4}{3}(k_1k_2+k_4^2) & & \\ & & & & \frac{4}{3}(k_1k_2+k_4^2) & \\ & & & & & \frac{4}{3}(k_2^2+k_6^2) \end{bmatrix} \tag{3.104}$$

Introducing the following substitution for k_1, k_2, k_4 and k_6

$$\frac{2}{9}k_1^2 = \frac{1}{k_z^2},\quad \frac{2}{9}k_2^2 = \frac{1}{k_x^2},\quad \frac{2}{3}(k_1k_2+k_4^2) = \frac{1}{k_{zx}^2},\quad \frac{2}{3}(k_2^2+k_6^2) = \frac{1}{k_{xy}^2}$$

we end up with format of the Voyiadjis and Thiagarajan condition analogous to (3.103) however 4–parameter, where not only k_x, k_z and k_{zx} but additionally k_{xy} are considered as independent (see doubly-underlined terms in (3.101) and (3.105))

$$\frac{\sigma_x^2+\sigma_y^2}{k_x^2}-\underline{\frac{\sigma_x\sigma_y}{k_x^2}}+\frac{\sigma_z^2}{k_z^2}-\frac{\sigma_y\sigma_z+\sigma_z\sigma_x}{k_zk_x}+\frac{\tau_{yz}^2+\tau_{zx}^2}{k_{zx}^2}+\underline{\underline{\frac{\tau_{xy}^2}{k_{xy}^2}}}=1 \tag{3.105}$$

Both transversely isotropic criteria: Hill's type of tetragonal symmetry (3.92) as well as Hu–Marin's type of hexagonal symmetry (3.101) describe cylindrical surfaces in space of principal stresses. However, Hill's type limit surface represents elliptical cylinder the axis of which coincides with the hydrostatic axis, in contrast to enhanced Hu–Marin's type limit surface that represents elliptic cylinder the axis of which forms a proportional stress/strength axis, different from the hydrostatic axis. It means that enhanced Hu–Marin's condition does not satisfy the deviatoricity property, which is a price for property of reducibility to the Huber–von Mises condition in the isotropy plane, with cylindricity ensured regardless of the magnitude of orthotropy degree.

A choice of appropriate transversely isotropic limit criterion, of either the tetragonal symmetry (3.92) or the hexagonal symmetry (3.101), depends on coincidence with experimental findings for real material. This may often lead to one of two above considered symmetry classes, but sometimes material limit response is different even from both of them. Note that the shape of limit curves in the transverse isotropy plane is the key to appropriate classification of real transversely isotropic material as exhibiting tetragonal symmetry or hexagonal or mixed symmetry properties.

3.4 Lecture—Implicit Formulation of Pressure Insensitive Yield Criteria

In this lecture another approach (implicit formulation) is discussed based on a series of papers developed by Barlat, Planckett, Cazacu and Khan to mention some names only. The implicit formulation involves the linear transformation of the Cauchy stress tensor $\boldsymbol{\sigma}$ to the transformed stress $\boldsymbol{\Sigma}=\boldsymbol{L}:\boldsymbol{\sigma}$ by use of transformation tensor $\boldsymbol{L}$ responsible for orthotropy. Such linear transformation concept of the stress tensor was first introduced by Sobotka (1969) and Boehler and Sawczuk (1970)

$$\widehat{\sigma}_{ij}=\widehat{A}_{ijkl}\sigma_{kl} \tag{3.106}$$

where $\widehat{A}_{ijkl}$ stands for a certain dimensionless tensor of anisotropy that satisfies general symmetry conditions $\widehat{A}_{ijkl}=\widehat{A}_{jikl}=\widehat{A}_{ijlk}=\widehat{A}_{klij}$ and the well known isotropic yield conditions to hold for anisotropic materials as well if σ_{ij} are replaced by $\widehat{\sigma}_{ij}$. This approach is not directly based on the theory of common invariants in sense of Sayir, Goldenblat, Kopnov, Spencer, Boehler, Betten, etc. formalism (explicit formulation). According to this implicit approach an extension of isotropic initial yield/failure criteria is performed to account for the tension/compression asymmetry property and to material anisotropy frame (usually orthotropy) by applying the linear transformation to the stress tensor and inserting this transformed stress tensor into the originally isotropic yield/failure criteria.

In Cazacu et al. (2006) the authors consider both the isotropic yield criterion for description of asymmetric yielding

$$f(J_{2s}, J_{3s}) = (|s_1| - \widehat{k}s_1)^a + (|s_2| - \widehat{k}s_2)^a + (|s_3| - \widehat{k}s_3)^a = 2k^a$$
$$\widehat{k} = \frac{1 - h(\frac{k_t}{k_c})}{1 + h(\frac{k_t}{k_c})} \qquad h(\tfrac{k_t}{k_c}) = \left[\frac{2^a - 2(\frac{k_t}{k_c})^a}{(2\frac{k_t}{k_c})^a - 2}\right]^{1/a} \tag{3.107}$$

where s_i, $i = 1, \ldots, 3$ are the principal values of the stress deviator and $\widehat{k}$ gives the size of the yield locus, as well as its extension to include orthotropy by the use of linear transformation of the Cauchy stress deviator $\boldsymbol{\Sigma} = \boldsymbol{C}' : \boldsymbol{s}$ through

$$\boldsymbol{C}' = \begin{bmatrix} C'_{11} & C'_{12} & C'_{13} & & & \\ C'_{12} & C'_{22} & C'_{23} & & & \\ C'_{13} & C'_{23} & C'_{33} & & & \\ & & & C'_{44} & & \\ & & & & C'_{55} & \\ & & & & & C'_{66} \end{bmatrix} \tag{3.108}$$

which lead to following anisotropic equation

$$(|\Sigma_1| - \widehat{k}\Sigma_1)^a + (|\Sigma_2| - \widehat{k}\Sigma_2)^a + (|\Sigma_3| - \widehat{k}\Sigma_3)^a = 2k^a \tag{3.109}$$

Authors proved convexity of the isotropic yield form (3.107) as well as pressure insensitivity of its orthotropic form (3.109) obtained through the linear transformation to the transformed stress frame. However the question of convexity of the orthotropic form (3.109) remains open in the light of discussion performed for Hill's (Fig. 3.6) and Hosford's (Fig. 3.8) extensions in case of a highly orthotropic materials.

The proposed yield function appears to be suitable for description of the strong asymmetry and anisotropy observed in textured Mg-Th and Mg-Li binary alloy sheets and for titanium 4Al-1/4O_2, see Cazacu et al. (2006). The orthotropic yield criterion proposed by Cazacu et al. (2006) was also investigated in a series of multiaxial loading experiments on Ti-6Al-4V titanium alloy by Khan et al. (2007).

Extension of Drucker's isotropic yield criterion to anisotropy by use of common invariants J_2^0 and J_3^0 is due to Cazacu and Barlat (2004), and investigated by Yoshida et al. (2013)

$$(J_2^0)^{3/2} - cJ_3^0 - k^3 = 0 \tag{3.110}$$

The constant c in the Eq. (3.110) accounts for the tension/compression asymmetry defined as

$$c = \frac{3\sqrt{3}(k_t^3 - k_c^3)}{2(k_t^3 + k_c^3)} \tag{3.111}$$

and belongs to two ranges

$$c \in \begin{cases} \left(0, \dfrac{3\sqrt{3}}{2}\right) & \text{for } k_\mathrm{t} > k_\mathrm{c} > 0 \\ \left(-\dfrac{3\sqrt{3}}{2}, 0\right) & \text{for } 0 < k_\mathrm{t} < k_\mathrm{c} \end{cases} \tag{3.112}$$

The second and third common invariants of orthotropy are defined as

$$\begin{aligned} J_2^0 &= \frac{1}{6}\left[a_1(\sigma_x-\sigma_y)^2 + a_2(\sigma_y-\sigma_z)^2 + a_3(\sigma_z-\sigma_x)^2\right] \\ &+ a_4\tau_{xy}^2 + a_5\tau_{xz}^2 + a_6\tau_{zy}^2 \\ J_3^0 &= \frac{1}{27}\left\{(b_1+b_2)\sigma_x^3 + (b_3+b_4)\sigma_y^3 + [2(b_1+b_4)-b_2-b_3]\sigma_z^3\right\} \\ &+ 2b_{11}\tau_{xy}\tau_{yz}\tau_{zx} + \frac{1}{9}\left\{2(b_1+b_2)\sigma_x\sigma_y\sigma_z - (b_1\sigma_y+b_2\sigma_z)\sigma_x^2\right. \\ &\left. - (b_3\sigma_z+b_2\sigma_x)\sigma_y^2 - [(b_1-b_2+b_4)\sigma_x + (b_1+b_3+b_4)\sigma_y]\sigma_z^2\right\} \\ &- \frac{1}{3}\left\{\tau_{yz}^2[(b_6+b_7)\sigma_x - b_6\sigma_y - b_7\sigma_z]\right. \\ &- \tau_{zx}^2[2b_9\sigma_y - b_8\sigma_z - (2b_9-b_8)\sigma_x] \\ &\left. - \tau_{xy}^2[2b_{10}\sigma_z - b_5\sigma_y - (2b_{10}-b_5)\sigma_x]\right\} \end{aligned} \tag{3.113}$$

The discussed anisotropic criterion was successfully verified for textured magnesium Mg-Th and Mg-Li alloy sheets. Authors proved convexity of the enhanced isotropic yield criterion only for $c(k_\mathrm{t}/k_\mathrm{c})$ belonging to the range $[-\frac{3\sqrt{3}}{2}, -\frac{3\sqrt{3}}{2}]$. In case of the anisotropic form of Cazacu and Barlat's criterion (3.110) the general proof of convexity for the wide class of highly tension/compression asymmetric and anisotropic materials may not be possible.

More complete representation of J_2^0 and J_3^0 common invariants as well as the extended model (3.110) verification for high-purity α-titanium is done by Nixon et al. (2010).

Korkolis and Kyriakides (2008) applied anisotropic extension of Hosford's isotropic criterion in terms of principal stress deviator s_1, s_2 in case of plane stress state

$$|s_1 - s_2|^n + |2s_1 + s_2|^n + |s_1 + 2s_2|^n = 2k^n \tag{3.114}$$

Following Barlat et al. (2003) they introduced anisotropy by use of a concept of two linear transformations $\boldsymbol{S}' = \boldsymbol{L}' : \boldsymbol{s}$ and $\boldsymbol{S}'' = \boldsymbol{L}'' : \boldsymbol{s}$ where $\boldsymbol{L}'$ and $\boldsymbol{L}''$ are transformation tensors introducing anisotropy

$$|S_1' - S_2'|^n + |2S_1'' + S_2''|^n + |S_1'' + 2S_2''|^n = 2k^n \tag{3.115}$$

Experimental validation of (3.115) is due to Korkolis and Kyriakides (2008) applied to Al-6260-T4 as well as due to Dunand et al. (2012); Luo et al. (2012) applied to AA6260-T6 alloys under classical tensile and butterfly shear tests.

3.5 Lecture–Yield/Failure Criteria for Hydrostatic Pressure Sensitive Materials

3.5.1 von Mises–Tsai–Wu Type Criteria

The linear term $\Pi_{ij}\sigma_{ij}$ in the Goldenblat–Kopnov criterion (3.37) plays essential role and cannot be omitted ($\alpha \neq 0$) when the pressure sensitive materials are considered.

As a rule it is convenient to reduce the general Goldenblat–Kopnov criterion (3.37) to the narrower format which exhibits dimensional homogeneity assuming $\alpha = 1, \beta = 1/2, \gamma = 1/3$ as follows

$$A_{ij}\sigma_{ij} + \left(A_{ijkl}\sigma_{ij}\sigma_{kl}\right)^{1/2} + \left(A_{ijklmn}\sigma_{ij}\sigma_{kl}\sigma_{mn}\right)^{1/3} - 1 = 0 \tag{3.116}$$

Limiting ourselves to the linear and the quadratic terms in the Eq. (3.116), in other words neglecting the third common invariant responsible for a distortion, we arrive at the particular sub-case of the Goldenblat–Kopnov criterion (Życzkowski 2001)

$$A_{ij}\sigma_{ij} + \sqrt{A_{ijkl}\sigma_{ij}\sigma_{kl}} - 1 = 0 \tag{3.117}$$

which, on the other hand, can be treated as an extension of isotropic Drucker–Prager's failure criterion to the case of anisotropy. Note however that due to material isotropy the Drucker–Prager criterion contains only stress invariants $J_{1\sigma}$ and J_{2s}, whereas in Eq. (3.117) to describe anisotropy the common invariants of stress and the structural anisotropy tensors A_{ij} and A_{ijkl} must be used. By contrast to the cases of the Huber–von Mises and the Hill yield conditions which represent the circular and the elliptic cylinders respectively (Fig. 3.12), in the considered case of the Drucker–Prager criterion and its anisotropic generalization (3.117) the respective failure surfaces can be recognized as the circular and the elliptic cones, respectively.

Another special sub-case of the Goldenblat–Kopnov criterion (3.37) is obtained if $\alpha = 1, \beta = 1$ and consecutive limitation of this format to the linear and the quadratic terms hold such that at the anisotropic extension of Burzyński's paraboloid is met, cf. Ganczarski and Lenczowski (1997)

$$A_{ij}\sigma_{ij} + A_{ijkl}\sigma_{ij}\sigma_{kl} - 1 = 0 \tag{3.118}$$

Note that the quadratic term $A_{ijkl}\sigma_{ij}\sigma_{kl}$ in Eq. (3.117) appears under square root whereas in Eq. (3.118) does not, hence the condition (3.117) can be interpreted as non circular cone whereas condition (3.118) non circular paraboloid.

Structural tensor of the second A_{ij} and fourth A_{ijkl} orders appearing in Eqs. (3.117)–(3.118) stand for two independent yield/failure anisotropy tensors the identification of which has to be performed on the basis of respective yield/failure tests in analogous way as that discussed in previous lecture. However in the present case two anisotropy tensors have to be calibrated hence the appropriate number of tests increases such that the difference between the tension and the compression uniaxial tests can be captured. Substituting for convenience Voigt's vector-matrix notation both above tensors can be represented as follows

$$[\boldsymbol{a}] = \begin{bmatrix} a_{11} & a_{12} & a_{13} \\ & a_{22} & a_{23} \\ & & a_{33} \end{bmatrix} = \left\{ a_{11}\ a_{22}\ a_{33}\ a_{23}\ a_{13}\ a_{12} \right\}^{\mathrm{T}} \tag{3.119}$$

and

$$[\boldsymbol{A}] = \begin{bmatrix} A_{11} & A_{12} & A_{13} & A_{14} & A_{15} & A_{16} \\ & A_{22} & A_{23} & A_{24} & A_{25} & A_{26} \\ & & A_{33} & A_{34} & A_{35} & A_{36} \\ & & & A_{44} & A_{45} & A_{46} \\ & & & & A_{55} & A_{56} \\ & & & & & A_{66} \end{bmatrix} \tag{3.120}$$

where yield/failure loci are determined by two yield/failure characteristic matrices $[\boldsymbol{a}]$ of the dimension (3×3) and $[\boldsymbol{A}]$ of the dimension (6×6). Hence in considered case of general anisotropy the number of modules defining yield/failure initiation is equal to $27 = 6 + 21$.

The condition of yield/failure initiation in anisotropic materials (3.118) takes in Voigt's notation the equivalent format

$$\{\boldsymbol{a}\}\{\boldsymbol{\sigma}\} + \{\boldsymbol{\sigma}\}^{\mathrm{T}}[\boldsymbol{A}]\{\boldsymbol{\sigma}\} - 1 = 0 \tag{3.121}$$

being an extension of anisotropic von Mises' criterion of plastic materials (3.43) however enriched by the additional term. From among general anisotropy number of modules 27 only $24 = 6 + 18$ are truly independent. However in practical application this number of material modules (24) can further be reduced assuming certain symmetry group.

In what follows an extension of anisotropic von Mises' criterion (3.42) enhanced by linear terms (3.118) is considered. Assume the deviatoric form of von Mises criterion (3.56) but enhanced by including linear terms

$$\begin{aligned} &-A_{12}\left(\sigma_x - \sigma_y\right)^2 - A_{13}\left(\sigma_x - \sigma_z\right)^2 - A_{23}\left(\sigma_y - \sigma_z\right)^2 \\ &+2\left\{\tau_{yz}\left[A_{24}\left(\sigma_y - \sigma_x\right) + A_{34}\left(\sigma_z - \sigma_x\right)\right] + \tau_{zx}\left[A_{15}\left(\sigma_x - \sigma_y\right)\right.\right. \\ &\left.+A_{35}\left(\sigma_z - \sigma_y\right)\right] + \tau_{xy}\left[A_{16}\left(\sigma_x - \sigma_z\right) + A_{26}\left(\sigma_y - \sigma_z\right)\right] \\ &\left.+A_{45}\tau_{yz}\tau_{zx} + A_{46}\tau_{xy}\tau_{yz} + A_{56}\tau_{zx}\tau_{xy}\right\} + A_{44}\tau_{yz}^2 + A_{55}\tau_{zx}^2 + A_{66}\tau_{xy}^2 \\ &+a_{11}\sigma_x + a_{22}\sigma_y + a_{33}\sigma_z + a_{12}\tau_{xy} + a_{13}\tau_{zx} + a_{23}\tau_{yz} = 1 \end{aligned} \tag{3.122}$$

Note that this form is strictly pressure insensitive only in the quadratic terms but it is pressure sensitive as far as the linear terms are concerned. In order to obtain form of Eq. (3.122) fully pressure insensitive the following additional constraint has to be satisfied

$$a_{11} + a_{22} + a_{33} = 0 \tag{3.123}$$

The condition (3.123) can be understood in one of the three following ways

$$a_{11} = -(a_{22} + a_{33}) \quad \text{or} \quad a_{22} = -(a_{11} + a_{33}) \quad \text{or} \quad a_{33} = -(a_{11} + a_{22}) \tag{3.124}$$

For instance, substituting the first of conditions (3.124) in the Eq. (3.122) we arrive at the first of three deviatoric forms of the von Mises–Tsai–Wu criterion, which is hydrostatic pressure insensitive

$$\begin{aligned}
&-A_{12}\left(\sigma_x - \sigma_y\right)^2 - A_{13}\left(\sigma_x - \sigma_z\right)^2 - A_{23}\left(\sigma_y - \sigma_z\right)^2 \\
&+2\left\{\tau_{yz}\left[A_{24}\left(\sigma_y - \sigma_x\right) + A_{34}\left(\sigma_z - \sigma_x\right)\right] + \tau_{zx}\left[A_{15}\left(\sigma_x - \sigma_y\right)\right.\right. \\
&\left.+A_{35}\left(\sigma_z - \sigma_y\right)\right] + \tau_{xy}\left[A_{16}\left(\sigma_x - \sigma_z\right) + A_{26}\left(\sigma_y - \sigma_z\right)\right] \\
&\left.+ A_{45}\tau_{yz}\tau_{zx} + A_{46}\tau_{xy}\tau_{yz} + A_{56}\tau_{zx}\tau_{xy}\right\} \\
&+A_{44}\tau_{yz}^2 + A_{55}\tau_{zx}^2 + A_{66}\tau_{xy}^2 \\
&+a_{22}\left(\sigma_y - \sigma_x\right) + a_{33}\left(\sigma_z - \sigma_x\right) + a_{12}\tau_{xy} + a_{13}\tau_{zx} + a_{23}\tau_{yz} = 1
\end{aligned} \tag{3.125}$$

The form analogous to (3.125) was considered by Szczepiński (1993) where the first of constraints (3.124_1) was chosen when calibrating anisotropic modules $A_{12}, A_{13}, A_{23}, \ldots, A_{66}$ (15 modules) and $a_{22}, a_{33}, \ldots, a_{23}$ (5 modules).

Experimental verification of Eq. (3.125) was done by Kowalewski and Śliwowski (1997) where the low carbon steel 18G2A specimens were used. In this experiment the cross-sections of limit surface (3.125) in the plane σ_x, τ_{xy}

$$-\left(A_{12} + A_{13}\right)\sigma_x^2 + 2A_{16}\tau_{xy}\sigma_x + A_{66}\tau_{xy}^2 - \left(a_{22} + a_{33}\right)\sigma_x + a_{12}\tau_{xy} = 1 \tag{3.126}$$

was considered. The Eq. (3.126) represents ellipse the center of which is shifted from the origin (σ_x, τ_{xy}), the axes are rotated with the axes ratio slightly different from that in isotropic material (Huber–von Mises' criterion), Fig. 3.11.

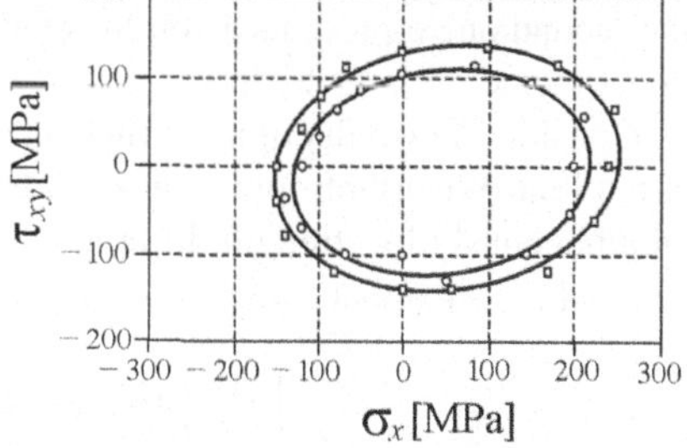

Fig. 3.11 Experimental verification of Eq. (3.126) in case of low carbon steel 18G2A subjected to monotonic prestrain ∘ − $\varepsilon_{\text{off}} = 1 \times 10^{-5}$, □ − $\varepsilon_{\text{off}} = 5 \times 10^{-5}$, after Kowalewski and Śliwowski (1997)

Limiting further considerations to orthotropic materials the both characteristic matrices $[_{\text{ort}}\boldsymbol{a}]$ and $[_{\text{ort}}\boldsymbol{A}]$ (3.120) and (3.121) take following forms valid for principal directions of orthotropy

$$
[_{\text{ort}}\boldsymbol{a}] = \begin{bmatrix} a_{11} & 0 & 0 \\ & a_{22} & 0 \\ & & a_{33} \end{bmatrix}
$$

$$
[_{\text{ort}}\boldsymbol{A}] = \begin{bmatrix} A_{11} & A_{12} & A_{13} & 0 & 0 & 0 \\ & A_{22} & A_{23} & 0 & 0 & 0 \\ & & A_{33} & 0 & 0 & 0 \\ & & & A_{44} & 0 & 0 \\ & & & & A_{55} & 0 \\ & & & & & A_{66} \end{bmatrix} \tag{3.127}
$$

The second rank matrix $[_{\text{ort}}\boldsymbol{a}]$ is of the dimension 3×3 whereas the fourth rank matrix $[_{\text{ort}}\boldsymbol{A}]$ dimension is 6×6. The matrix $[_{\text{ort}}\boldsymbol{a}]$ has diagonal form and the matrix $[_{\text{ort}}\boldsymbol{A}]$ is of the identical symmetry as the von Mises plastic orthotropy matrix (3.57). Both matrices (3.127) are defined by $12 = 3 + 9$ modules.

Therefore the condition of yield/failure initiation for anisotropic materials (3.121) takes a form typical for the rotationally symmetric group

$$
\{_{\text{ort}}\boldsymbol{a}\}\{\boldsymbol{\sigma}\} + \{\boldsymbol{\sigma}\}^{\mathrm{T}}[_{\text{ort}}\boldsymbol{A}]\{\boldsymbol{\sigma}\} - 1 = 0 \tag{3.128}
$$

being an extension of the von Mises orthotropic yield condition (3.117) for pressure sensitive materials. The von Mises orthotropic yield/failure initiation criterion (3.128) can be written down in the following extended form

$$
\begin{aligned}
&A_{11}\sigma_x^2 + A_{22}\sigma_y^2 + A_{33}\sigma_z^2 + 2(A_{12}\sigma_x\sigma_y + A_{23}\sigma_y\sigma_z + A_{31}\sigma_z\sigma_x) + \\
&A_{44}\tau_{yz}^2 + A_{55}\tau_{zx}^2 + A_{66}\tau_{xy}^2 + a_{11}\sigma_x + a_{22}\sigma_y + a_{33}\sigma_z - 1 = 0
\end{aligned} \tag{3.129}
$$

Note that above equation represents fully tensorial form of the orthotropic yield/failure criterion contrary to the deviatoric form which is characteristic for the Hill yield criterion. This means that $12 = 3 + 9$ material modules defining yield/failure material characteristic tensors $_{\text{ort}}\boldsymbol{a}$ and $_{\text{ort}}\boldsymbol{A}$ are required for its identification. The first term in Eq. (3.128) refers to the strength differential effect whereas the second one represents a von Mises-type surface the axis of which generally does not coincide with the hydrostatic axis.

Consider now reduction of criterion (3.128) to narrower form known in literature as the Tsai–Wu orthotropic criterion of failure. The Tsai–Wu criterion is characterized simultaneously by strength differential effect and pressure insensitivity of $[_{\text{ort}}\boldsymbol{A}]$ in Eq. (3.127) such that $[_{\text{ort}}\boldsymbol{A}] \longrightarrow [\boldsymbol{A}^{\mathrm{TW}}]$ (see (3.122))

$$
\{\boldsymbol{a}^{\mathrm{TW}}\}\{\boldsymbol{\sigma}\} + \{s\}^{\mathrm{T}}[\boldsymbol{A}^{\mathrm{TW}}]\{s\} - 1 = 0 \tag{3.130}
$$

This leads to the following representation of both characteristic matrices (see (3.123))

$$\left[\boldsymbol{a}^{\mathrm{TW}}\right]=\begin{bmatrix} a_{11} & 0 & 0 \\ & a_{22} & 0 \\ & & a_{33} \end{bmatrix} \tag{3.131}$$

$$\left[\boldsymbol{A}^{\mathrm{TW}}\right]=\begin{bmatrix} -A_{12}-A_{13} & A_{12} & A_{13} & 0 & 0 & 0 \\ & -A_{12}-A_{23} & A_{23} & 0 & 0 & 0 \\ & & -A_{13}-A_{23} & 0 & 0 & 0 \\ & & & A_{44} & 0 & 0 \\ & & & & A_{55} & 0 \\ & & & & & A_{66} \end{bmatrix} \tag{3.132}$$

Form of Eq. (3.130) and its representation (3.132) reflects "hybrid notation" in the following sense: the first term represents linear common invariant of the stress tensor $\boldsymbol{\sigma}$ and the structural tensor $\boldsymbol{a}^{\mathrm{TW}}$ (analogy to the pressure sensitivity in case of isotropic material) whereas the second term represents quadratic common invariant of the stress deviator $\boldsymbol{s}$ and the structural tensor $\boldsymbol{A}^{\mathrm{TW}}$ (defining shape and orientation of surface in the stress space). The criterion (3.130) takes therefore explicit form of 9–parameter Tsai–Wu's criterion (Tsai and Wu 1971)

$$\begin{gathered}-\left[A_{23}\left(\sigma_y-\sigma_z\right)^2+A_{13}\left(\sigma_z-\sigma_x\right)^2+A_{12}\left(\sigma_x-\sigma_y\right)^2\right] \\ +A_{44}\tau_{yz}^2+A_{55}\tau_{zx}^2+A_{66}\tau_{xy}^2+a_{11}\sigma_x+a_{22}\sigma_y+a_{33}\sigma_z-1=0\end{gathered} \tag{3.133}$$

As a matter of fact, any addition of a hydrostatic pressure to all normal stresses $\sigma_x \rightarrow \sigma_x \pm \sigma_{\mathrm{h}}$ does not change magnitude of quadratic terms in condition (3.133) but simultaneously causes the linear terms still dependent on σ_{h}. Hence, finally the Tsai–Wu criterion in the format given by (3.133) remains the pressure sensitive one through the linear terms.

3.5.2 *Transversely Isotropic Case of Tsai–Wu Type Criteria*

Similarly to previous lecture a reduction of 9–parameter yield/failure orthotropic Tsai–Wu's criterion (3.133) to narrower case of the transverse isotropy requires precise distinction between the tetragonal and hexagonal symmetry classes. Assuming after Chen and Han (1995) plane of transverse isotropy xy, the 4th-rank orthotropy matrix $[\boldsymbol{A}^{\mathrm{TW}}]$ (3.132) reduces to the transversely isotropic format $[{}_{\mathrm{tris}}\boldsymbol{A}^{\mathrm{TW}}]$ analogously to the transversely isotropic Hill criterion (3.153) or (3.154) possessing only four independent material constants whereas the 2nd-rank transversely isotropic matrix $[{}_{\mathrm{tris}}\boldsymbol{a}^{\mathrm{TW}}]$ reduces to a form possessing only two independent material constants. Finally, assuming $A_{23}=A_{13}$, $A_{44}=A_{55}$, $a_{11}=a_{22}$ in (3.132) instead of the 9–parameter form (3.132) we arrive at two 6–parameter forms of the transversely isotropic yield/failure criterion of tetragonal symmetry that directly refer to formulations (3.122) or (3.123), namely

$$\left[{}_{\mathrm{tris}}\boldsymbol{a}^{\mathrm{TW}}\right] = \begin{bmatrix} a_{11} & 0 & 0 \\ & a_{11} & 0 \\ & & a_{33} \end{bmatrix} \tag{3.134}$$

and

$$\left[{}_{\mathrm{tris}}\boldsymbol{A}^{\mathrm{TW}}\right] = \begin{bmatrix} -A_{12}-A_{13} & A_{12} & A_{13} & 0 & 0 & 0 \\ & -A_{12}-A_{13} & A_{13} & 0 & 0 & 0 \\ & & -2A_{13} & 0 & 0 & 0 \\ & & & A_{44} & 0 & 0 \\ & & & & A_{44} & 0 \\ & & & & & A_{66} \end{bmatrix} \tag{3.135}$$

or alternatively

$$\left[{}_{\mathrm{tris}}\boldsymbol{A}^{\mathrm{TW}}\right] = \begin{bmatrix} A_{11} & \dfrac{A_{33}-2A_{11}}{2} & -\dfrac{A_{33}}{2} & 0 & 0 & 0 \\ & A_{11} & -\dfrac{A_{33}}{2} & 0 & 0 & 0 \\ & & A_{33} & 0 & 0 & 0 \\ & & & A_{44} & 0 & 0 \\ & & & & A_{44} & 0 \\ & & & & & A_{66} \end{bmatrix} \tag{3.136}$$

In case when non abbreviated notation is used the 6–parameter transversely isotropic Tsai–Wu yield/failure criterion of tetragonal symmetry takes the following form

$$\begin{aligned} &-A_{13}\left[\left(\sigma_y-\sigma_z\right)^2+\left(\sigma_z-\sigma_x\right)^2\right]-A_{12}\left(\sigma_x-\sigma_y\right)^2 \\ &+A_{44}\left(\tau_{yz}^2+\tau_{zx}^2\right)+A_{66}\tau_{xy}^2+a_{11}\left(\sigma_x+\sigma_y\right)+a_{33}\sigma_z-1=0 \end{aligned} \tag{3.137}$$

or alternatively

$$\begin{aligned} &A_{11}\left(\sigma_x^2+\sigma_y^2\right)+A_{33}\sigma_z^2+\left(A_{33}-2A_{11}\right)\sigma_x\sigma_y-A_{33}\left(\sigma_x\sigma_z+\sigma_y\sigma_z\right) \\ &+A_{44}\left(\tau_{yz}^2+\tau_{zx}^2\right)+A_{66}\tau_{xy}^2+a_{11}\left(\sigma_x+\sigma_y\right)+a_{33}\sigma_z-1=0 \end{aligned} \tag{3.138}$$

The above transversely isotropic limit equations are expressed in terms of six material anisotropy modules: $A_{12}, A_{13}, A_{44}, A_{66}, a_{11}, a_{33}$ or $A_{11}, A_{33}, A_{44}, A_{66}, a_{11}, a_{33}$ if corresponding matrix representations (3.134)–(3.135) or (3.134)–(3.136) are implemented.

However, the number of independent modules can further be reduced to five, since the sixth diagonal modulus A_{66} has to satisfy the relationships (cf. Chen and Han 1995; Ganczarski and Skrzypek 2009))

$$A_{66} = -2(A_{13} + 2A_{12}) \quad \text{or} \quad A_{66} = 4A_{11} - A_{33} \tag{3.139}$$

if the corresponding formats (3.135) or (3.136) are used. The conditions (3.139) satisfy reducibility of the criteria (3.135) or (3.136) to the forms invariant with respect to two equivalent stress states $\tau_{xy} = \sigma$ and $\sigma_x = \sigma$, $\sigma_y = -\sigma$ in the transverse isotropy plane.

Taking above conditions into account Eqs. (3.137) and (3.138) contain only five independent material coefficients referring to appropriate tensile and compressive strengths $k_{tx}, k_{cx}, k_{tz}, k_{cz}$ and shear strength k_{zx}. Hence, in order to calibrate them following tests have to be performed if, for instance, the format (3.137) is used:

$$\begin{aligned}
&\sigma_x = k_{tx}, \quad \sigma_y = \ldots = \tau_{zx} = 0 \longrightarrow (-A_{13} - A_{12})\, k_{tx}^2 + a_{11} k_{tx} = 1 \\
&\sigma_x = -k_{cx}, \sigma_y = \ldots = \tau_{zx} = 0 \longrightarrow (-A_{13} - A_{12})\, k_{cx}^2 - a_{11} k_{cx} = 1 \\
&\sigma_z = k_{tz}, \quad \sigma_x = \ldots = \tau_{zx} = 0 \longrightarrow -2A_{13} k_{tz}^2 + a_{33} k_{tz} = 1 \\
&\sigma_z = -k_{cz}, \sigma_x = \ldots = \tau_{zx} = 0 \longrightarrow -2A_{13} k_{cz}^2 - a_{33} k_{cz} = 1 \\
&\tau_{zx} = k_{zx}, \quad \sigma_x = \ldots = \tau_{yz} = 0 \longrightarrow A_{44} k_{zx}^2 = 1
\end{aligned} \tag{3.140}$$

Solution of Eqs. (3.140) with respect to A_{13}, A_{12}, A_{44}, a_{11} and a_{33} takes the form

$$\begin{aligned}
&-A_{13} = \frac{1}{2k_{tz}k_{cz}}, \quad -A_{12} = \frac{1}{k_{tx}k_{cx}} - \frac{1}{2k_{tz}k_{cz}}, \quad A_{44} = \frac{1}{2k_{zx}^2} \\
&a_{11} = \frac{1}{k_{tx}} - \frac{1}{k_{cx}}, \quad a_{33} = \frac{1}{k_{tz}} - \frac{1}{k_{cz}},
\end{aligned} \tag{3.141}$$

Magnitude of material modules A_{66}, referring to shear strength in the plane of transverse isotropy is not independent but given by Eq. (3.139), hence

$$A_{66} = \frac{4}{k_{tx}k_{cx}} - \frac{1}{k_{tz}k_{cz}} \tag{3.142}$$

Note that both formats in Eq. (3.139) lead to the same calibration for A_{66} (3.142). Hence, after substitution of Eqs. (3.141)–(3.142) to Eq. (3.137) one can get final form of the hexagonal transversely isotropic Tsai–Wu criterion in terms of 5 independent constants $k_{tx}, k_{cx}, k_{tz}, k_{cz}$ and k_{zx}

$$\begin{aligned}
&\frac{\sigma_x^2+\sigma_y^2}{k_{tx}k_{cx}}+\frac{\sigma_z^2}{k_{tz}k_{cz}}-\left(\frac{2}{k_{tx}k_{cx}}-\frac{1}{k_{tz}k_{cz}}\right)\sigma_x\sigma_y-\frac{\sigma_y\sigma_z+\sigma_x\sigma_z}{k_{tz}k_{cz}}\\
&+\frac{\tau_{yz}^2+\tau_{zx}^2}{k_{zx}^2}+\underline{\left(\frac{4}{k_{tx}k_{cx}}-\frac{1}{k_{tz}k_{cz}}\right)}\tau_{xy}^2\\
&+\left(\frac{1}{k_{tx}}-\frac{1}{k_{cx}}\right)\left(\sigma_x+\sigma_y\right)+\left(\frac{1}{k_{tz}}-\frac{1}{k_{cz}}\right)\sigma_z=1
\end{aligned} \tag{3.143}$$

Inspection of the transversely isotropic format of the Tsai–Wu criterion (3.143) reveals that underlined coefficient preceding τ_{xy} differs in format from the analogous term in the transversely isotropic Hill criterion (3.93) since independent shear limit in the transverse isotropy plane k_{xy} is used. Obviously the transition from the Tsai–Wu criterion (3.143) to the Hill criterion (3.93) requires to ignore the tension/compression asymmetry effect $k_{tx}=k_{cx}$ and $k_{tz}=k_{cz}$ which leads simultaneously to vanishing of linear terms. In other words in this case the Tsai–Wu transversely isotropic criterion reducible to the Hill criterion becomes a pressure insensitive, by contrast to the Eq. (3.143) in which pressure sensitivity $k_{tx}\neq k_{cx}$ and $k_{tz}\neq k_{cz}$ plays essential role.

It is seen that material coefficients in the xy-plane of transverse isotropy that precede terms $\sigma_x\sigma_y$ and τ_{xy}^2 are not fully independent since they contain not only the in-plane tensile and compressive limits k_{tx}, k_{cx} but also the out-of-plane tensile and compressive limits k_{tz} and k_{cz}. Consequently, Eq. (3.143) can be classified as the hexagonal transversely isotropic Tsai–Wu criterion of initial yield/failure.

Applicability range of the Tsai–Wu orthotropic criterion (3.143) to properly describe initiation of failure in some engineering materials that exhibit high orthotropy degree, is bounded by a possible ellipticity loss of the limit surface, see Ganczarski and Adamski (2015). In other words, a physically inadmissible degeneration of a single convex and simply connected elliptic limit surface into two concave hyperbolic surfaces occurs.

The following inequality bounds the range of applicability of the transversely isotropic Tsai–Wu criterion to ensure convexity

$$\frac{1}{k_{tz}k_{cz}}\left(\frac{4}{k_{tx}k_{cx}}-\frac{1}{k_{tz}k_{cz}}\right)>0 \tag{3.144}$$

which can easily be recognized as an extension of the relevant bounding inequality for Hill's criterion (3.77). Substitution of the dimensionless parameter

$$\overline{R}=2\left(\frac{k_{tz}k_{cz}}{k_{tx}k_{cx}}\right)-1,$$

leads to the simplified restriction

$$\overline{R}>-0.5 \tag{3.145}$$

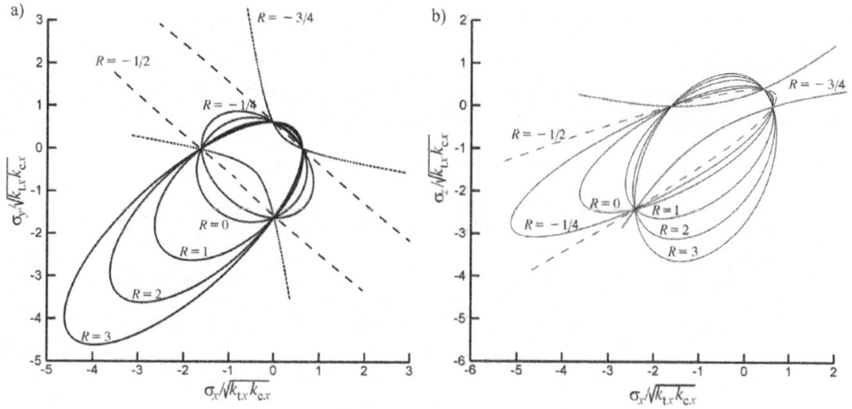

Fig. 3.12 Degeneration of the Tsai–Wu limit surface with respect to parameter $\overline{R}$: **a** transverse isotropy plane, **b** orthotropy plane, after Ganczarski and Adamski (2015)

If the above inequalities (3.144)–(3.145) do not hold, elliptic cross sections of the limit surface degenerate to two hyperbolic branches and the lack of convexity occurs. To illustrate this limitation, the yield curves in two planes:

- the transverse isotropy plane (σ_x, σ_y)

$$\sigma_x^2 - \frac{2\overline{R}}{1+\overline{R}}\sigma_x\sigma_y + \sigma_y^2 + (k_{cx} - k_{tx})\left(\sigma_x + \sigma_y\right) = k_{tx}k_{cx} \tag{3.146}$$

- and the orthotropy plane (σ_x, σ_z)

$$\sigma_x^2 - \frac{2}{1+\overline{R}}\sigma_x\sigma_z + \frac{2}{1+\overline{R}}\sigma_z^2 + (k_{cx} - k_{tx})\,\sigma_x + k_{tz}k_{cz}\sigma_z = k_{tx}k_{cx} \tag{3.147}$$

for various $\overline{R}$-values, are sketched in Fig. 3.12a, b respectively. It is observed that when $\overline{R}$, starting from $\overline{R} = 3$, approaches the limit $\overline{R} = -0.5$, the limit curves change from closed ellipses to two parallel lines, whereas for $\overline{R} < -0.5$, concave hyperbolas appear.

Except the hexagonal transversely isotropic Tsai–Wu criterion Eq. (3.143) one can introduce the other hexagonal transversely isotropic Tsai–Wu failure criterion, see Ganczarski and Adamski (2015). In order to do this let us consider the more general transverse isotropic von Mises–Tsai–Wu criterion of the format

$$\begin{aligned} &A_{11}\left(\sigma_x^2 + \sigma_y^2\right) + A_{33}\sigma_z^2 + 2\Pi_{12}\sigma_x\sigma_y + 2A_{13}\left(\sigma_x + \sigma_y\right)\sigma_z \\ &+A_{44}\left(\tau_{yz}^2 + \tau_{zx}^2\right) + A_{66}\tau_{xy}^2 + a_{11}\left(\sigma_x + \sigma_y\right) + a_{33}\sigma_z = 1 \end{aligned} \tag{3.148}$$

Equation (3.148) contains $8 = 6 + 2$ independent modules and it is straightforward simplification of the orthotropic von Mises–Tsai–Wu criterion (3.129) by introducing obvious symmetry conditions $A_{11} = A_{22}$, $A_{23} = A_{31}$, $A_{44} = A_{55}$ and $a_{11} = a_{22}$. For calibration of it following tests are to be performed:

- six uniaxial tension/compression and shear conditions

$$
\begin{array}{lll}
\sigma_x = k_{tx}, & \sigma_y = \ldots = \tau_{zx} = 0 \longrightarrow & A_{11}k_{tx}^2 + a_{11}k_{tx} = 1 \\
\sigma_x = -k_{cx}, & \sigma_y = \ldots = \tau_{zx} = 0 \longrightarrow & A_{11}k_{cx}^2 - a_{11}k_{cx} = 1 \\
\sigma_z = k_{tz}, & \sigma_x = \ldots = \tau_{zx} = 0 \longrightarrow & A_{33}k_{tz}^2 + a_{33}k_{tz} = 1 \\
\sigma_z = -k_{cz}, & \sigma_x = \ldots = \tau_{zx} = 0 \longrightarrow & A_{33}k_{cz}^2 - a_{33}k_{cz} = 1 \\
\tau_{zx} = k_{zx}, & \sigma_x = \ldots = \tau_{yz} = 0 \longrightarrow & A_{44}k_{zx}^2 = 1 \\
\tau_{xy} = \sqrt{k_{tx}k_{cx}/3}, & \sigma_x = \ldots = \tau_{yz} = 0 \longrightarrow & A_{66}k_{tx}k_{cx}/3 = 1
\end{array}
\qquad (3.149)
$$

- and two biaxial conditions that allows to capture magnitudes of A_{12} and A_{13}

$$
\sigma_x = \sigma_y = k_{(xy)} = -(k_{cx} - k_{tx}) \mp \sqrt{\Delta_1}, \sigma_z = \ldots = \tau_{yz} = 0
$$

$$
\longrightarrow \frac{2k_{(xy)}^2}{k_{tx}k_{cx}} - 2A_{12}k_{(xy)}^2 + \frac{k_{cx} - k_{tx}}{k_{tx}k_{cx}}k_{(xy)} = 1
$$

$$
\sigma_x = \sigma_z = k_{(xz)} = -\frac{1}{2}\left[(k_{cx} - k_{tx}) + (k_{cz} - k_{tz})\frac{k_{tx}k_{cx}}{k_{tz}k_{cz}} \pm \sqrt{\Delta_2}\right],
$$

$$
\sigma_y = \ldots = \tau_{yz} = 0 \longrightarrow \frac{k_{(xz)}^2}{k_{tx}k_{cx}} + \frac{k_{(xz)}^2}{k_{tz}k_{cz}} - 2A_{13}k_{(xz)}^2 \qquad (3.150)
$$

$$
+\left(\frac{1}{k_{tx}} - \frac{1}{k_{cx}}\right)k_{(xz)} + \left(\frac{1}{k_{tz}} - \frac{1}{k_{cz}}\right)k_{(xz)} = 1
$$

Symbols Δ_1 and Δ_2 used for brevity denote: $\Delta_1 = (k_{cx} - k_{tx})^2 + k_{tx}k_{cx}$ and $\Delta_2 = [(k_{cx} - k_{tx}) + (k_{cz} - k_{tz})\frac{k_{tx}k_{cx}}{k_{tz}k_{cz}}]^2 + 4k_{tx}k_{cx}$. Solution of Eqs. (3.149) and (3.150) with respect to A_{11}, A_{12}, A_{13}, A_{33}, A_{44}, A_{66}, a_{11} and a_{33} yields

$$
\begin{array}{lll}
A_{11} = 1/k_{tx}k_{cx} & A_{12} = -1/2k_{tx}k_{cx} & A_{13} = -1/2k_{tz}k_{cz} \\
A_{33} = 1/k_{tz}k_{cz} & A_{44} = 1/k_{zx}^2, & A_{66} = 3/k_{tx}k_{cx} \\
a_{11} = 1/k_{tx} - 1/k_{cx} & a_{33} = 1/k_{tz} - 1/k_{cz} &
\end{array}
\qquad (3.151)
$$

which finally leads to the new hexagonal transversely isotropic von Mises–Tsai–Wu failure criterion also in terms of 5 independent constants k_{tx}, k_{cx}, k_{tz}, k_{cz} and k_{zx}, but different from (3.143)

Table 3.6 Experimental data for columnar ice after Ralston (1977)

Tensile strength		Compressive strength	
k_{tx}	1.01 MPa	k_{cx}	7.11 MPa
k_{tz}	1.21 MPa	k_{cz}	13.5 MPa

$$\frac{\sigma_x^2+\sigma_y^2}{k_{tx}k_{cx}}+\frac{\sigma_z^2}{k_{tz}k_{cz}}-\underline{\frac{\sigma_x\sigma_y}{k_{tx}k_{cx}}}-\frac{\sigma_y\sigma_z+\sigma_x\sigma_z}{k_{tz}k_{cz}}+\frac{\tau_{yz}^2+\tau_{zx}^2}{k_{zx}^2}+\underline{\frac{3}{k_{tx}k_{cx}}\tau_{xy}^2}$$
$$+\left(\frac{1}{k_{tx}}-\frac{1}{k_{cx}}\right)(\sigma_x+\sigma_y)+\left(\frac{1}{k_{tz}}-\frac{1}{k_{cz}}\right)\sigma_z=1 \tag{3.152}$$

Note that the coefficients preceding $\sigma_x\sigma_y$ and τ_{xy}^2, underlined terms in (3.152) are always positive by contrast to analogous terms in (3.143) that can change sign. These prevent elliptic form of failure curves from loss of ellipticity and reduce Eq. (3.152) to the "shifted" Huber–von Mises ellipse from the origin of co-ordinate system in case of transverse isotropy plane. In other words this new hexagonal format of Tsai–Wu's failure criterion is unconditionally stable and preserves reducibility to isotropic Huber–von Mises ellipse but shifted in the isotropy plane.

Both the Tsai–Wu transversely isotropic initial failure criteria: hexagonal Eq. (3.143) and new hexagonal type Eq. (3.152) are compared for columnar ice, the experimental data of which was established by Ralston (1977) in Table 3.6 in plane of transverse isotropy (σ_x, σ_y), shear plane (σ_x, τ_{xy}) and in plane of orthotropy (σ_x, σ_z), see Fig. 3.13. Subsequent cross sections of the limit surface are ellipses, that exhibit strong oblateness in case tetragonal symmetry, the centers of which are shifted outside the origin of co-ordinate system towards the quarter referring to compressive stresses. In case of cross section by plane of transverse isotropy (see Fig. 3.13a) the symmetry axis has obviously inclination equal 45° to the axes of co-ordinate system, in other words it overlaps projection of hydrostatic axis at the transverse isotropy plane (σ_x, σ_y), contrary to the cross section by plane of orthotropy (see Fig. 3.13b) the main semi-axis of ellipse is inclined by 71.1°. It has to be emphasize that in case of columnar ice compressive strength along othotropy axis k_{cz} is over 10 times greater than tensile strength k_{tz}, whereas analogous ratio k_{cx}/k_{tx} is approximately equal to 7 in case of transverse isotropy plane. Moreover, ratio of semi-axes for Tsai–Wu tetragonal ellipse in (σ_x, σ_y) plane essentially exceeds analogous ratio for Huber–von Mises ellipse, contrary to the case of Tsai–Wu hexagonal ellipse. It is also worth to emphasize that although the tetragonal transversely isotropic Tsai–Wu failure criterion Eq. (3.143) and the hexagonal transversely isotropic Tsai–Wu failure criterion Eq. (3.152) contain the same number of 5 independent strengths k_{tx}, k_{cx}, k_{tz}, k_{cz} and k_{zx}, only criterion (3.152) is free from convexity loss and simultaneously truly transversely isotropic in sense of hexagonal class of symmetry.

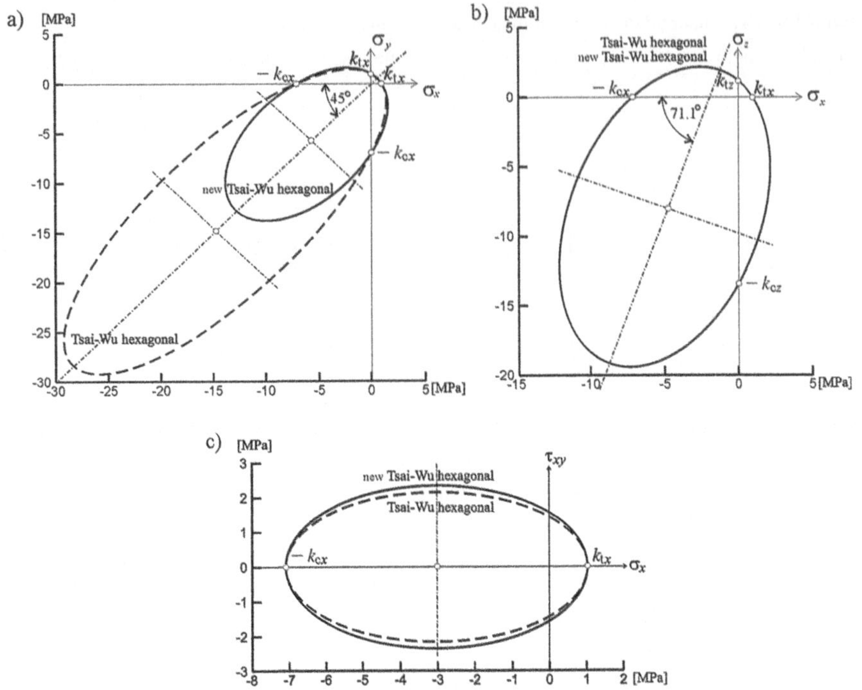

Fig. 3.13 Comparison of transversely isotropic Tsai–Wu's initial failure criteria of hexagonal and new hexagonal types for columnar ice: **a** plane of transverse isotropy (σ_x, σ_y), **b** plane of orthotropy (σ_x, σ_z), **c** shear plane (σ_x, τ_{xy}), after Ganczarski and Adamski (2015)

3.6 Lecture—Implicit Formulation of Pressure Sensitive Anisotropic Initial Failure Criteria

In this lecture selected examples of implementation of the implicit approach to the broader class accounting for anisotropy, tension/compression asymmetry and pressure sensitivity are thoroughly considered.

Khan and Liu (2012) applied the following extension of the 9–parameter orthotropic von Mises criterion (3.116) to describe the ductile fracture of the Ti-6Al-4V alloy accounting for hydrostatic pressure sensitivity, anisotropy and significant tension/compression asymmetry effect

$$\begin{aligned}&\sqrt{\exp[C(\zeta+1)]\left(F\sigma_1^2+G\sigma_2^2+H\sigma_3^2+L\sigma_1\sigma_2+M\sigma_2\sigma_3+N\sigma_1\sigma_3\right.}\\&\overline{\left.+P\sigma_{12}^2+Q\sigma_{13}^2+R\sigma_{23}^2\right)}=\exp\left(c_1\frac{I_1}{\sqrt{3}}\right)\end{aligned} \tag{3.153}$$

Both the hydrostatic pressure dependence I_1 and the tension/compression asymmetry J_3 are included in an implicit fashion as arguments of two exponential functions appearing as multipliers at the right- and the left-hand sides of orthotropic von Mises' equation. According to authors interpretation the main advantage of such formulation is that the anisotropy and tension/compression asymmetry are uncoupled into separate multiplicative terms which allow the anisotropic parameters and tension/compression asymmetry coefficient to be determined independently. The following definitions hold: F, G, H, L, M, N, P, Q and R are anisotropic parameters, C is the tension/compression asymmetry coefficient, ζ denotes the Lode parameter $\zeta = \cos 3\theta = \frac{27}{2}\frac{J_3}{(\sqrt{3J_2})^3}$, where θ is the Lode angle. Although the general form of limit criterion (3.153) which accounts for all three features: anisotropy, tension/compression asymmetry and hydrostatic pressure dependence, in fact its calibration performed by authors leads the form capturing only the tension/compression asymmetry and hydrostatic pressure dependence. As a consequence the limit curve of Al2024-T351 alloy exhibits only one axis of symmetry which means that this corresponds to the case of partly distorted limit surface. By the use of above formula authors succeeded with fitting experimental data in rolling direction (RD), transverse to rolling direction (TD) and the thickness direction (ND). However, hydrostatic pressure dependence introduced by the use of right-hand side exponential function leads to loss of convexity of the fracture surface along the meridian direction in the Haigh–Westergaard space as it was shown by Khan and Liu (2012) in Fig. 3.14b. The convexity loss discussed in this case is significant only from theoretical point of view because in such a case Drucker's postulate is violated. However, for the data cited by authors the concave meridian effect is very small such that it can probably be ignored from engineering point of view for the considered material data. Nevertheless in spite of possible convexity loss along meridian none convexity loss along circumference is observed although there exists second exponential function dependent on J_2 and J_3 being a multiplier of the Hill form on the left hand side of Eq. (3.153).

In another paper by Khan et al. (2012) the direct hydrostatic pressure dependence is dropped however both significant anisotropy and tension/compression asymmetry are saved.

$$\begin{aligned} \exp[-C(\zeta+1)]\,\big(&F\sigma_1^2 + G\sigma_2^2 + H\sigma_3^2 + L\sigma_1\sigma_2 + M\sigma_2\sigma_3 + N\sigma_1\sigma_3 \\ &+ P\sigma_{12}^2 + Q\sigma_{13}^2 + R\sigma_{23}^2\big) = 1 \end{aligned} \tag{3.154}$$

Although the general form of limit criterion (3.154) accounts for nine independent anisotropy parameters, in example considered by authors in Khan et al. (2012), due to calibration the material constant G is determined from the equi-biaxial compression test so it depends on three compression limits like in case of Hill's criterion. Under assumption of plane stress state it reduces to 4–parameter orthotropic Hill's condition (3.70). Fitting of experimental data for Ti-6Al-4V alloy at different strain rates and temperatures shows excellent coincidence between the experimental findings and simulation. By contrast to the previous formulation (3.153) the symmetry of the limit curve is lost completely as shown in Fig. 3.14.

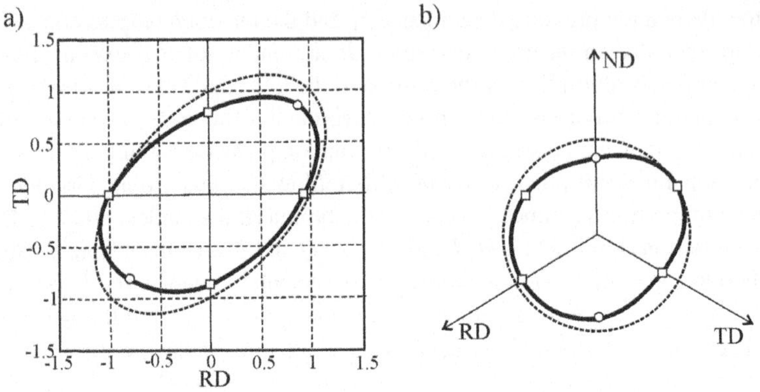

Fig. 3.14 Correlation of yield loci of Ti-6Al-4V alloy (○—experimental data points, □—points calculated from ND experimental data) with the yield function by Khan et al. (solid line) and Huber–von Mises criterion (dashed line): **a** RD-TD plane, **b** deviatoric plane

Orthotropic yield criterion proposed by Yoon et al. (2014) being anisotropic extension of the isotropic criterion by Cazacu and Barlat (2004)

$$\widetilde{I}_1 + \sqrt[3]{J_2'^{3/2} - J_3''} = 1$$
$$\widetilde{I}_1 = h_x \sigma_{xx} + h_y \sigma_{yy} + h_z \sigma_{zz}, \; J_2' = \frac{1}{2} \boldsymbol{s}' : \boldsymbol{s}', \; J_3'' = \det(\boldsymbol{s}'') \tag{3.155}$$

The stress tensors $\boldsymbol{s}'$ and $\boldsymbol{s}''$ are transformed from the stress tensor $\boldsymbol{\sigma}$ to the transformed space by two fourth-order linear transformation tensors $\boldsymbol{L}'$ and $\boldsymbol{L}''$ as follows $\boldsymbol{s}' = \boldsymbol{L}' : \boldsymbol{\sigma}$ and $\boldsymbol{s}'' = \boldsymbol{L}'' : \boldsymbol{\sigma}$ with

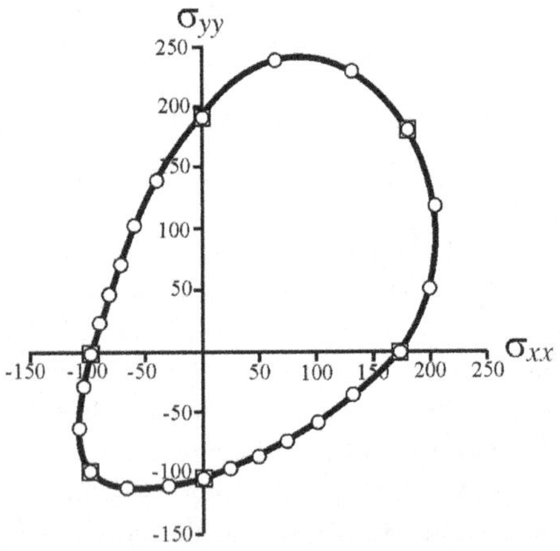

Fig. 3.15 Comparison of the yield surface of AZ31 magnesium alloy: ○—experimental data points, □—yield function proposed by Yoon et al. (2014)

$$
\boldsymbol{L}^{(i)} = \begin{bmatrix} \frac{1}{3}(C_2^{(i)} + C_3^{(i)}) & -\frac{1}{3}C_3^{(i)} & -\frac{1}{3}C_2^{(i)} & & & \\ & \frac{1}{3}(C_3^{(i)} + C_1^{(i)}) & -\frac{1}{3}C_1^{(i)} & & & \\ & & \frac{1}{3}(C_1^{(i)} + C_2^{(i)}) & & & \\ & & & C_4^{(i)} & & \\ & & & & C_5^{(i)} & \\ & & & & & C_6^{(i)} \end{bmatrix} \tag{3.156}
$$

where superscript $(i) = {}^{\prime}$ or ${}^{\prime\prime}$.

This highly extended yield criterion is capable of capturing all three features: anisotropy, tension/compression asymmetry and hydrostatic pressure sensitivity of various metals like AA 2008-T4, high purity α-Titanium and AZ31 Magnesium alloy. Excellent fitting of proposed yield criterion and experimental data of AZ31 is shown in Fig. 3.15.

References

Barlat, F., Brem, J. C., Yoon, J. W., Chung, K., Dick, R. E., Lege, D. J., et al. (2003). Plane stress function for aluminium alloy sheets—Part I: Theory. *International Journal of Plasticity, 19,* 1297–1319.

Betten, J. (1988). Applications of tensor functions to the formulation of yield criteria for anisotropic materials. *International Journal of Plasticity, 4,* 29–46.

Boehler, J. P., & Sawczuk, A. (1970). Equilibre limite des sols anisotropes. *J. Mécanique, 9,* 5–33.

Cazacu, O., & Barlat, F. (2004). A criterion for description of anisotropy and yield differential effects in pressure-insensitive materials. *International Journal of Plasticity, 20,* 2027–2045.

Cazacu, O., Planckett, B., & Barlat, F. (2006). Orthotropic yield criterion for hexagonal close packed metals. *International Journal of Plastics, 22,* 1171–1194.

Chen, W. F., & Han, D. J. (1995). *Plasticity for Structural Engineers*. Berlin, Heidelberg: Springer.

Dunand, M., Maertens, A. P., Luo, M., & Mohr, D. (2012). Experiments and modeling of anisotropic aluminum extrusions under multi-axial loading—Part I: Plasticity. *International Journal of Plasticity, 36,* 34–49.

Ganczarski, A., & Lenczowski, J. (1997). On the convexity of the Goldenblatt-Kopnov yield condition. *Archives of Mechanics, 49*(3), 461–475.

Ganczarski, A., & Skrzypek, J. (2009). *Plasticity of Engineering Materials* (in Polish), Issue of Cracow University Technology.

Ganczarski, A., & Skrzypek, J. (2011). Modeling of limit surfaces for transversely isotropic composite SCS-6/Ti-15-3 (in Polish). *Acta Mechanica et Automatica, 5*(3), 24–30.

Ganczarski, A., & Skrzypek, J. (2013). *Mechanics of Novel Materials* (in Polish). Wyd. Politechniki Krakowskie

Ganczarski, A., & Skrzypek, J. (2014). Constraints on the applicability range of Hill's criterion: Strong orthotropy or transverse isotropy. *Acta Mechanica, 225,* 2568–2582.

Ganczarski, A., & Adamski, M. (2015). Tetragonal or hexagonal symmetry in modeling of yield criteria for transversely isotropic materials. *Acta Mechanica et Automatica, 29,* 125–128.

Goldenblat, I. I. (1995). *Some Problems of Mechanics of Deformable Media*. Moskva: Gostekhizdat (in Russian).

И.И. Гольденблат, В.А. Копнов, Обобщенная теория пластического течения анизотропных сред, Сборник Строительная Механика, Стройиздат, Москва, pages 307–319, 1966.

Haigh, B. F. (1920). The strain-energy function and the elastic limit. *Engineering, London, 109*, 158–160.

Herakovich, C. T., & Aboudi, J. (1999). Thermal effects in composites. In *Thermal Stresses V* (pp. 1–142). Lastran Corp. Publ. Division.

Hill, R. (1948). A theory of the yielding and plastic flow of anisotropic metals. *Proceedings of The Royal Society London, A193*, 281–297.

Hill, R. (1950). *The Mathematical Theory of Plasticity*. Oxford: Clarendon Press.

Hosford, W. F., & Backhofen, W. A. (1964). Strength and plasticity of textured metals. In *Fundamentals of Deformation Processing* (pp. 259–298). Syracuse University Press.

Hosford, W. F. (1972). A generalized isotropic yield criterion. *Transactions of the ASME, E39*(2), 607–609.

Hu, Z. W., & Marin, J. (1956). Anisotropic loading functions for combined stresses in the plastic range. *The Journal of Applied Mechanics, 22*, 1.

Jackson, L. R., Smith, K. F., & Lankford, W. T. (1948). Plastic flow in anisotropic steel sheet. *American Institute of Mining and Metallurgical Engineers, 2440*, 1–15.

Khan, A. S., Kazmi, R., & Farrokh, B. (2007). Multiaxial and non-proportional loading responses, anisotropy and modeling of Ti-6Al-4V titanium alloy over wide ranges of strain rates and temperatures. *International Journal of Plasticity, 23*, 931–950.

Khan, A. S., & Liu, H. (2012). Strain rate and temperature dependent fracture criteria for isotropic and anisotropic metals. *International Journal of Plasticity, 37*, 1–15.

Khan, A. S., Yu, S., & Liu, H. (2012). Deformation enhanced anisotropic responses of Ti-6Al-4V alloy, Part II: A stress rate and temperature dependent anisotropic yield criterion. *International Journal of Plasticity, 38*, 14–26.

Korkolis, Y. P., Kyriakides, S. (2008). An advanced yield function including deformation-induced anisotropy. Inflation and burst of aluminum tubes Part II. *International Journal of Plastics, 24*, 1625–1637.

Kowalewski, Z. L., & Śliwowski, M. (1997). Effect of cyclic loading on the yield surface evolution of 18G2A low-alloy steel. *International Journal of Mechanical Sciences, 39*(1), 51–68.

Kowalsky, U. K., Ahrens, H., & Dinkler, D. (1999). Distorted yield surfaces—Modeling by higher order anisotropic hardening tensors. *Computational Materials Science, 16*, 81–88.

Lankford, W. T., Low, J.R., & Gensamer, M. (1947). The plastic flow of aluminium alloy sheet under combined loads. *Transactions of the AIME, 171*, 574; TP 2238, Met. Techn.

Love, A. E. H. (1944). *A Treatise on the Mathematical Theory of Elasticity*. New York: Dover Publication.

Luo, M., Dunand, M., & Moth, D. (2012). Experiments and modeling of anisotropic aluminum extrusions under multi-axial loading—Part II: Ductile fracture. *International Journal of Plasticity, 32–33*, 36–58.

Luo, X. Y., Li, M., Boger, R. K., Agnew, S. R., & Wagoner, R. H. (2007). Hardening evolution of AZ31B Mg sheet. *International Journal of Plasticity, 23*, 44–86.

Malinin, N. N., & Rżysko, J. (1981). *Mechanika materiałów*. Warszawa: PWN.

Nixon, M. E., Cazacu, O., & Lebensohn, R. A. (2010). Anisotropic response of high-purity α-titanium: Experimental characterization and constitutive modeling. *International Journal of Plasticity, 26*, 516–532.

Nye, J. F. (1957). *Physical Properties of Crystals their Representations by Tensor and Matrices*. Oxford: Clarendon Press.

Ottosen, N. S., & Ristinmaa, M. (2005). *The Mechanics of Constitutive Modeling*. Amsterdam: Elsevier.

Plunkett, B., Cazacu, O., & Barlat, F. (2008). Orthotropic yield criteria for description of the anisotropy in tension and compression of sheet metal. *International Journal of Plasticity, 24*, 847–866.

Ralston, T. D. (1977). *Yield and Plastic Deformation in ICE Crushing Failure*. Seatle, Washington: ICSI. AIDJEX Symposium on Sea Ice-Processes and Models.
Rogers, T. G. (1990). Yield criteria, flow rules, and hardening in anisotropic plasticity. In *Yielding, Damage and Failure of Anisotropic Solids* (pp. 53–79). London: Mechanical Engineering Publication.
Rymarz, C. Z. (1993). *Continuum Mechanics*. Warszawa: PWN.
Sayir, M. (1970). Zur Fließbedingung der Plastiztätstheorie. *Ingenierarchiv, 39,* 414–432.
Skrzypek, J., & Ganczarski, A. (2013). Anisotropic initial yield and failure criteria including temperature effect. In *Encyclopedia of Thermal Stresses*. Springer Science+Business Media Dordrecht.
Sobotka, Z. (1969). Theorie des plastischen Fliessens von anisotropen Körpern. *Z. Angew. Math. Mechanik, 49,* 25–32.
Spencer, A. J. M. (1971). Theory of invariants. In *Continuum Physics* (pp. 239–353). Academic Press.
Sun, C. T., & Vaidya, R. S. (1996). Prediction of composite properties from a representative volume element. *Composites Science Technology, 56,* 171–179.
Szczepiński, W. (1993). On deformation-induced plastic anisotropy of sheet metals. *Archives of Mechanics, 45*(1), 3–38.
Tsai, S. T., & Wu, E. M. (1971). A general theory of strength for anisotropic materials. *International Journal for Numerical Methods in Engineering, 38,* 2083–2088.
von Mises, R. (1913). Mechanik der festen Körper im plastisch deformablen Zustand, Götingen Nachrichten. *Mathematical Plasticity, 4*(1), 582–592.
von Mises, R. (1928). Mechanik der plastischen Formänderung von Kristallen. *ZAMM, 8*(13), 161–185.
Voyiadjis, G. Z., & Thiagarajan, G. (1995). An anisotropic yield surface model for directionally reinforced metal-matrix composites. *International Journal of Plasticity, 11,* 867–894.
Westergaard, H. M. (1920). On the resistance of ductile materials to combined stresses in two and three directions perpendicular to one another. *Journal of the Franklin Institute, 189*, 627–640.
Yoon, J. W., Lou, Y., Yoon, J., & Glazoff, M. V. (2014) Asymmetric yield function based on stress invariants for pressure sensitive metals. *International Journal of Plasticity*.
Yoshida, F., Hamasaki, H. M., & Uemori, T. (2013). A user-friendly 3D yield function to describe anisotropy of steel sheets. *International Journal of Plasticity, 45,* 119–139.
Życzkowski, M. (2001). Anisotropic yield conditions. In *Handbook of Materials Behavior Models* (pp. 155–165). San Diego: Academic Press.

Chapter 4
Time-dependent and Time-independent Models of Cyclic Plasticity for Low-cycle and Thermomechanical Fatigue Life Assessment

Thomas Seifert

Abstract In this work, time-independent and time-dependent plasticity models are presented that are well suited for the calculation of stresses and strains with the finite-element method to assess the low-cycle and thermomechanical fatigue life of engineering components. The focus are plasticity models that are available in finite-element programs nowadays as standard material models and describe isotropic and kinematic hardening, strain-rate dependency as well as static recovery of hardening. For the presented models, aspects relevant for the application of the models are addressed as the determination of the material properties and the numerical implementation. Nevertheless, the plasticity models are also embedded in the thermodynamic framework used for the derivation of thermodynamically consistent plasticity models. Only uniaxial formulations are used to achieve a good readability and preventing the use of tensors.

4.1 Introduction

4.1.1 Motivation

Many engineering components are subjected to cyclic mechanical loadings. In notched regions, local cyclic plastic deformations can occur that have a significant effect on the lifetime and can result in low-cycle fatigue (LCF) of the material. In case of additional thermal transients, the yield stress of the material decreases with increasing material temperature and plastic deformations are even more likely to occur. Moreover, the material changes from almost time-independent plastic behavior at lower temperatures to time-dependent plastic behavior at higher temperatures. Hence, creep may contribute to additional damage of the material and thermomechanical fatigue (TMF) of the material needs to be assessed to find the appropriate

T. Seifert (✉)
Offenburg University of Applied Sciences, Badstraße 24, 77652 Offenburg, Germany
e-mail: thomas.seifert@hs-offenburg.de

H. Altenbach and A. Ganczarski (eds.), *Advanced Theories for Deformation, Damage and Failure in Materials*, CISM International Centre for Mechanical Sciences 605, https://doi.org/10.1007/978-3-031-04354-3_4

design and material that ensure the integrity of the component for a whole product life.

Since component testing in bench or field tests is expensive and time-consuming, computational approaches are used and developed allowing the calculation of the lifetime of the components and the optimization of the components via computer simulations. In this context, the finite-element method is a powerful tool, since it allows the computation of the transient stress and strain fields during cyclic mechanical or thermomechanical loading. The computed local stresses and strains can then be used together with an appropriate lifetime model to predict the fatigue life of the component.

To obtain reliable stress and strain fields in the finite-element calculations, constitutive equations are needed for the description of time-independent and time-dependent cyclic plasticity. Over the last years, commercial finite element programs as ABAQUS and ANSYS have improved their capabilities of modelling plasticity and have implemented isotropic and kinematic hardening law considering rate-effects and recovery of hardening at higher temperatures. Hence, they allow to describe the essential phenomena, namely strain-rate dependency, creep and stress relaxation, as well as the Bauschinger effect and cyclic hardening and softening. These models contain a larger number of material properties that must be determined on the basis of experimental data. On the one hand, the determination of the material properties requires deeper understanding of the models and, on the other hand, appropriate experimental data is often not available since it is expensive and time-consuming to generate. Both aspects are major barriers that prevent the introduction of better constitutive models into industrial application and their usage in research projects.

4.1.2 Aims of This Work

It is the aim of this work to present time-independent and time-dependent plasticity models that can be used for LCF and TMF life assessment of components. The models are presented from the user's perspective, e.g. users of finite-element programs as ABAQUS or ANSYS. Hence, the focus are models that are available in these finite-element programs nowadays as standard material models. Besides the mathematical description of the material behavior, rheological models are presented that help in understanding the interrelationships of different mechanisms in the plasticity models. Finally, with the understanding of the models, the material properties are introduced and practical ways to determine the material properties from uniaxial material tests are described. Only uniaxial formulations of the plasticity models are used to achieve a good readability and prevent the use of tensors.

Even though the focus is on the user's perspective, the thermodynamic framework for the derivation of thermodynamically consistent plasticity models is addressed as well since it gives the link to energy-based approaches used e.g. in fracture mechanics where the stored energy density plays an important role. Fracture mechanics-based models are very well suited for the assessment of the LCF and TMF life, where fatigue crack growth is the dominant damage mechanism. Fracture mechanics-based models for fatigue life prediction will not be addressed in this work. They are presented and applied e.g. in Nissley (1995), Heitmann et al. (1984), Hoffmeyer et al. (2001), Hoffmeyer et al. (2006), Seifert et al. (2010), Seifert and Riedel (2010), Schlesinger et al. (2017).

Textbooks including the presented concepts and models of time-independent and time-dependent plasticity are e.g. Lemaitre and Chaboche (1990) and Simo and Hughes (1998). Well-known works on plasticity models are due to J.L. Chaboche: e.g. Chaboche (1986, 1989, 1993, 1997, 2008). Further works dealing with the modeling of the cyclic plastic behavior, in particular with the sophisticated modeling under thermomechanical loading, are e.g. Yaguchi et al. (2002), Seifert (2006), Aktaa and Petersen (2009), Seifert et al. (2010), Oesterlin and Maier (2014), Ohmenhäuser et al. (2014), Wilhelm et al. (2014), Mao et al. (2015), Rekun and Jörg (2016), Seifert et al. (2019), Jilg and Seifert (2018).

4.1.3 Structure of This Work

The work is structured as follows: The thermodynamic framework with internal variables and normality rules is described in Sect. 4.2. Then, time-independent plasticity models are presented in Sect. 4.3 and time-dependent plasticity models are addressed in Sect. 4.4. In both sections on plasticity models, the rheological and mathematical models as well as the relevant material properties are introduced and embedded in the thermodynamic framework. Moreover, numerical and analytical solutions are derived for the plasticity models. Some aspects related to the application of the models to thermomechanical loadings are considered in Sect. 4.5, before a short conclusion is given in Sect. 4.6.

4.2 Concept of Internal Variables and Normality Rules

This section gives an introduction into the framework of thermodynamically consistent plasticity models resulting in the concept of internal variables and normality rules. To this end, we consider a non-isothermal mechanical problem where temperature is a prescribed field to determine the current temperature-dependent material properties and thermal strains. The presented framework might give the impression of being very abstract. Therefore, practical examples of plasticity models follow in Sects. 4.3 and 4.4 on time-independent and time-dependent formulations.

4.2.1 Helmholtz Free Energy and Internal Variables

Starting point for the following derivations is the Clausius-Planck inequality that is a result of the second law of thermodynamics requiring a positive entropy production. The Clausius-Planck inequality describes the non-negative local dissipation $\mathcal{D}_{loc}$ defined by

$$\mathcal{D}_{loc} = \sigma\dot{\varepsilon} - s\dot{T} - \dot{\psi} \geq 0. \tag{4.1}$$

σ and ε are the stress and the strain, respectively. T and s are the temperature and the entropy per unit volume, respectively. The Helmholtz free energy per unit volume ψ characterizes the current state of the material. A point above a quantity denotes its time derivative. For elastic materials, the Helmholtz free energy is a function of the elastic strain and temperature alone, while for irreversible processes during plastic deformation a set of $\alpha = 1, ..., n_\xi$ internal variables ξ_α enters the free energy. Hence, the functional form of the Helmholtz free energy is $\psi = \hat{\psi}\,(\varepsilon^e, T, \xi_\alpha)$. Using this functional form in Eq. (4.1) yields

$$\mathcal{D}_{loc} = \sigma\dot{\varepsilon} - s\dot{T} - \frac{\partial\psi}{\partial\varepsilon^e}\dot{\varepsilon}^e - \frac{\partial\psi}{\partial T}\dot{T} - \sum_{\alpha=1}^{n_\xi}\frac{\partial\psi}{\partial\xi_\alpha}\dot{\xi}_\alpha \geq 0. \tag{4.2}$$

Using an additive decomposition of the strains into reversible thermal strains ε^t, reversible elastic strains ε^e and irreversible plastic strains ε^p, i.e.

$$\varepsilon = \varepsilon^t + \varepsilon^e + \varepsilon^p, \tag{4.3}$$

one obtains

$$\mathcal{D}_{loc} = \left(\sigma - \frac{\partial\psi}{\partial\varepsilon^e}\right)\dot{\varepsilon}^e + \sigma\dot{\varepsilon}^p - \left(s + \frac{\partial\psi}{\partial T} - \sigma\frac{\partial\varepsilon^t}{\partial T}\right)\dot{T} - \sum_{\alpha=1}^{n_\xi}\frac{\partial\psi}{\partial\xi_\alpha}\dot{\xi}_\alpha \geq 0. \tag{4.4}$$

In case of reversible deformation processes with no plastic deformations ($\dot{\varepsilon}^p = 0$) and no change of the internal variables ($\dot{\xi}_\alpha = 0$), there is no local dissipation. Thus,

$$\mathcal{D}_{loc} = \left(\sigma - \frac{\partial\psi}{\partial\varepsilon^e}\right)\dot{\varepsilon}^e - \left(s + \frac{\partial\psi}{\partial T} - \sigma\frac{\partial\varepsilon^t}{\partial T}\right)\dot{T} = 0, \tag{4.5}$$

which is only fulfilled for arbitrary strain rates $\dot{\varepsilon}^e$ and temperature rates $\dot{T}$ if the terms in the brackets are equal to zero (Coleman and Gurtin 1967). This results in the relations

$$\sigma = \frac{\partial \psi}{\partial \varepsilon^e} \tag{4.6}$$

$$s = -\frac{\partial \psi}{\partial T} + \sigma \frac{\partial \varepsilon^t}{\partial T}, \tag{4.7}$$

so that the Helmholtz free energy is a thermodynamic potential from which stress and entropy can be derived. With these results and introducing thermodynamic forces f_α conjugate to the internal variables ξ_α as

$$f_\alpha = \frac{\partial \psi}{\partial \xi_\alpha}, \tag{4.8}$$

the Clausius-Planck inequality reduces to

$$\mathcal{D}_{loc} = \sigma \dot{\varepsilon}^p - \sum_{\alpha=1}^{n_\xi} f_\alpha \dot{\xi}_\alpha \geq 0. \tag{4.9}$$

Evolution laws for the internal variables must fulfill the reduced Clausius-Planck inequality so that they are thermodynamically consistent, i.e. do not violate the laws of thermodynamics. For purely mechanical problems, where temperature is a prescribed field, the entropy s is no longer necessary as a physical quantity.

4.2.2 *Flow Potential and Normality Rule*

A potential relation for the plastic flow rule defining $\dot{\varepsilon}^p$ and for the evolution laws for each internal variable ξ_α is obtained assuming that the evolution of the plastic strain is only dependent on the stress and each internal variable is only dependent on its conjugate force. With the flow potential defined by

$$\Omega = \int_0^\sigma \dot{\varepsilon}^p \mathrm{d}\sigma - \sum_{\alpha=1}^{n_\xi} \int_0^{f_\alpha} \dot{\xi}_\alpha \mathrm{d} f_\alpha, \tag{4.10}$$

one obtains the normality rules for the flow rule and the evolution equations

$$\dot{\varepsilon}^p = \frac{\partial \Omega}{\partial \sigma} \tag{4.11}$$

$$\dot{\xi}_\alpha = -\frac{\partial \Omega}{\partial f_\alpha}. \tag{4.12}$$

For the derivation of the flow rule and the evolution equations for time-independent and time-dependent plasticity models in Sects. 4.3 and 4.4, respectively, the principle of maximum plastic dissipation is used. It postulates that for all admissible choices for the stress $\sigma^\star$ and the thermodynamic forces $f_\alpha^\star$, the relevant stress σ and the relevant thermodynamic forces f_α maximize the local dissipation:

$$\underbrace{\sigma\dot{\varepsilon}^p - \sum_{\alpha=1}^{n_\xi} f_\alpha \dot{\xi}_\alpha}_{\mathcal{D}_{loc}} \geq \underbrace{\sigma^\star\dot{\varepsilon}^p - \sum_{\alpha=1}^{n_\xi} f_\alpha^\star \dot{\xi}_\alpha}_{\mathcal{D}_{loc}^\star} . \tag{4.13}$$

It can be shown that the principle of maximum dissipation requires that the flow potential Ω is a convex function with respect to stress σ and the thermodynamic forces f_α. Besides the used principle of maximum plastic dissipation, other approaches were proposed or discussed to derive the normality rules, e.g. (Hill 1968; Rice 1971; Fischer and Svoboda 2006; Hackl and Fischer 2008).

4.3 Time-independent Cyclic Plasticity

In this section, time-independent plasticity models are introduced for ductile materials that describe two different hardening mechanisms: isotropic hardening and kinematic hardening. Linear elastic behavior is assumed as typical for metals. The respective material behavior is shown in terms of stress-strain curves and rheological models and the mathematical models are presented. In this section, isothermal conditions are assumed. Hence, no thermal strains occur: $\varepsilon^t = 0$.

To describe the non-smooth transition from elastic to plastic loading, an elastic domain is assumed in which stresses only result in elastic loading increments. The boundary of this domain is described by the yield function $\phi_y = \hat{\phi}_y(\sigma, f_\alpha) = 0$ and defines stresses σ and thermodynamic forces f_α that result in plastic yielding. For stresses σ and thermodynamic forces f_α resulting in $\phi_y < 0$, no yielding occurs. The elastic domain can, thus, be expressed by

$$\mathbb{E} = \left\{\sigma, f_\alpha \middle| \hat{\phi}_y(\sigma, f_\alpha) \leq 0\right\} . \tag{4.14}$$

The elastic domain defines all admissible stresses and thermodynamic forces and is an inequality constraint in maximizing the local dissipation $\mathcal{D}_{loc}$, i.e. $\left\{\sigma^\star, f_\alpha^\star\right\} \in \mathbb{E}$ in Eq. (4.13). The constraint maximization problem can be solved using a Lagrange multiplier $\dot{\lambda}$ which enforces the inequality constraint and by changing the sign of the dissipation to convert the maximization to a minimization problem. The corresponding Lagrange function to be minimized reads

$$\hat{\mathcal{L}}\left(\sigma, f_\alpha, \dot{\lambda}\right) = -\sigma\dot{\varepsilon}^p + \sum_{\alpha=1}^{n_\xi} f_\alpha \dot{\xi}_\alpha + \dot{\lambda}\phi_y \longrightarrow \text{min.} \tag{4.15}$$

The necessary conditions for the solution are

$$\frac{\partial \mathcal{L}}{\partial \sigma} = -\dot{\varepsilon}^p + \dot{\lambda}\frac{\partial \phi_y}{\partial \sigma} = 0 \tag{4.16}$$

$$\frac{\partial \mathcal{L}}{\partial f_\alpha} = \dot{\xi}_\alpha + \dot{\lambda}\frac{\partial \phi_y}{\partial f_\alpha} = 0 \tag{4.17}$$

and the Karush-Kuhn-Tucker optimality conditions, also known as the loading/unloading conditions,

$$\dot{\lambda} \geq 0, \qquad \phi_y \leq 0, \dot{\lambda}\phi_y = 0. \tag{4.18}$$

Hence, the flow rule and the evolution equations for the internal variables are governed by normality rules using ϕ_y as flow potential:

$$\dot{\varepsilon}^p = \dot{\lambda}\frac{\partial \phi_y}{\partial \sigma} \tag{4.19}$$

$$\dot{\xi}_\alpha = -\dot{\lambda}\frac{\partial \phi_y}{\partial f_\alpha}. \tag{4.20}$$

Convexity of ϕ_y with respect to σ and f_α follows from the principle of maximum dissipation.

4.3.1 Isotropic Hardening

Plastic deformation of metals is a result of the slip of dislocations. During plastic deformation, new dislocations are continuously generated at dislocation sources, so that the dislocation density increases. At lower temperatures the dislocations can not annihilate by annealing. Instead, the increasing number of dislocations results in an increasing interaction of dislocations: The dislocations serve as pinning points or obstacles that impede the motion of other dislocations. As a result, the material hardens during plastic deformation.

Such strain hardening is usually described by isotropic hardening. Isotropic hardening is a hardening mechanism that does not depend on the loading direction. For uniaxial loadings, as considered in this work, this means that a strength achieved under tensile load is also present when the load is reversed under compressive load and vice versa. In Fig. 4.1 the stress-strain curve of an isotropic hardening material is shown for a monotonic load (dashed line) and a cyclic load (solid line).

The elastic domain is introduced that bounds the stresses to the current strength of the material. Initially, for the plastically undeformed material, the boundary of the elastic domain is defined by the initial yield stress R_e. Besides the initial elastic domain, the elastic domain is shown in Fig. 4.1 for the current states at the points of load reversal for the cyclic curve. During plastic deformation the strength increases by the isotropic hardening variable R and, hence, the elastic domain increases keeping it center at zero stress.

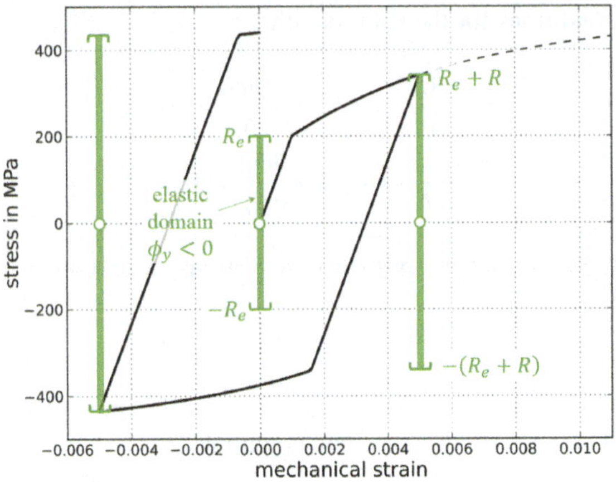

Fig. 4.1 Stress-strain curves of isotropic hardening material using the combined linear and exponential hardening law from Eq. (4.23) with the following material properties: $E = 200000$ MPa, $R_e = 200$ MPa, $H = 5000$ MPa, $Q_\infty = 200$ MPa and $b = 300$

4.3.1.1 Rheological Model

A rheological model representing the isotropic hardening elastic-plastic material is shown in Fig. 4.2. Two basic rheological elements are connected in series: the spring element and the friction element. Elastic strain represented by the spring and plastic strain represented by the friction element add up to the mechanical strain resulting in the additive decomposition of the mechanical strain

$$\varepsilon = \varepsilon^e + \varepsilon^p. \tag{4.21}$$

The stiffness of the spring is defined by Young's modulus E that corresponds to the slope of linear elastic loadings and unloadings in the stress-strain curves. The stress required to overcome friction in the friction element is defined by the current strength $R_e + R$.

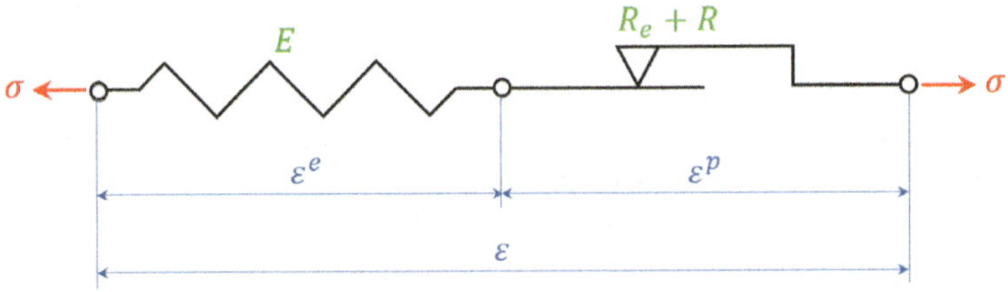

Fig. 4.2 Rheological model of isotropic hardening elastic-plastic material

4.3.1.2 Mathematical Model

The isotropic hardening variable must increase independent of the loading direction. This is achieved by defining R to be a function of the so-called accumulated plastic strain p, i.e. $R = \hat{R}(p)$. The accumulated or equivalent plastic strain p is defined by

$$p = \int_t \dot{p} \mathrm{d}t \text{ where } \dot{p} = \left|\dot{\varepsilon}^p\right|. \tag{4.22}$$

With $\dot{p} \geq 0$, p sums up all strain increments regardless of their sign and measures the "total plastic strain" that the material has undergone. A typical isotropic hardening function is a combination of a linear and an exponential function:

$$R = Hp + Q_\infty \left(1 - e^{-bp}\right). \tag{4.23}$$

H, Q_∞ and b are material properties: The hardening modulus H describes the slope of the linear portion. Q_∞ is the saturation value of the exponential function and b the transition constant that controls the hardening rate to reach saturation. This hardening law is used to compute the stress-strain curves shown in Fig. 4.1 and is e.g. available as standard hardening model in the finite-element program ANSYS. In the finite-element program ABAQUS, the exponential function is available to define isotropic hardening via material properties.

Stresses only arise if the spring is strained, i.e. for non-zero elastic strains, so that the law of elasticity

$$\sigma = E\varepsilon^e = E\left(\varepsilon - \varepsilon^p\right) \tag{4.24}$$

is obtained. The plastic strain only can change if the friction element is sliding. To this end, the yield function

$$\phi_y = |\sigma| - (R_e + R) \leq 0 \tag{4.25}$$

is introduced that describes with $\phi_y < 0$ the elastic domain of the material, in which stresses only result in elastic loading increments and, thus, $\dot{\lambda} = 0$ in the flow rule in Eq. (4.19). If stresses attain values that give $\phi_y = 0$, plastic loading increments additionally occur, i.e. the friction element slides and $\dot{\lambda} > 0$ in the flow rule. The stresses can, however, not exceed the current strength, so that $\phi_y > 0$ is not possible. These situations are expressed by the loading/unloading conditions in Eq. (4.18). ϕ_y is a convex function of σ and R as required by the principle of maximum dissipation.

In the following, the model is embedded in the framework of thermodynamically consistent plasticity models. To this end, the Helmholtz free energy is additively decomposed in to an elastic part ψ^e and a plastic part ψ^p:

$$\psi = \underbrace{\int_{\varepsilon^e} \sigma \mathrm{d}\varepsilon^e}_{\psi^e} + \underbrace{\int_{p} R \mathrm{d}p + C}_{\psi^p} . \tag{4.26}$$

The elastic part describes the elastically stored energy in the spring element of the rheological model (Fig. 4.2). In the plastic part the accumulated plastic strain p is interpreted as internal variable. The isotropic hardening variable R is its conjugate thermodynamic force. The integration constant C must be determined such that $\psi^p = 0$ for $p = 0$. For linear elasticity, Eq. (4.24), and the isotropic hardening law given in Eq. (4.23), the Helmholtz free energy is

$$\psi = \underbrace{\frac{1}{2} E \varepsilon^{e2}}_{\psi^e} + \underbrace{\frac{1}{2} H p^2 + Q_\infty \left(p + \frac{1}{b} e^{-bp} - \frac{1}{b} \right)}_{\psi^p} . \tag{4.27}$$

Using Eqs. (4.6) and (4.8), this yields the desired relations

$$\sigma = \frac{\partial \psi}{\partial \varepsilon^e} = E\varepsilon^e = E\left(\varepsilon - \varepsilon^p\right) \tag{4.28}$$

$$R = \frac{\partial \psi}{\partial p} = Hp + Q_\infty \left(1 - e^{-bp}\right) . \tag{4.29}$$

The flow rule and the evolution equation for the internal variable p is obtained from the normality rules

$$\dot{\varepsilon}^p = \dot{\lambda} \frac{\partial \phi_y}{\partial \sigma} = \dot{\lambda} \underbrace{\frac{\partial \phi_y}{\partial |\sigma|}}_{1} \underbrace{\frac{\partial |\sigma|}{\partial \sigma}}_{N = \frac{\sigma}{|\sigma|}} = \dot{\lambda} N = \dot{p} N \tag{4.30}$$

$$\dot{p} = -\dot{\lambda} \underbrace{\frac{\partial \phi_y}{\partial R}}_{-1} = \dot{\lambda} = \left|\dot{\varepsilon}^p\right| . \tag{4.31}$$

N is the flow direction that is determined by the sign of the stress: N is 1 in case of tensile stress and -1 in case of compressive stress.

With the potential functions ψ from Eq. (4.27) and ϕ_y from Eq. (4.25), the plasticity model with isotropic hardening is defined resulting in the constitutive Eqs. (4.28) to (4.31). The last step is to determine the multiplier $\dot{\lambda}$. The loading/unloading conditions (4.18) state that the stress is forced to lie on the boundary of the elastic domain during plastic yielding, i.e. $\phi_y = 0$ when $\dot{\lambda} > 0$. From this it can be concluded that ϕ_y must stay zero during yielding and, thus, $\dot{\phi}_y = 0$ holds as well. The latter is the so-called consistency condition that gives one equations from which the multiplier can be determined:

$$\dot{\phi}_y = \underbrace{\frac{\partial \phi_y}{\partial \sigma}}_{N} \dot{\sigma} + \underbrace{\frac{\partial \phi_y}{\partial R}}_{-1} \frac{\partial R}{\partial p} \dot{p} = 0. \tag{4.32}$$

With Eqs. (4.24), (4.30) and (4.31) one finds

$$\begin{aligned} \dot{\phi}_y &= NE\left(\dot{\varepsilon} - \dot{\lambda}N\right) - \frac{\partial R}{\partial p}\dot{\lambda} \\ &= NE\dot{\varepsilon} - \underbrace{N^2}_{1} E\dot{\lambda} - \frac{\partial R}{\partial p}\dot{\lambda} = 0, \end{aligned} \tag{4.33}$$

which can be solved for the multiplier:

$$\dot{\lambda} = \frac{NE\dot{\varepsilon}}{E + \dfrac{\partial R}{\partial p}}. \tag{4.34}$$

The higher the hardening rate $\partial R/\partial p$, the lower the multiplier $\dot{\lambda}$ and, thus, the lower the plastic strain rate.

4.3.1.3 Thermodynamic Consistency

With the internal variable p and its conjugate force R, the local dissipation is given by

$$\begin{aligned} \mathcal{D}_{loc} &= \sigma \dot{\varepsilon}^p - R\dot{p} \\ &= \dot{\lambda} \underbrace{\sigma N}_{\frac{\sigma^2}{|\sigma|} = |\sigma|} - \dot{\lambda}R \\ &= \dot{\lambda}\left(|\sigma| - R\right) \geq 0. \end{aligned} \tag{4.35}$$

With $\dot{\lambda} \geq 0$ and $\phi_y = |\sigma| - (R_0 + R) = 0$ during plastic yielding, it follows that

$$\begin{aligned} |\sigma| - R &= R_e \geq 0 \\ |\sigma| &= R_e + R \geq 0. \end{aligned} \tag{4.36}$$

Hence, thermodynamic consistency requires $R_e \geq 0$, since $R = 0$ initially when $p = 0$. Moreover, $R_e \geq -R$ must hold. This is only strictly the case in the combined linear and exponential hardening if $H \geq 0$ and $R_e \geq -Q_\infty$.

4.3.1.4 Numerical Solution with Predictor/Corrector Algorithm

An efficient way to solve the coupled differential equations with the inequality constraints of the loading/unloading conditions are so-called strain-controlled predictor-corrector algorithms. A prediction-corrector algorithm for time-independent plasticity model with isotropic and kinematic hardening is shown in Sect. 4.3.3.

4.3.1.5 Analytical Solution for Determination of Material Properties

If experimental data from uniaxial material tests is available from which the material properties should be determined, then the plastic strains can be computed from the data and do not need to be computed by integration of the flow rule. In this case, an analytical solution for the stresses can be obtained that can be used to determine the material properties of the model such that the computed stresses give a good description of the experimentally measured stresses. Assuming plastic yielding,

$$\phi_y = |\sigma| - (R_e + R) = 0 \tag{4.37}$$

holds. Solving for $|\sigma|$ and multiplication of both sides with $\sigma / |\sigma|$ yields

$$\underbrace{\frac{\sigma}{|\sigma|}|\sigma|}_{\sigma} = \underbrace{\frac{\sigma}{|\sigma|}}_{N}(R_e + R)\,. \tag{4.38}$$

For the linear and exponential isotropic hardening law of Eq. (4.23), one finally obtains

$$\sigma = N\left(R_e + Hp + Q_\infty\left(1 - e^{-bp}\right)\right). \tag{4.39}$$

Examples for the determination of material properties for a plasticity model with isotropic and kinematic hardening based on analytical solutions for the stress are presented in Sect. 4.3.3.

4.3.2 Kinematic Hardening

Isotropic hardening alone does not describe the cyclic material behavior well. Under cyclic loading conditions, the so-called Bauschinger effect is observed where plastic yielding under reversed loading occurs for lower stresses than attained before load reversal. This is shown with the stress-strain curve for cyclic loadings (solid line) in Fig. 4.3. The Bauschinger effect can be explained by the pile-up of dislocations in front of obstacles as grain boundaries or second phase particles during plastic deformation. The residual stress fields of the piled-up dislocation superimpose and

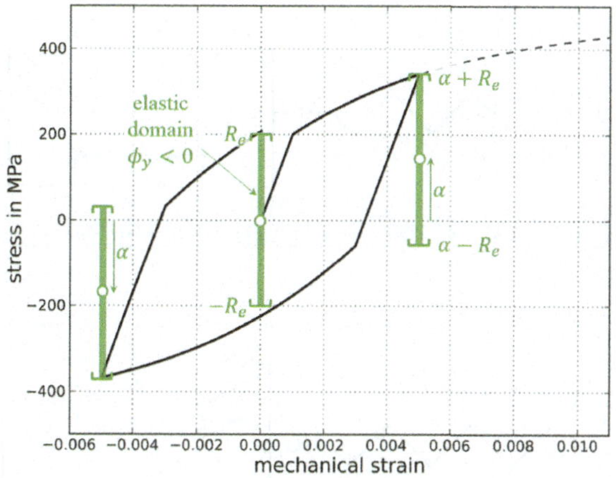

Fig. 4.3 Stress-strain curves of kinematic hardening material using two backstresses, the first with a linear and the second with a non-linear evolution equation, and the following material properties: $E = 200000$ MPa, $R_e = 200$ MPa, $C_1 = 5000$ MPa, $\gamma_1 = 0$, $C_2 = 60000$ MPa and $\gamma_2 = 300$

generate a backstress that promotes the slip of the dislocations in the opposite direction.

The Bauschinger effect can be described in plasticity models by kinematic hardening. In contrast to isotropic hardening, where the elastic domain increases keeping the center at zero stress (Fig. 4.1), kinematic hardening results in a translation of the elastic domain in direction of plastic loading (Fig. 4.3). The translation is described by the backstress α, that represents the superimposed residual stress field of dislocation pile-ups in a macroscopic sense and can be positive or negative.

4.3.2.1 Rheological Model

To account for the generation of backstresses during plastic deformation, a spring element is connected in parallel to the friction element in the rheological model for elastic-plastic behavior with kinematic hardening (Fig. 4.4a). During plastic deformation, the spring with the stiffness C stores energy to promote yielding in the opposite direction. Often, several springs are used in parallel in the plastic part to describe the total backstress acting on the friction element as shown in the rheological model of the kinematic hardening elastic-plastic material with several backstresses (Fig. 4.4b). The reason is that short- and long-range residual stresses exist whose effect can be treated using several springs.

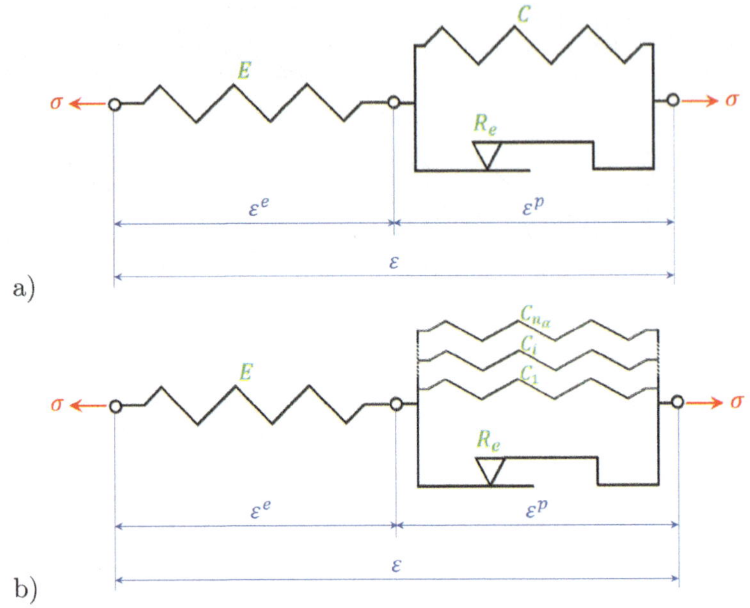

Fig. 4.4 Rheological model of kinematic hardening elastic-plastic material with **a** one backstress and **b** $i = 1, ..., n_\alpha$ backstresses

4.3.2.2 Mathematical Model

The rheological model implies the law of elasticity, where stresses in the materials are generated by elastic straining:

$$\sigma = E\varepsilon^e = E\left(\varepsilon - \varepsilon^p\right). \tag{4.40}$$

In the parallel connection controlling the plastic strain, however, the applied stress is divided between the spring element and the friction element. Hence, the stress acting on the friction element is $\sigma - \alpha$. Due to residual stresses described by the backstress, there can be load to the friction element, even if no external stress σ is applied. If $|\sigma - \alpha|$ attains the yield stress R_e of the friction element, plastic yielding occurs. This results in the yield function

$$\phi_y = |\sigma - \alpha| - R_e \leq 0. \tag{4.41}$$

With $\phi_y < 0$, the current stresses are in the elastic domain of the material, so that $\dot{\lambda} = 0$. If stresses attain values that give $\phi_y = 0$, plastic loading increments additionally occur and $\dot{\lambda} > 0$ in the flow rule, as expressed by the loading/unloading conditions in Eq. (4.18). ϕ_y is a convex function of σ and α as required by the principle of maximum dissipation.

A frequently used model to describe the evolution of the backstress during plastic loading is the Armstrong-Frederick kinematic hardening model (Armstrong and Frederick 1966; Frederick and Armstrong 2007). It is available as standard plasticity model in the finite-element programs ABAQUS and ANSYS. The Armstrong-Frederick kinematic hardening model describes the evolution of backstress by

$$\dot{\alpha} = \underbrace{C\dot{\varepsilon}^p}_{\text{hardening}} - \underbrace{\gamma\dot{p}\alpha}_{\text{dynamic recovery}} . \tag{4.42}$$

C and γ are material properties: C represents the stiffness of the spring and gives the initial hardening modulus. γ controls the decreasing hardening rate with increasing hardening and, thus, is responsible for the non-linearity of the hardening law. For $\gamma = 0$, linear hardening is obtained. The second term is called dynamic recovery term since recovery of hardening only occurs with plastic deformation. In Sect. 4.4, static recovery is considered as well.

The Armstrong-Frederick hardening law describes an exponential hardening curve as shown later in this section when deriving analytical solutions. To increase the flexibility of the model when describing experimental data, the backstress can be additively decomposed in $i = 1, ..., n_\alpha$ backstresses α_i (Fig. 4.4b), where each backstress follows the Armstrong-Frederick model and has the corresponding material properties C_i and γ_i:

$$\alpha = \sum_{i=1}^{n_\alpha} \alpha_i \text{ where } \dot{\alpha}_i = C_i\dot{\varepsilon}^p - \gamma_i\dot{p}\alpha_i . \tag{4.43}$$

The use of several backstresses can be motivated by long- and short-range residual stresses of dislocations and dislocation pile-ups as described earlier in this section.

In the following, the Armstrong-Frederick kinematic hardening model is embedded in the thermodynamic framework of plasticity models. Therefore, the backstresses α_i are interpreted as thermodynamic forces that drive the evolution of internal variables a_i. The internal variables a_i enter the Helmholtz free energy as follows:

$$\psi = \underbrace{\frac{1}{2}E\varepsilon^{e2}}_{\psi^e} + \underbrace{\sum_{i=1}^{n_\alpha}\frac{1}{2}C_i a_i^2}_{\psi^p} . \tag{4.44}$$

The elastic part describes the elastically stored energy in the spring element controlled by the elastic strain in the rheological model (Fig. 4.4). In the plastic part, the energy stored in the springs with stiffness C_i is considered.

Applying Eqs. (4.6) and (4.8) yields

$$\sigma = \frac{\partial \psi}{\partial \varepsilon^e} = E\varepsilon^e = E\left(\varepsilon - \varepsilon^p\right) \tag{4.45}$$

$$\alpha_i = \frac{\partial \psi}{\partial a_i} = C_i a_i . \tag{4.46}$$

To apply the normality rules to the flow rule and the evolution equations for the internal variables a_i, the following flow potential convex with respect to σ and α_i is formulated:

$$\phi = \phi_y + \sum_{i=1}^{n_\alpha} \left(\frac{1}{2} \frac{\gamma_i}{C_i} \left(\alpha_i^2 - C_i^2 a_i^2 \right) \right) . \tag{4.47}$$

The second term on the right hand side is written such that with Eq. (4.46) zero results. However, a proper thermodynamic formulation is obtained by formally adding the term. One obtains:

$$\dot{\varepsilon}^p = \dot{\lambda} \frac{\partial \phi}{\partial \sigma} = \dot{\lambda} \underbrace{\frac{\partial \phi_y}{\partial \sigma}}_{N = \frac{\sigma - \alpha}{|\sigma - \alpha|}} = \dot{\lambda} N = \dot{p} N \tag{4.48}$$

$$\dot{a}_i = -\dot{\lambda} \frac{\partial \phi}{\partial \alpha_i} = -\dot{\lambda} \underbrace{\frac{\partial \phi_y}{\partial \alpha_i}}_{-N} - \dot{\lambda} \frac{\gamma_i}{C_i} \alpha_i = \dot{\varepsilon}^p - \frac{\gamma_i}{C_i} \dot{p} \alpha_i . \tag{4.49}$$

N is the flow direction that is determined by the sign of the effective stress $\sigma - \alpha$ acting on the friction element in the rheological model. $\dot{p} = |\dot{\varepsilon}^p|$ is the equivalent plastic strain rate as defined in Eq. (4.22). The evolution equation of the Armstrong-Frederick kinematic hardening law (4.42) is obtained by taking the time derivative of Eq. (4.46), followed by insertion of Eq. (4.49), i.e.

$$\dot{\alpha}_i = C_i \dot{a}_i = C_i \left(\dot{\varepsilon}^p - \frac{\gamma_i}{C_i} \dot{p} \alpha_i \right) = C_i \dot{\varepsilon}^p - \gamma_i \dot{p} \alpha_i . \tag{4.50}$$

With the potential functions ψ given in Eq. (4.44) and ϕ_y given in Eq. (4.47), the plasticity model with kinematic hardening is defined by the resulting constitutive Eqs. (4.45), (4.46), (4.48) and (4.49). The last step is to determine the multiplier $\dot{\lambda}$. Since the stress is forced to lie on the boundary of the elastic domain during plastic yielding, i.e. $\phi_y = 0$ when $\dot{\lambda} > 0$, we use the consistency condition (as for the case of the isotropic hardening plasticity model in Sect. 4.3.1):

$$\dot{\phi}_y = \underbrace{\frac{\partial \phi_y}{\partial \sigma}}_{N} \dot{\sigma} + \sum_{i=1}^{n_\alpha} \underbrace{\frac{\partial \phi_y}{\partial \alpha_i}}_{-N} \dot{\alpha}_i = 0. \tag{4.51}$$

With Eqs. (4.40), (4.48) and (4.50) one finds

$$\begin{aligned}\dot{\phi}_y &= NE\left(\dot{\varepsilon} - \dot{\lambda}N\right) - \sum_{i=1}^{n_\alpha} N\left(C_i\dot{\lambda}N - \gamma_i\dot{\lambda}\alpha_i\right)\\ &= NE\dot{\varepsilon} - N^2E\dot{\lambda} - \dot{\lambda}\sum_{i=1}^{n_\alpha}\left(N^2C_i - N\gamma_i\alpha_i\right) = 0,\end{aligned} \tag{4.52}$$

which can be solved for the multiplier:

$$\dot{\lambda} = \frac{NE\dot{\varepsilon}}{E + \sum_{i=1}^{n_\alpha}\left(C_i - N\gamma_i\alpha_i\right)}. \tag{4.53}$$

4.3.2.3 Thermodynamic Consistency

With the internal variables a_i and their conjugate forces α_i, the local dissipation is given by

$$\begin{aligned}\mathcal{D}_{loc} &= \sigma\dot{\varepsilon}^p - \sum_{i=1}^{n_\alpha}\alpha_i\dot{a}_i\\ &= \sigma\dot{\lambda}N - \sum_{i=1}^{n_\alpha}\alpha_i\left(\dot{\lambda}N - \frac{\gamma_i}{C_i}\dot{\lambda}\alpha_i\right)\\ &= \dot{\lambda}\quad\underbrace{(\sigma-\alpha)\,N}_{\frac{(\sigma-\alpha)^2}{|\sigma-\alpha|} = |\sigma-\alpha|}\quad + \sum_{i=1}^{n_\alpha}\frac{\gamma_i}{C_i}\alpha_i^2\\ &= \dot{\lambda}\,|\sigma-\alpha| + \dot{\lambda}\sum_{i=1}^{n_\alpha}\frac{\gamma_i}{C_i}\alpha_i^2 \geq 0.\end{aligned} \tag{4.54}$$

With $\dot{\lambda} \geq 0$ and $\phi_y = |\sigma - \alpha| - R_e = 0$ during plastic yielding, it follows for the first term on the left hand side that

$$|\sigma - \alpha| = R_e \geq 0. \tag{4.55}$$

Imposing the strong restrictions $R_e \geq 0$ and $\gamma_i/C_i \geq 0$ from the second term on the material properties, thermodynamic consistency is satisfied a priori.

4.3.2.4 Numerical Solution with Predictor/Corrector Algorithm

An efficient way to solve the coupled differential equations with the inequality constraints of the loading/unloading conditions are so-called strain-controlled predictor-corrector algorithms. A prediction-corrector algorithm for time-independent plasticity model with isotropic and kinematic hardening is shown in Sect. 4.3.3.

4.3.2.5 Analytical Solution for Determination of Material Properties

For the determination of the material properties of the plasticity model with kinematic hardening based on uniaxial experimental data, analytical solutions of the Armstrong-Frederick law can be derived. For the evaluation of the analytical solutions, the plastic strains are known from the experimental data. In the following, Eq. (4.50) is integrated from a load reversal point, where the current backstress has the value α_i^0 and the current accumulated plastic strain is p^0. From Eq. (4.50) one obtains:

$$\begin{aligned} \dot{\alpha}_i &= (C_i N - \gamma_i \alpha_i)\, \dot{p} \\ \mathrm{d}\alpha_i &= \gamma_i \left(\frac{C_i}{\gamma_i} N - \alpha_i \right) \mathrm{d}p \end{aligned} \tag{4.56}$$

Separation of the variables and integration starting from the point of load reversal yields

$$\begin{aligned} \int_{\alpha_i^0}^{\alpha_i} \frac{1}{\frac{C_i}{\gamma_i} N - \bar{\alpha}_i} \mathrm{d}\bar{\alpha}_i &= \int_{p^0}^{p} \gamma_i \mathrm{d}\bar{p} \\ \left[-\ln \left(\frac{C_i}{\gamma_i} N - \bar{\alpha}_i \right) \right]_{\alpha_i^0}^{\alpha_i} &= \left[\gamma_i \bar{p} \right]_{p^0}^{p} \\ \ln \frac{\frac{C_i}{\gamma_i} N - \alpha_i}{\frac{C_i}{\gamma_i} N - \alpha_i^0} &= -\gamma_i \left(p - p^0 \right) \\ \frac{\frac{C_i}{\gamma_i} N - \alpha_i}{\frac{C_i}{\gamma_i} N - \alpha_i^0} &= e^{-\gamma_i (p - p^0)}. \end{aligned} \tag{4.57}$$

By solving for the backstress α_i, the analytical solution

$$\begin{aligned}\alpha_i &= \frac{C_i}{\gamma_i}N - \left(\frac{C_i}{\gamma_i}N - \alpha_i^0\right)e^{-\gamma_i(p-p^0)} \\ &= \alpha_i^0 e^{-\gamma_i(p-p^0)} + \frac{C_i}{\gamma_i}N\left(1 - e^{-\gamma_i(p-p^0)}\right)\end{aligned} \tag{4.58}$$

is obtained which can be used for the integration from one point of load reversal to the next, respectively. Assuming plastic yielding,

$$\phi_y = |\sigma - \alpha| - R_e = 0, \tag{4.59}$$

one obtains

$$\underbrace{\frac{\sigma - \alpha}{|\sigma - \alpha|}|\sigma - \alpha|}_{\sigma - \alpha} = \underbrace{\frac{\sigma - \alpha}{|\sigma - \alpha|}}_{N} R_e, \tag{4.60}$$

where both sides are multiplied by $(\sigma - \alpha) / |\sigma - \alpha|$. This gives with Eq. (4.58) the following analytical expression for the stress:

$$\sigma = \underbrace{NR_e}_{\substack{\text{stress}\\ \text{in friction element}}} + \underbrace{\sum_{i=1}^{n_\alpha}\left(\alpha_i^0 e^{-\gamma_i(p-p^0)} + \frac{C_i}{\gamma_i}N\left(1 - e^{-\gamma_i(p-p^0)}\right)\right)}_{\text{stress in springs}}. \tag{4.61}$$

This result shows the additive decomposition of the stress in the parallel connection of the rheological elements in the plastic part of the rheological model in Fig. 4.4.

In case of a linear kinematic hardening law, i.e. $\gamma_i = 0$ in Eq. (4.56), integration yields

$$\begin{aligned}\int_{\alpha_i^0}^{\alpha_i} \mathrm{d}\bar{\alpha}_i &= \int_{p^0}^{p} C_i N \mathrm{d}\bar{p} \\ [\bar{\alpha}_i]_{\alpha_i^0}^{\alpha_i} &= [C_i N \bar{p}]_{p^0}^{p} \\ \alpha_i &= \alpha_i^0 + C_i N\left(p - p^0\right).\end{aligned} \tag{4.62}$$

Assuming $n_\alpha = 2$ backstresses, with the first one having a linear evolution, then

$$\sigma = NR_e + \alpha_1^0 + C_1 N\left(p - p^0\right) + \alpha_2^0 e^{-\gamma_2(p-p^0)} + \frac{C_2}{\gamma_2}N\left(1 - e^{-\gamma_2(p-p^0)}\right). \tag{4.63}$$

In case of monotonic loading of a previously undeformed material with $\alpha_i^0 = 0$ and $p^0 = 0$, this equation gives

$$\sigma = N\left(R_e + C_1 p + \frac{C_2}{\gamma_2}\left(1 - e^{-\gamma_2 p}\right)\right) \tag{4.64}$$

and is exactly the same compared to Eq. (4.39) that describes the isotropic hardening material: If one chooses the material properties $\gamma_2 = b$, $C_1 = H$ and $C_2/\gamma_2 = Q_\infty$ the identical behavior is obtained for isotropic and kinematic hardening under monotonic loading. This is shown in Figs. 4.1 and 4.3 where the values of the material properties are chosen exactly this way (see caption of the figures). The monotonic curves are identical for isotropic and kinematic hardening. The difference of both hardening mechanisms becomes evident only if a plastic load reversal is applied. This insight is important for the determination of the respective material properties, as will be shown in examples with a plasticity model with isotropic and kinematic hardening in Sect. 4.3.3.

4.3.3 *Combined Hardening*

Now, both hardening mechanisms, isotropic hardening and kinematic hardening, are combined in a plasticity model. Hence, the elastic domain can increase due to isotropic hardening and it can translate in direction of plastic loading due to kinematic hardening (Fig. 4.5).

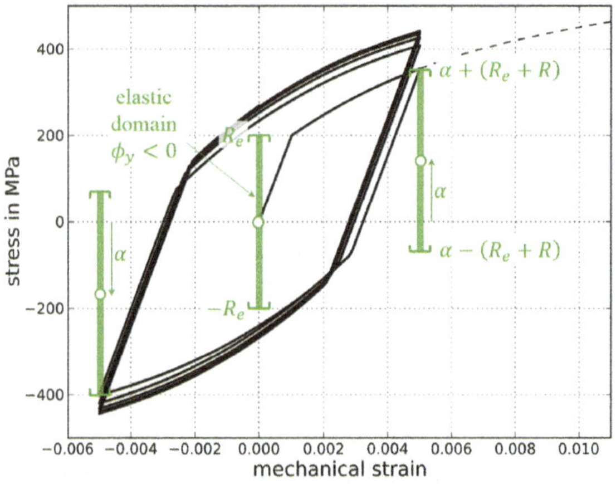

Fig. 4.5 Stress-strain curves of isotropic and kinematic hardening material using two backstresses and the following material properties: $E = 200000$ MPa, $R_e = 200$ MPa, $H = 0$ MPa, $Q_\infty = 100$ MPa and $b = 50$, $C_1 = 5000$ MPa, $\gamma_1 = 0$, $C_2 = 60000$ MPa and $\gamma_2 = 300$

4.3.3.1 Rheological Model

In the rheological model of the combined hardening plasticity model, besides the springs generating backstresses α_i for kinematic hardening in parallel to the friction element, the friction element can increase its strength by the isotropic hardening variable R (Fig. 4.6).

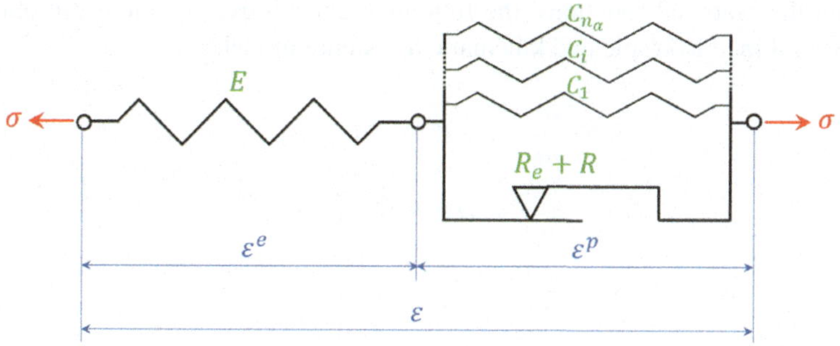

Fig. 4.6 Rheological model of isotropic and kinematic hardening elastic-plastic material

4.3.3.2 Mathematical Model

Before the constitutive equations of the combined hardening model are presented, the potential functions, namely the Helmholtz free energy and the flow potential, are given that fully govern the constitutive equations. By combining the Helmholtz free energies of the isotropic hardening model, Eq. (4.27) giving a linear and exponential hardening curve, and the kinematic hardening model, Eq. (4.44) resulting in the Armstrong-Frederick kinematic hardening law with 1, ..., n_α backstresses, the Helmholtz free energy of the combined hardening model is obtained:

$$\psi = \underbrace{\frac{1}{2}E\varepsilon^{e2}}_{\psi^e} + \underbrace{\frac{1}{2}Hp^2 + Q_\infty\left(p + \frac{1}{b}e^{-bp} - \frac{1}{b}\right) + \sum_{i=1}^{n_\alpha}\frac{1}{2}C_i a_i^2}_{\psi^p}. \tag{4.65}$$

In analogy to Eq. (4.47), the convex flow potential is

$$\phi = \phi_y + \sum_{i=1}^{n_\alpha}\left(\frac{1}{2}\frac{\gamma_i}{C_i}\left(\alpha_i^2 - C_i^2 a_i^2\right)\right) \tag{4.66}$$

and contains the yield function including isotropic and kinematic hardening:

$$\phi_y = |\sigma - \alpha| - (R_e + R) \leq 0 \text{ with } \alpha = \sum_{i=1}^{n_\alpha} \alpha_i. \tag{4.67}$$

With $\phi_y < 0$, the current stresses are in the elastic domain of the material, so that $\dot{\lambda} = 0$. If stresses attain values that give $\phi_y = 0$, plastic loading increments additionally occur and $\dot{\lambda} > 0$ in the flow rule, as expressed by the loading/unloading conditions in Eq. (4.18).

From the potential functions, the following constitutive equations are obtained for the combined isotropic and kinematic hardening model:

$$\sigma = \frac{\partial \psi}{\partial \varepsilon^e} = E\varepsilon^e = E\left(\varepsilon - \varepsilon^p\right) \tag{4.68}$$

$$R = \frac{\partial \psi}{\partial p} = Hp + Q_\infty \left(1 - e^{-bp}\right) \tag{4.69}$$

$$\alpha_i = \frac{\partial \psi}{\partial a_i} = C_i a_i. \tag{4.70}$$

The flow rule and the evolution equations for the internal variables p and a_i are obtained from the normality rules

$$\dot{\varepsilon}^p = \dot{\lambda}\frac{\partial \phi}{\partial \sigma} = \dot{\lambda} \underbrace{\frac{\partial \phi_y}{\partial \sigma}}_{N = \frac{\sigma - \alpha}{|\sigma - \alpha|}} = \dot{\lambda} N = \dot{p} N \tag{4.71}$$

$$\dot{p} = -\dot{\lambda} \underbrace{\frac{\partial \phi_y}{\partial R}}_{-1} = \dot{\lambda} = \left|\dot{\varepsilon}^p\right|. \tag{4.72}$$

$$\dot{a}_i = -\dot{\lambda}\frac{\partial \phi}{\partial \alpha_i} = -\dot{\lambda} \underbrace{\frac{\partial \phi_y}{\partial \alpha_i}}_{-N} - \dot{\lambda}\frac{\gamma_i}{C_i}\alpha_i = \dot{\varepsilon}^p - \frac{\gamma_i}{C_i}\dot{p}\alpha_i. \tag{4.73}$$

N is the flow direction that determines the sign of the plastic strain rate. Usually, the kinematic hardening law is expressed in terms of the rate of the backstresses:

$$\dot{\alpha}_i = C_i \dot{a}_i = C_i \dot{\varepsilon}^p - \gamma_i \dot{p} \alpha_i. \tag{4.74}$$

Finally, the multiplier $\dot{\lambda}$ is obtained from the consistency condition $\dot{\phi}_y = 0$, requiring the stresses to be bounded by the elastic domain:

$$\dot{\phi}_y = \underbrace{\frac{\partial \phi_y}{\partial \sigma}}_{N} \dot{\sigma} + \sum_{i=1}^{n_\alpha} \underbrace{\frac{\partial \phi_y}{\partial \alpha_i}}_{-N} \dot{\alpha}_i + \underbrace{\frac{\partial \phi_y}{\partial R}}_{-1} \frac{\partial R}{\partial p} \dot{p} = 0. \tag{4.75}$$

With Eqs. (4.68), (4.71) and (4.74) one finds

$$
\begin{aligned}
\dot{\phi}_y &= NE\left(\dot{\varepsilon} - \dot{\lambda}N\right) - \sum_{i=1}^{n_\alpha} N\left(C_i\dot{\lambda}N - \gamma_i\dot{\lambda}\alpha_i\right) - \frac{\partial R}{\partial p}\dot{\lambda} \\
&= NE\dot{\varepsilon} - N^2E\dot{\lambda} - \dot{\lambda}\sum_{i=1}^{n_\alpha}\left(N^2C_i - N\gamma_i\alpha_i\right) - \frac{\partial R}{\partial p}\dot{\lambda} = 0.
\end{aligned}
\tag{4.76}
$$

Hence, $\dot{\lambda}$ is given by

$$
\dot{\lambda} = \frac{NE\dot{\varepsilon}}{E + \sum_{i=1}^{n_\alpha}(C_i - N\gamma_i\alpha_i) + \frac{\partial R}{\partial p}}. \tag{4.77}
$$

4.3.3.3 Thermodynamic Consistency

With the internal variables a_i and p and their conjugate forces α_i and R the local dissipation is given by

$$
\begin{aligned}
\mathcal{D}_{loc} &= \sigma\dot{\varepsilon}^p - \sum_{i=1}^{n_\alpha}\alpha_i\dot{a}_i - R\dot{p} \\
&= \sigma\dot{\lambda}N - \sum_{i=1}^{n_\alpha}\alpha_i\left(\dot{\lambda}N - \frac{\gamma_i}{C_i}\dot{\lambda}\alpha_i\right) - R\dot{\lambda} \\
&= \dot{\lambda}\underbrace{(\sigma - \alpha)N}_{\frac{(\sigma-\alpha)^2}{|\sigma-\alpha|} = |\sigma-\alpha|} + \dot{\lambda}\sum_{i=1}^{n_\alpha}\frac{\gamma_i}{C_i}\alpha_i^2 - R\dot{\lambda} \\
&= \dot{\lambda}\left(|\sigma - \alpha| - R\right) + \dot{\lambda}\sum_{i=1}^{n_\alpha}\frac{\gamma_i}{C_i}\alpha_i^2 \geq 0.
\end{aligned}
\tag{4.78}
$$

Requiring the strong restrictions that both terms should be ≥ 0, one obtains for the first term with $\dot{\lambda} \geq 0$ and $\phi_y = 0$ during plastic yielding

$$
\begin{aligned}
|\sigma - \alpha| - R &= R_e \geq 0 \\
|\sigma - \alpha| &= R_e + R \geq 0.
\end{aligned}
\tag{4.79}
$$

Hence, thermodynamic consistency is satisfied a priori if the material properties are chosen such that $R_e + R \geq 0$ and, from the second term, $\gamma_i/C_i \geq 0$. $R_e \geq 0$, since

$R = 0$ initially when $p = 0$. Moreover, $R_e \geq -R$ must hold. This is only strictly the case in the combined linear and exponential hardening if $H \geq 0$ and $R_e \geq -Q_\infty$.

4.3.3.4 Numerical Solution with Predictor/Corrector Algorithm

An efficient way to solve the coupled differential equations with the inequality constraints of the loading/unloading conditions are so-called strain-controlled predictor-corrector algorithms. To this end, the flow rule and the evolution equations are discretized in time by an unconditionally stable implicit integration scheme as the backward Euler method. In each time increment, the strain increment is first assumed being fully elastic (elastic predictor step). If the resulting stresses violate the loading/unloading conditions, the multiplier $\dot{\lambda}$ is found by enforcing the incremental consistency condition to hold (plastic corrector step). This step is often referred to as return mapping algorithm.

For the numerical treatment of the constitutive equations, a time interval $\left[t_n, t_{n+1}\right]$, $\Delta t = t_{n+1} - t_n$, is considered where all variables at time t_n are assumed to be known. A deformation driven context is assumed, where the strain increment $\Delta\varepsilon$ and the temperature increment ΔT in Δt is given and the update of the stresses σ_{n+1} and the updates of the hardening variables are computed.

The flow rule (4.71) is integrated using the fully implicit backward Euler method:

$$\varepsilon^p_{n+1} = \varepsilon^p_n + \Delta t \dot{\varepsilon}^p_{n+1} = \varepsilon^p_n + \underbrace{\Delta t \dot{\lambda}_{n+1}}_{\Delta\lambda} \underbrace{N_{n+1}}_{\frac{\sigma_{n+1} - \alpha_{n+1}}{|\sigma_{n+1} - \alpha_{n+1}|}} . \tag{4.80}$$

Using the law of elasticity (4.68), the update of the stresses is

$$\begin{aligned} \sigma_{n+1} &= E\left(\varepsilon_{n+1} - \varepsilon^p_{n+1}\right) \\ &= E\left(\varepsilon_{n+1} - \varepsilon^p_n - \Delta\lambda N_{n+1}\right) \\ &= \underbrace{E\left(\varepsilon_{n+1} - \varepsilon^p_n\right)}_{\sigma_\star} - E\Delta\lambda N_{n+1}, \end{aligned} \tag{4.81}$$

where the strain is computed using the prescribed strain increment:

$$\varepsilon_{n+1} = \varepsilon_n + \Delta\varepsilon.$$

$\sigma_\star$ is the stress that would be computed if no plastic deformation takes place in the time increment, i.e. if $\Delta\lambda = 0$. Often, $\sigma_\star$ is called trial stress.

The evolution equation for the accumulated plastic strain, Eq. (4.72), is also integrated using the fully implicit backward Euler method, yielding

$$p_{n+1} = p_n + \Delta t \dot{p}_{n+1} = p_n + \Delta\lambda, \tag{4.82}$$

as well as the evolution equation of the backstresses, Eq. (4.74):

$$\begin{aligned}\alpha_{i,n+1} &= \alpha_{i,n} + \Delta t \dot{\alpha}_{i,n+1} \\ &= \alpha_{i,n} + C_i \Delta\lambda N_{n+1} - \gamma_i \Delta\lambda \alpha_{i,n+1} \\ &= \frac{1}{1+\gamma_i \Delta\lambda} \left(\alpha_{i,n} + C_i \Delta\lambda N_{n+1}\right).\end{aligned} \tag{4.83}$$

The yield function at time t_{n+1} is

$$\phi_{y,n+1} = |\sigma_{n+1} - \alpha_{n+1}| - (R_0 + R_{n+1}) \tag{4.84}$$

$$\text{where } \alpha_{n+1} = \sum_{i=1}^{n_\alpha} \alpha_{i,n+1} \tag{4.85}$$

$$R_{n+1} = H p_{n+1} + Q_\infty \left(1 - e^{-b p_{n+1}}\right). \tag{4.86}$$

Using Eqs. (4.81) and (4.83), the effective stress can be written as

$$\sigma_{n+1} - \alpha_{n+1} = \sigma_\star - \underbrace{\sum_{i=1}^{n_\alpha} \frac{1}{1+\gamma_i \Delta\lambda} \alpha_{i,n}}_{\alpha_\star} - \Delta\lambda \left(E + \sum_{i=1}^{n_\alpha} \frac{C_i}{1+\gamma_i \Delta\lambda}\right) N_{n+1}. \tag{4.87}$$

Insertion of the identities

$$\sigma_{n+1} - \alpha_{n+1} = |\sigma_{n+1} - \alpha_{n+1}| \underbrace{\frac{\sigma_{n+1} - \alpha_{n+1}}{|\sigma_{n+1} - \alpha_{n+1}|}}_{N_{n+1}} \tag{4.88}$$

$$\sigma_\star - \alpha_\star = |\sigma_\star - \alpha_\star| \underbrace{\frac{\sigma_\star - \alpha_\star}{|\sigma_\star - \alpha_\star|}}_{N_\star} \tag{4.89}$$

yields

$$|\sigma_{n+1} - \alpha_{n+1}| \, N_{n+1} = |\sigma_\star - \alpha_\star| \, N_\star - \Delta\lambda \left(E + \sum_{i=1}^{n_\alpha} \frac{C_i}{1+\gamma_i \Delta\lambda}\right) N_{n+1}. \tag{4.90}$$

This equations reveals that $N_\star = N_{n+1}$ since the sum of two directions (on the right hand side) can only give one of these directions if both directions are the same. Hence, the equation can be simplified to

$$|\sigma_{n+1} - \alpha_{n+1}| = |\sigma_\star - \alpha_\star| - \Delta\lambda \left(E + \sum_{i=1}^{n_\alpha} \frac{C_i}{1+\gamma_i \Delta\lambda}\right) \tag{4.91}$$

and inserted into the yield function (4.86):

$$\begin{aligned}\phi_{y,n+1} &= |\sigma_\star - \alpha_\star| - \Delta\lambda \left(E + \sum_{i=1}^{n_\alpha} \frac{C_i}{1+\gamma_i \Delta\lambda} \right) \\ &\quad - \left(R_0 + H\left(p_n + \Delta\lambda\right) + Q_\infty \left(1 - e^{-b(p_n+\Delta\lambda)}\right)\right).\end{aligned} \tag{4.92}$$

In the following, the predictor/corrector algorithm is applied. During the elastic predictor step, it is assumed that no plastic deformation occurs, i.e. if $\Delta\lambda = 0$. This results in the so-called trial yield function

$$\phi_{y,\star} = |\sigma_\star - \alpha_\star| - \left(R_0 + H p_n + Q_\infty \left(1 - e^{-bp_n}\right)\right). \tag{4.93}$$

If the trial yield function is less or equal than zero, then the stresses are in the elastic domain, so that no plasticity needs to be considered in the time increment and $\Delta\lambda = 0$ as assumed. If, however, the trial yield function attains values larger than zero, inadmissible stresses outside the elastic domain are computed in the elastic predictor step. Hence, the plastic corrector step must be applied to bring the stresses to the boundary of the elastic domain:

$$\phi_{y,\star} = \begin{cases} \leq 0, \ \Delta\lambda = 0 \text{ (elastic loading step)} \\ > 0, \ \Delta\lambda > 0 \text{ (elastic-plastic loading step)} \end{cases} \tag{4.94}$$

In the plastic corrector step, the incremental consistency condition is enforced to find the multiplier $\Delta\lambda$, i.e. $\Delta\lambda$ must be determined such that $\phi_{y,n+1} = 0$ in Eq. (4.92). Since non-linear hardening laws are considered, $\phi_{y,n+1}$ is a non-linear function of $\Delta\lambda$ and can not be solved directly for $\Delta\lambda$. A solution is obtained iteratively using Newton's method that yields the iteration scheme

$$\Delta\lambda_{j+1} = \Delta\lambda_j - \left(\left.\frac{\partial \phi_{y,n+1}}{\partial \Delta\lambda}\right|_{\Delta\lambda_j} \right)^{-1} \left.\phi_{y,n+1}\right|_{\Delta\lambda_j} \tag{4.95}$$

$$\begin{aligned}\text{where } \frac{\partial \phi_{y,n+1}}{\partial \Delta\lambda} &= N_\star \sum_{i=1}^{n_\alpha} \frac{\gamma_i}{(1+\gamma_i\Delta\lambda)^2} \alpha_{i,n} \\ &\quad - \left(E + \sum_{i=1}^{n_\alpha} \frac{C_i}{1+\gamma_i\Delta\lambda} \right) + \Delta\lambda \sum_{i=1}^{n_\alpha} \frac{C_i \gamma_i}{(1+\gamma_i\Delta\lambda)^2} \\ &\quad - H - Q_\infty b e^{-b(p_n+\Delta\lambda)}.\end{aligned} \tag{4.96}$$

Convergence in iteration $j+1$ is obtained if the absolute value of $\left.\phi_{y,n+1}\right|_{\Delta\lambda_{j+1}}$ gets smaller than a defined tolerance. The updates of the plastic strain, the stress, the accumulated plastic strain and the backstresses are obtained by insertion of the determined multiplier $\Delta\lambda$ in Eqs. (4.80) to (4.83). In no stage of the integration algorithm, the time increment Δt is explicitly necessary because time is not relevant for time-independent plasticity models. The integration algorithm for time-dependent plasticity models will include Δt explicitly (see Sect. 4.4.1).

4.3.3.5 Analytical Solution for Determination of Material Properties

For the determination of the material properties on the basis of results from uniaxial material tests, the already available analytical solutions for the kinematic hardening law from Eq. (4.58) can be used to derive an analytical expression for calculation of the stresses based on the experimentally measured plastic strains. To this end,

$$\phi_y = |\sigma - \alpha| - (R_e + R) = 0 \tag{4.97}$$

is used that holds for plastic yielding. Rearranging to

$$\underbrace{\frac{\sigma - \alpha}{|\sigma - \alpha|} |\sigma - \alpha|}_{\sigma - \alpha} = \underbrace{\frac{\sigma - \alpha}{|\sigma - \alpha|}}_{N} (R_e + R)\,, \tag{4.98}$$

where both sides are multiplied by $(\sigma - \alpha) / |\sigma - \alpha|$, and solving for the stress yields:

$$\begin{aligned} \sigma = {} & \underbrace{N\left(R_e + Hp + Q_\infty\left(1 - e^{-bp}\right)\right)}_{\text{stress in friction element}} \\ & + \underbrace{\sum_{i=1}^{n_\alpha}\left(\alpha_i^0 e^{-\gamma_i(p-p^0)} + \frac{C_i}{\gamma_i} N\left(1 - e^{-\gamma_i(p-p^0)}\right)\right)}_{\text{stress in springs}}. \end{aligned} \tag{4.99}$$

This result shows the additive decomposition of the stress in the parallel connection of the rheological elements in the plastic part of the rheological model in Fig. 4.6. This equation will be used in the following example regarding the determination of the material properties for the combined isotropic and kinematic hardening model.

4.3.3.6 Example: Determination of Material Properties

In this section the determination of the material properties of the combined isotropic and kinematic hardening model is addressed. To this end, experimental data for uniaxial loading conditions is assumed that is available as stress-strain data is terms of stress-strain data pairs $\{\sigma_k^{\text{exp}}, \varepsilon_k^{\text{exp}}\}$ at $k = 1, \ldots, n_{\text{data}}$ discrete time point t_k. From the data, the elastic strains and plastic strains are obtained from

$$\varepsilon_k^{e,\exp} = \sigma_k^{\exp}/E \tag{4.100}$$
$$\varepsilon_k^{p,\exp} = \varepsilon_k^{\exp} - \varepsilon_k^{e,\exp}, \tag{4.101}$$

respectively. The accumulated plastic strain and the flow direction are computed from plastic strain increments according to

$$p_k^{\exp} = \sum_{i=2}^{k} \left|\varepsilon_i^{p,\exp} - \varepsilon_{i-1}^{p,\exp}\right| \tag{4.102}$$
$$N_k^{\exp} = \begin{cases} +1, & \varepsilon_k^{p,\exp} - \varepsilon_{k-1}^{p,\exp} \geq 0 \\ -1, & \varepsilon_k^{p,\exp} - \varepsilon_{k-1}^{p,\exp} < 0 \end{cases} \tag{4.103}$$

for $k \geq 2$. $p_1^{\exp} = 0$ and $N_1^{\exp} = 1$ are used as initial conditions.

For the determination of the material properties, the analytical expression for the stress from Eq. (4.99) is used, so that

$$\sigma_k^{\text{sim}} = N_k^{\exp}\left(R_e + H p_k^{\exp} + Q_\infty\left(1 - e^{-b p_k^{\exp}}\right)\right) + \sum_{i=1}^{n_\alpha}\left(\alpha_i^0 e^{-\gamma_i\left(p_k^{\exp} - p^0\right)} + \frac{C_i}{\gamma_i} N_k^{\exp}\left(1 - e^{-\gamma_i\left(p_k^{\exp} - p^0\right)}\right)\right). \tag{4.104}$$

The values α_i^0 and p^0 are determined at points of load reversal where the flow direction changes its sign. The computed stress is a function of a set of material properties $\boldsymbol{p}$, i.e. $\sigma_k^{\text{sim}} = \hat{\sigma}_k^{\text{sim}}(\boldsymbol{p})$. $n_\alpha = 2$ backstresses are used with the first backstress having a linear evolution equation with $\gamma_1 = 0$, so that $\boldsymbol{p} = \left[R_e \; H \; Q_\infty \; b \; C_1 \; C_2 \; \gamma_2\right]$ are the material properties to be determined.

An experience-based approach is to find the material properties by trial and error and choose their values so that visually a good description of the data is obtained. A visual inspection of the results is depending on the person that is engaged with the determination of the material properties. One person might believe that one set of material properties provides a good description, while another person might believe that the description with other values is better. Hence, non-objective material properties can result.

An objective criterion for the quality of the description of the data with some chosen material properties is e.g. the least-square function

$$f = \hat{f}(\boldsymbol{p}) = \sum_{k=1}^{n_{\text{data}}} \left(\hat{\sigma}_k^{\text{sim}}(\boldsymbol{p}) - \sigma_k^{\exp}\right)^2. \tag{4.105}$$

The lower the function value f, the better the description of the data. But still one person might already be satisfied with the description and another person might continue looking for a better solution with a lower f. Hence, non-objective material properties are the result.

Instead of finding the material properties with the experience-based approach that gives non-objective material properties, optimization-based approaches can be used. To this end, the least-squares function is minimized using numerical optimization methods as e.g. gradient-based optimization methods that find a better set of material properties in $j = 1, ...$ iterations on the basis of the negative gradient of the least-squares function that shows in the direction of steepest descent:

$$\boldsymbol{p}_{j+1} = \boldsymbol{p}_j - \boldsymbol{H} \left. \frac{\partial f}{\partial \boldsymbol{p}} \right|_j . \tag{4.106}$$

$\boldsymbol{H}$ is some iteration matrix that defines the selected method, e.g. the method of steepest descent, the conjugate-gradient method or Newton's method, e.g. Seifert et al. (2007). Some initial values $\boldsymbol{p}_0$ must be supplied. The optimal material properties $\boldsymbol{p}_\star$ are obtained after convergence of the iterative solution. Convergence can be defined if the increments $\hat{f}\left(\boldsymbol{p}_j\right) - \hat{f}\left(\boldsymbol{p}_{j+1}\right)$ or $\left|\boldsymbol{p}_{j+1} - \boldsymbol{p}_j\right|$ or the gradient $\partial f / \partial \boldsymbol{p}|_{j+1}$ fall below a specified tolerance.

For the following examples, MICROSOFT EXCEL is used to process the experimental data that acts as experimental data $\left\{\sigma_k^{\text{exp}}, \varepsilon_k^{\text{exp}}\right\}$ according Eqs. (4.101) to (4.103) and to implement the plasticity model according to Eq. (4.104). Moreover, the least-square function (4.105) is computed and the MICROSOFT EXCEL built-in gradient-based solver is used with the standard tolerances to minimize the least-square sum.

In the following two examples, the experimental data is generated synthetically using prescribed values for the material properties, the master values $\boldsymbol{p}_M$. It is the aim to re-determine the material properties from the synthetically generated data. For the generation of the data, the plasticity model is numerically integrated using the predictor/corrector algorithm previously described in this section, so that the results are available in terms of the stress-strain data pairs $\left\{\sigma_k^{\text{exp}}, \varepsilon_k^{\text{exp}}\right\}$ for the $k = 1, ..., n_{\text{data}}$ calculated time points t_k. For the re-determination of the material properties, three different sets of initial values, namely $\boldsymbol{p}_0^{\text{I}}$, $\boldsymbol{p}_0^{\text{II}}$ and $\boldsymbol{p}_0^{\text{III}}$ are used and a gradient-based optimization is performed for each set of initial values. The initial values are given together with the master values in Table 4.1.

In the first example, data from a tensile test is considered. The data, generated with the master material properties $\boldsymbol{p}_M$, is shown in Fig. 4.7. In this figure, also the results of the model are shown obtained with the initial values $\boldsymbol{p}_0^{\text{I}}$, $\boldsymbol{p}_0^{\text{II}}$ and $\boldsymbol{p}_0^{\text{III}}$. After applying the MICROSOFT EXCEL solver, optimal material properties $\boldsymbol{p}_\star^{\text{I}}$, $\boldsymbol{p}_\star^{\text{II}}$ and $\boldsymbol{p}_\star^{\text{III}}$ corresponding to each set of initial values are found and compiled in Table 4.2. The resulting stress-strain curves for the optimal material properties are shown in Fig. 4.8 together with the data underlying the optimization. For all optimizations, a very good description is obtained visually and in terms of the least-square sum. However, the optimized material properties are quite different for the varying initial values. The mean value and the standard deviation is also shown in Table 4.2. Hence, the material properties are non-objective: Different persons using different initial values but the same material data, the same material model and the same optimization method find

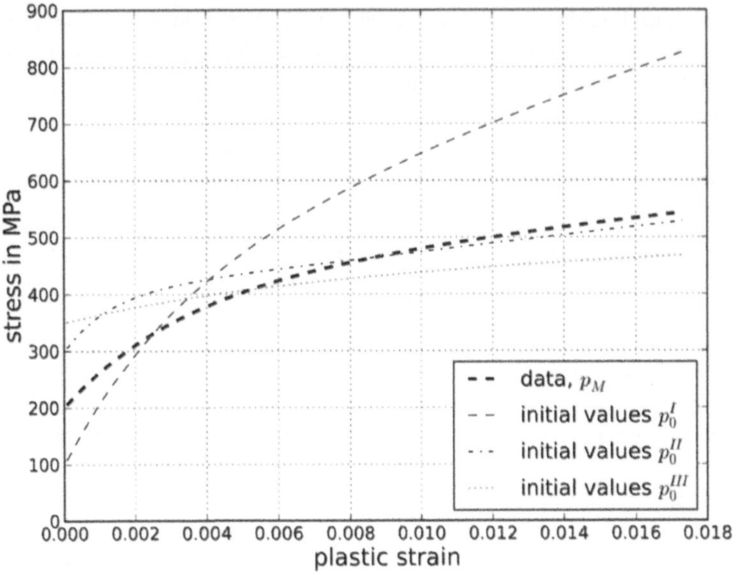

Fig. 4.7 Stress-strain curves of tensile test obtained with the master material properties $\boldsymbol{p}_M$ and the initial values $\boldsymbol{p}_0^{\mathrm{I}}$, $\boldsymbol{p}_0^{\mathrm{II}}$ and $\boldsymbol{p}_0^{\mathrm{III}}$ given in Table 4.1

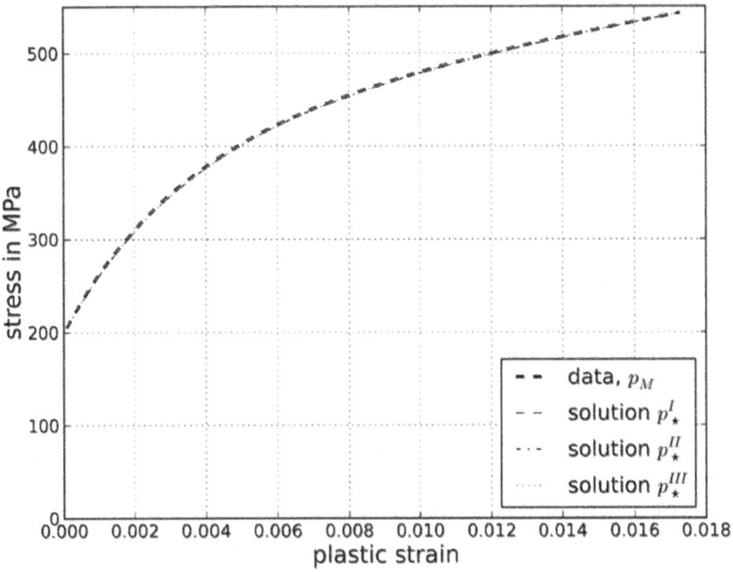

Fig. 4.8 Stress-strain curves of tensile test obtained with the master material properties $\boldsymbol{p}_M$ and the optimal material properties $\boldsymbol{p}_\star^{\mathrm{I}}$, $\boldsymbol{p}_\star^{\mathrm{II}}$ and $\boldsymbol{p}_\star^{\mathrm{III}}$ given in Table 4.2

Table 4.1 Master material properties $\boldsymbol{p}_M$ for the generation of synthetic experimental data and three sets of initial values $\boldsymbol{p}_0^{\mathrm{I}}$, $\boldsymbol{p}_0^{\mathrm{II}}$ and $\boldsymbol{p}_0^{\mathrm{III}}$ for the re-determination of the material properties from the synthetic experimental data

$\boldsymbol{p}$	$\boldsymbol{p}_M$	$\boldsymbol{p}_0^{\mathrm{I}}$	$\boldsymbol{p}_0^{\mathrm{II}}$	$\boldsymbol{p}_0^{\mathrm{II}}$
R_e in MPa	200	100	300	150
H in MPa	0	2000	5000	10000
Q_∞ in MPa	100	200	50	300
b	50	100	10	70
C_1 in MPa	5000	10000	2000	15000
C_2 in MPa	60000	30000	80000	20000
γ_2	5000	10000	80000	20000

Table 4.2 Master material properties $\boldsymbol{p}_M$ for the generation of synthetic experimental data and the optimal material properties $\boldsymbol{p}_\star^{\mathrm{I}}$, $\boldsymbol{p}_\star^{\mathrm{II}}$ and $\boldsymbol{p}_\star^{\mathrm{III}}$ obtained from the re-determination from the tensile test with three different sets of initial values as well as resulting least-square sum f for all material properties

$\boldsymbol{p}$	$\boldsymbol{p}_M$	$\boldsymbol{p}_\star^{\mathrm{I}}$	$\boldsymbol{p}_\star^{\mathrm{II}}$	$\boldsymbol{p}_\star^{\mathrm{II}}$	mean value	standard deviation
R_e in MPa	200	200	200	200	200	0
H in MPa	0	3247	4840	2772	3620	884
Q_∞ in MPa	100	164	50	194	136	62
b	50	316	10	301	209	525
C_1 in MPa	5000	3247	1974	2773	2665	68
C_2 in MPa	60000	11224	61174	5472	25957	25013
γ_2	300	154	279	72	168	85
f in MPa^2		0.0476	3.78	0.0168		

different optimal material properties. The reason is that under monotonic loading isotropic and kinematic hardening can not be distinguished. It requires at least one plastic load reversal so that the increase and the translation of the elastic domain can be observed and isotropic and kinematic hardening can be uniquely identified.

In the second example, data from a cyclic test is considered. The cyclic data is again generated with the master material properties $\boldsymbol{p}_M$ given in Table 4.1. The data is shown in Fig. 4.9 together with the results of the model obtained with the initial values $\boldsymbol{p}_0^{\mathrm{I}}$, $\boldsymbol{p}_0^{\mathrm{II}}$ and $\boldsymbol{p}_0^{\mathrm{III}}$ given in Table 4.1. After applying the MICROSOFT EXCEL solver, optimal material properties $\boldsymbol{p}_\star^{\mathrm{I}}$, $\boldsymbol{p}_\star^{\mathrm{II}}$ and $\boldsymbol{p}_\star^{\mathrm{III}}$ corresponding to each set of initial values are found and compiled in Table 4.3. The resulting stress-strain curves for the optimal material properties are shown in Fig. 4.10 together with the data underlying the optimization. For the cyclic tests, also a good visual agreement and very low least-square sums are obtained for all initial values. For the cyclic test, however, also the optimal material properties for all initial values agree well with the master material properties and show a very low standard deviation (Table 4.3).

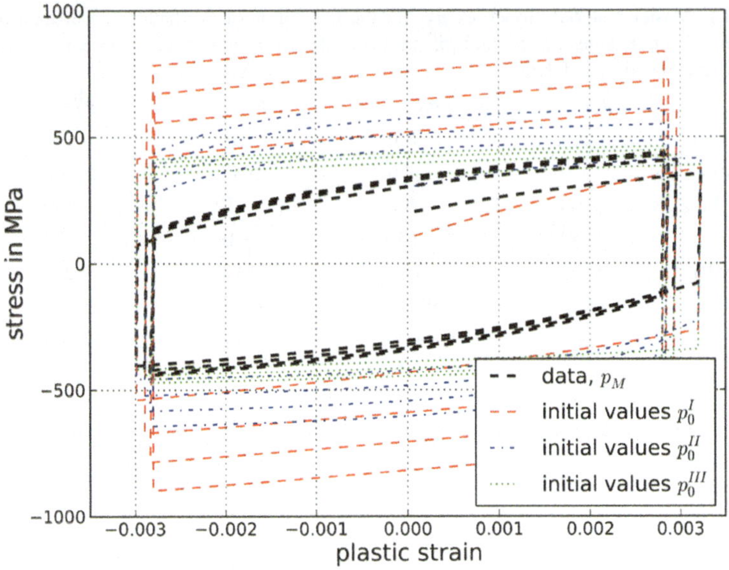

Fig. 4.9 Stress-strain curves of cyclic test obtained with the master material properties $\boldsymbol{p}_M$ and the initial values $\boldsymbol{p}_0^{\mathrm{I}}$, $\boldsymbol{p}_0^{\mathrm{II}}$ and $\boldsymbol{p}_0^{\mathrm{III}}$ given in Table 4.1

Table 4.3 Master material properties $\boldsymbol{p}_M$ for the generation of synthetic experimental data and the optimal material properties $\boldsymbol{p}_\star^{\mathrm{I}}$, $\boldsymbol{p}_\star^{\mathrm{II}}$ and $\boldsymbol{p}_\star^{\mathrm{III}}$ obtained from the re-determination from the cyclic test with three different sets of initial values as well as resulting least-square sum f for all material properties

p	p_M	$p_\star^{\mathrm{I}}$	$p_\star^{\mathrm{II}}$	$p_\star^{\mathrm{II}}$	mean value	standard deviation
R_e in MPa	200	200	200	200	200	0
H in MPa	0	4	4	5	4	0.7
Q_∞ in MPa	100	100	100	100	100	0
b	50	50	50	50	50	0
C_1 in MPa	5000	4990	4989	4993	4990	1.4
C_2 in MPa	60000	59346	59347	59344	59346	1.2
γ_2	300	297	297	297	297	0
f in MPa^2		0.0602	0.0597	0.0621		

Due to the plastic load reversals, isotropic and kinematic hardening can be uniquely identified, so that objective material properties are determined. The material data is sensitive for the plasticity model.

In both examples, the synthetically generated data was ideal with no scattering in the data as usually the case for experimental data. Hence, the examples investigate the stability of the material properties. By introducing random scattering to the data

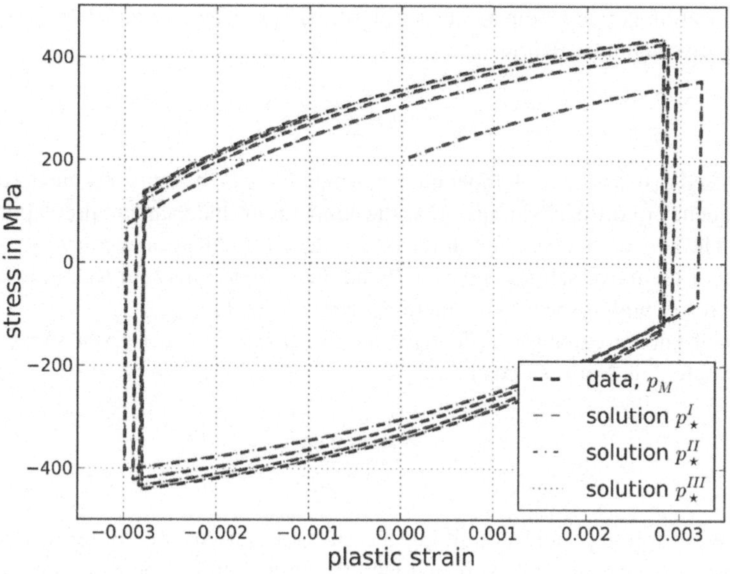

Fig. 4.10 Stress-strain curves of cyclic test obtained with the master material properties $\boldsymbol{p}_M$ and the optimal material properties $\boldsymbol{p}^{\mathrm{I}}_\star$, $\boldsymbol{p}^{\mathrm{II}}_\star$ and $\boldsymbol{p}^{\mathrm{III}}_\star$ given in Table 4.3

and using different data sets with random scattering to re-determine the material properties, the robustness of the material properties can be assessed to get further insights into the objectivity of the material properties. Moreover, analytical methods can be used to assess the robustness. An assessment of the robustness of material properties is found in e.g. Seifert (2006), Rekun and Jörg (2016).

4.4 Time-dependent Cyclic Plasticity

In this section, plasticity models are introduced for ductile materials that are exposed to higher temperature. Isotropic hardening and kinematic hardening will be considered, linear elastic behavior and isothermal loading conditions are assumed. In material tests at higher temperature, materials show time-dependent plastic behavior. Materials can creep under stress loading or the stresses in the material can relax under strain loading. Moreover, increasing the strain rate at higher temperature also increases the stress level. Responsible for the time-dependent behavior are typically thermally activated processes, in particular stress-controlled diffusion processes. Climbing of dislocations over obstacles, which act as barrier to dislocation movement at lower temperature, is a result of the residual stress fields around the dislocations: Vacancies diffuse to the compressive residual stress field so that the dislocation climbs over several slip planes until the obstacle is overcome and the dislocation is free to slip again. Since diffusion processes control the creep rate, often the Arrhenius equation

$$k = A_0 e^{-\frac{Q}{RT}} \tag{4.107}$$

can be used to describe the temperature-dependency. k is the rate constant describing the frequency of diffusive jumps. A_0 is the temperature-independent pre-exponential factor. The activation energy Q characterizes the basic diffusion mechanism and the product of the universal gas constant R and the temperature T in Kelvin is used as scaling for Q with respect to the thermal energy available.

A well-known equation to describe the creep rate $\dot{\varepsilon}^c$ for a given stress σ and temperature T is Norton's creep law:

$$\dot{\varepsilon}^c = \underbrace{A_0 e^{-\frac{Q}{RT}}}_{A} \sigma^n. \tag{4.108}$$

n is the Norton exponent describing the stress sensitivity of the creep rate. The Arrhenius term can also be expressed with the temperature-dependent pre-factor A. Norton's creep law does not consider isotropic or/and kinematic hardening of the material and, hence, does only allow to describe secondary creep. This is why in the next sections different approaches are presented to model rate-dependent plasticity with isotropic and kinematic hardening and to include the effect of static recovery of hardening, i.e. a recovery of hardening with time observable at higher temperature. The thermodynamic framework as well as rheological models are addressed.

4.4.1 Rate-dependent Yielding: Unified Models

Stress-strain curves for a cyclically loaded isotropic and kinematic hardening material are shown in Fig. 4.11 where rate-dependency is assumed. The higher the strain rate, the higher the resistance of the material against plastic yielding and, hence, the higher the observable yield stress limiting the elastic domain. The yield stress R_0 is defined in the figure for infinitely low strain rate, which gives the rate-independent limit. By increasing the strain rate, an additional overstress σ_v increases the elastic domain. For rate-dependent plasticity, the stress is not bound anymore by the condition $\phi_y \leq 0$ as it is the case for the rate-independent material.

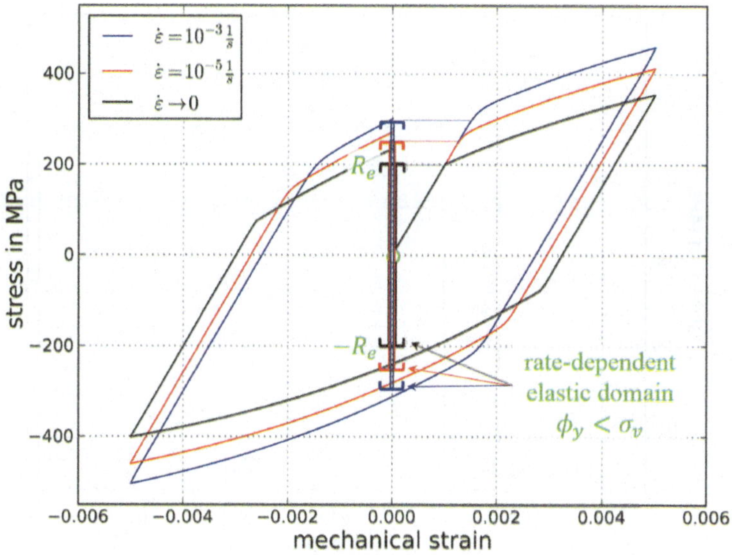

Fig. 4.11 Stress-strain curves of rate-dependent isotropic and kinematic hardening material with the following material properties: $E = 200000$ MPa, $R_e = 200$ MPa, $H = 0$ MPa, $Q_\infty = 100$ MPa and $b = 50$, $C_1 = 5000$ MPa, $\gamma_1 = 0$, $C_2 = 60000$ MPa, $\gamma_2 = 300$, $K = 300$ MPa and $n = 8$

4.4.1.1 Rheological Model

In the rheological model, the rate-effect can be included introducing a damper element in parallel within the time-dependent plastic part (Fig. 4.12) that determines the viscoplastic strains ε^{vp}. The damper generates a viscous overstress σ_v against

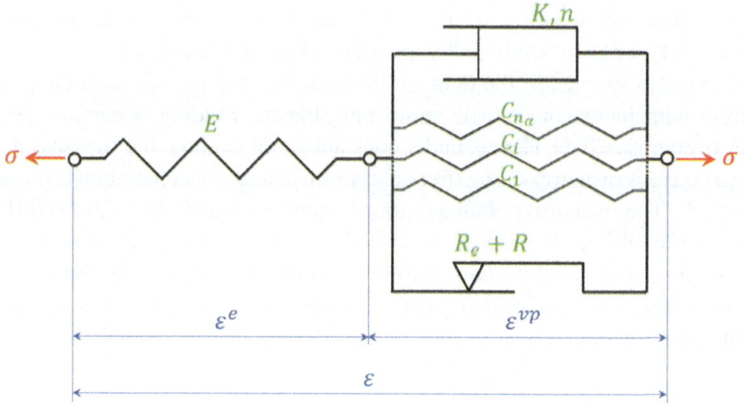

Fig. 4.12 Rheological model of the unified rate-dependent isotropic and kinematic hardening elastic-plastic material

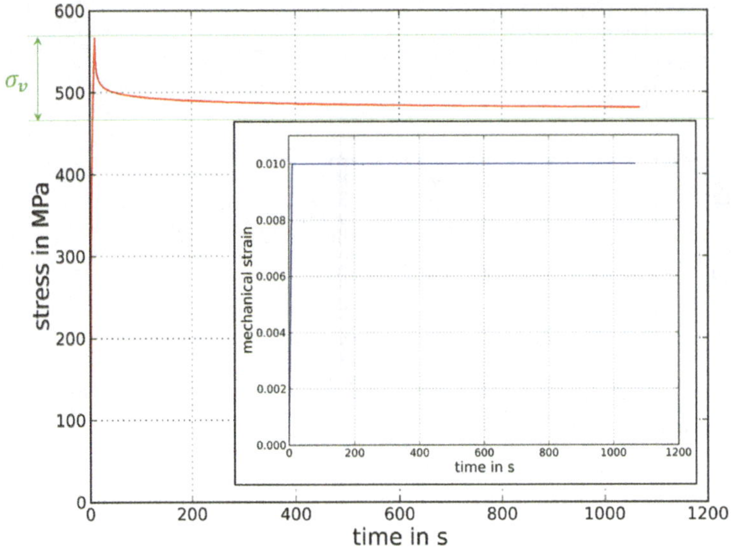

Fig. 4.13 Prescribed strain history and resulting stress history in a stress relaxation test calculated with rate-dependent isotropic and kinematic hardening material with the following material properties: $E = 200000$ MPa, $R_e = 200$ MPa, $H = 0$ MPa, $Q_\infty = 100$ MPa and $b = 50$, $C_1 = 5000$ MPa, $\gamma_1 = 0$, $C_2 = 60000$ MPa, $\gamma_2 = 300$, $K = 300$ MPa and $n = 8$

plastic yielding that depends on the equivalent viscoplastic strain rate $\dot{p} = |\dot{\varepsilon}^{vp}|$, i.e. $\sigma_v = \hat{\sigma}_v(\dot{p})$. Hence, the effect of the damper is independent on the loading direction (tension/compression going loading). The functional dependency also includes material properties. The material properties K and n shown in Fig. 4.12 are introduced later in this section. Since there is one viscoplastic strain describing the rate-dependent behavior of the material, it is called a unified model. In Sect. 4.4.2, non-unified models will be addressed as well, where the viscoplastic strain is split into a rate-independent plastic strain and a rate-dependent viscous strain.

Due to the properties of the damper element, the material shows higher viscous overstress with increasing plastic strain rate. Plastic yielding occurs as long as a viscous overstress exists. Hence, under constant strain loading, the stresses can relax as long as there is overstress. The stress and strain history of a relaxation test is shown in Fig. 4.13. The overstress obtained after loading with a strain rate of 0.001 1/s is indicated in the figure. In case of creep loading, where a constant stress is applied, an evolution of the plastic strains is only possible if there is still overstress. The overstress reduces due to hardening so that primary creep is found. Figure 4.14 shows a calculated primary creep curve, where the load was applied with 40 MPa/s.

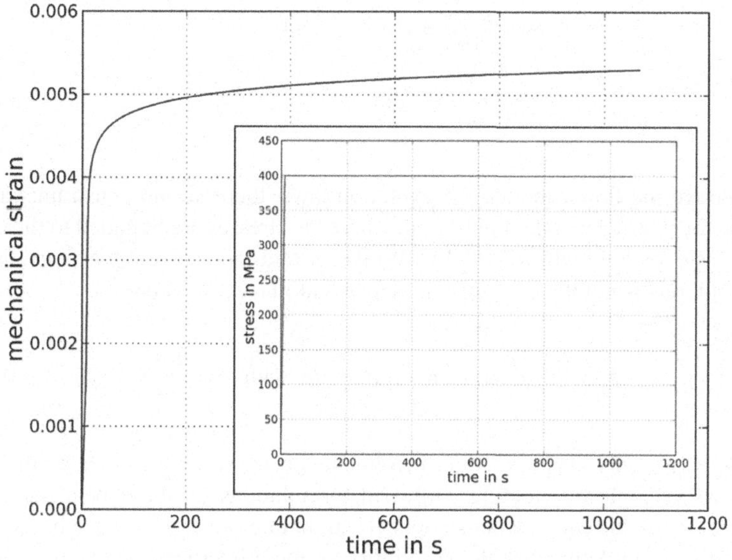

Fig. 4.14 Prescribed stress history and resulting strain history in a creep test calculated with rate-dependent isotropic and kinematic hardening material with the following material properties: $E = 200000$ MPa, $R_e = 200$ MPa, $H = 0$ MPa, $Q_\infty = 100$ MPa and $b = 50$, $C_1 = 5000$ MPa, $\gamma_1 = 0$, $C_2 = 60000$ MPa, $\gamma_2 = 300$, $K = 300$ MPa and $n = 8$

4.4.1.2 Mathematical Model

In the following, the constitutive equations for a rate-dependent plasticity model with isotropic and kinematic hardening according to Sect. 4.4 are derived. The additional damper in the rheological model does not contribute to energy storage in the material since it is a purely dissipative element. Hence, the stored Helmholtz free energy is the same as used for the rate-independent plasticity model given in Eq. (4.65) for combined linear and exponential isotropic hardening and kinematic hardening with $1, .., n_\alpha$ backstresses, namely:

$$\psi = \underbrace{\frac{1}{2}E\varepsilon^{e2}}_{\psi^e} + \underbrace{\frac{1}{2}Hp^2 + Q_\infty\left(p + \frac{1}{b}e^{-bp} - \frac{1}{b}\right) + \sum_{i=1}^{n_\alpha}\frac{1}{2}C_i a_i^2}_{\psi^p}. \tag{4.109}$$

This gives the definition of the stress σ and the thermodynamic forces related to isotropic hardening, R, and related to kinematic hardening, the backstresses α_i:

$$\sigma = \frac{\partial\psi}{\partial\varepsilon^e} = E\varepsilon^e = E\left(\varepsilon - \varepsilon^{vp}\right) \tag{4.110}$$

$$R = \frac{\partial \psi}{\partial p} = Hp + Q_\infty \left(1 - e^{-bp}\right) \tag{4.111}$$

$$\alpha_i = \frac{\partial \psi}{\partial a_i} = C_i a_i . \tag{4.112}$$

To derive the flow rule and the evolution laws, there are no constraints as it is the case for rate-independent plasticity, where the stresses are bounded to the elastic domain as expressed with Eq. (4.14): While for rate-independent plasticity, $\phi_y \leq 0$ must hold, this is not the case for rate-dependent plasticity. There,

$$\phi_y = |\sigma - \alpha| - (R_e + R) \leq \sigma_v \text{ with } \alpha = \sum_{i=1}^{n_\alpha} \alpha_i . \tag{4.113}$$

With $\phi_y < \sigma_v$, only elastic loading increments occur, while for $\phi_y = \sigma_v$ the material shows viscoplastic yielding. Different formulations for the dependency of the overstress σ_v on the equivalent viscoplastic strain rate $\dot{p}$, i.e. $\hat{\sigma_v}\left(\dot{p}\right)$ can be assumed. Frequently used relations for the description of the properties of the damper during yielding are the Chaboche model (Chaboche 1989, 2008),

$$\sigma_v = K \dot{p}^{\frac{1}{n}} \tag{4.114}$$

or the Perzyna model (Perzyna 1966),

$$\sigma_v = (R_0 + R) \left(\frac{\dot{p}}{D}\right)^{\frac{1}{n}} . \tag{4.115}$$

Both models are available as standard models in the finite-element programs ABAQUS and ANSYS. In the Perzyna model, the overstress depends on the strength induced by isotropic hardening, i.e. $(R_0 + R)$, in addition to the viscoplastic strain rate. The exponent n is a material property controlling the rate sensitivity. K and D are also material properties, respectively. While D has unit time, K is usually given in unit strength, e.g. MPa. With this interpretation, Eq. (4.114) is not consistent in units. One should theoretically add the unit time to compensate time from the strain rate. $K = 0$ gives the rate-independent limit case, where no viscous overstresses are allowed. In the rheological model shown in Fig. 4.12, the damper is assumed behave according to the Chaboche model with the material properties K and n.

By Eqs. (4.114) and (4.115), the equivalent viscoplastic strain rate is directly defined, while for rate-independent plasticity the consistency condition $\dot{\phi}_y = 0$ is required to determine $\dot{p}$ via the multiplier $\dot{\lambda}$, e.g. Eq. (4.75). Equations (4.114) and (4.115) can be directly solved for the equivalent viscoplastic strain rate and the Macaulay (Föppl) brackets

$$\langle x\rangle = \begin{cases} 0, \ x \le 0 \\ x, \ x > 0 \end{cases} \tag{4.116}$$

can be introduced to describe the non-smooth transition from elastic to viscoplastic loading. Using $\phi_y = \sigma_v$ during plastic yielding, one obtains for the Chaboche model

$$\dot{p} = \left\langle \frac{\phi_y}{K} \right\rangle^n = \left\langle \frac{|\sigma - \alpha| - (R_e + R)}{K} \right\rangle^n \tag{4.117}$$

and for the Perzyna model

$$\dot{p} = D\left\langle \frac{\phi_y}{R_0 + R} \right\rangle^n = D\left\langle \frac{|\sigma - \alpha|}{R_0 + R} - 1 \right\rangle^n . \tag{4.118}$$

Hence, a consistency condition is not required to determine the equivalent viscoplastic strain in case of rate-dependent plasticity.

For the principle of maximum dissipation to hold, the flow potential Ω, Eq. (4.10), must be a convex function of the stress and the thermodynamic forces. For the Chaboche model, Eq. (4.117), the flow potential is

$$\Omega = \frac{K}{n+1} \left\langle \frac{\phi}{K} \right\rangle^{n+1} , \tag{4.119}$$

where ϕ is defined as for the rate-independent case, Eq. (4.66), namely

$$\phi = \phi_y + \sum_{i=1}^{n_\alpha} \left(\frac{1}{2} \frac{\gamma_i}{C_i} \left(\alpha_i^2 - C_i^2 a_i^2 \right) \right) . \tag{4.120}$$

The second term on the right hand side vanishes with the result of Eq. (4.112). Hence, $\phi_y = \phi$. It is formally introduced in that way to obtain the evolution equations from the flow potential. The following flow rule and evolution equations are obtained:

$$\dot{\varepsilon}^{vp} = \frac{\partial \Omega}{\partial \sigma} = \underbrace{\frac{\partial \Omega}{\partial \phi} \frac{\partial \phi}{\partial \phi_y}}_{\dot{p} = \left\langle \frac{\phi_y}{K} \right\rangle^n} \underbrace{\frac{\partial \phi_y}{\partial \sigma}}_{N = \frac{\sigma - \alpha}{|\sigma - \alpha|}} = \dot{p} N \tag{4.121}$$

$$\dot{p} = -\frac{\partial \Omega}{\partial R} = -\left\langle \frac{\phi_y}{K} \right\rangle^n \underbrace{\frac{\partial \phi_y}{\partial R}}_{-1} = \left\langle \frac{\phi_y}{K} \right\rangle^n \tag{4.122}$$

$$\dot{a}_i = -\frac{\partial \Omega}{\partial \alpha_i} = -\left\langle \frac{\phi_y}{K} \right\rangle^n \underbrace{\frac{\partial \phi_y}{\partial \alpha_i}}_{-N} + \left\langle \frac{\phi_y}{K} \right\rangle^n \frac{\gamma_i}{C_i} \alpha_i = \dot{\varepsilon}^{vp} - \frac{\gamma_i}{C_i} \dot{p} \alpha_i . \tag{4.123}$$

N is the flow direction that takes the value 1 or -1 and determines the sign of the viscoplastic strain rate. Usually, the kinematic hardening law is expressed in terms of the rate of the backstresses. Using Eqs. (4.112) and (4.123),

$$\dot{\alpha}_i = C_i \dot{a}_i = C_i \dot{\varepsilon}^{vp} - \gamma_i \dot{p} \alpha_i . \tag{4.124}$$

With the law of elasticity (4.110), the isotropic hardening law (4.111), the flow rule (4.121), the equivalent viscoplastic strain rate (4.122) and the evolution equation for the backstresses (4.124), a coupled system of differential equations is obtained that describes rate-dependent plasticity with isotropic and kinematic hardening.

4.4.1.3 Thermodynamic Consistency

With the internal variables a_i and p and their conjugate forces α_i and R, the local dissipation is given by

$$\begin{aligned} \mathcal{D}_{loc} &= \sigma \dot{\varepsilon}^{vp} - \sum_{i=1}^{n_\alpha} \alpha_i \dot{a}_i - R\dot{p} \\ &= \sigma \dot{p} N - \sum_{i=1}^{n_\alpha} \alpha_i \left(\dot{p} N - \frac{\gamma_i}{C_i} \dot{p} \alpha_i \right) - R\dot{p} \\ &= \dot{p} \underbrace{(\sigma - \alpha)\, N}_{\frac{(\sigma-\alpha)^2}{|\sigma-\alpha|} = |\sigma - \alpha|} + \dot{p} \sum_{i=1}^{n_\alpha} \frac{\gamma_i}{C_i} \alpha_i^2 - R\dot{p} \\ &= \dot{p} \left(|\sigma - \alpha| - R \right) + \dot{p} \sum_{i=1}^{n_\alpha} \frac{\gamma_i}{C_i} \alpha_i^2 \geq 0. \end{aligned} \tag{4.125}$$

The strong restriction is imposed that both terms should be ≥ 0. Since

$$\dot{p} = \left\langle \frac{\phi_y}{K} \right\rangle^n \geq 0 \tag{4.126}$$

where $K > 0$, the first term with $\phi_y = \sigma_v$ during plastic yielding results in

$$\begin{aligned} |\sigma - \alpha| - R &= R_e + \sigma_v \geq 0 \\ |\sigma - \alpha| &= R_e + R + \sigma_v \geq 0. \end{aligned} \tag{4.127}$$

This equation has to be fulfilled for arbitrary strain rates, i.e. also very low strain rates with $\dot{p} \to 0$ and, thus, $\sigma_v \to 0$. Hence, the material properties must be chosen such that $R_e + R \geq 0$. $R_e \geq 0$, since $R = 0$ initially when $p = 0$. Moreover, $R_e \geq -R$

must hold. This is only strictly the case in the combined linear and exponential hardening if $H \geq 0$ and $R_e \geq -Q_\infty$. From the second term, one finds $\gamma_i/C_i \geq 0$.

4.4.1.4 Numerical Solution with Predictor/Corrector Algorithm

An efficient way to solve the coupled differential equations for the rate-dependent plasticity model are predictor-corrector algorithms. The integration of the flow rule and the evolution equations and the calculation of the yield function is the same as for the rate-independent model, see Sect. 4.3.3, Eqs. (4.80) to (4.92).

The elastic predictor step assumes that only elastic loading is occurring in the time increment Δt. Hence, the yield function at the new time t_{n+1}, $\phi_{y,n+1}$ in Eq. (4.92), is computed assuming $\Delta\lambda = 0$:

$$\phi_{y,\star} = |\sigma_\star - \alpha_\star| - \left(R_0 + Hp_n + Q_\infty\left(1 - e^{-bp_n}\right)\right). \tag{4.128}$$

With the trial yield function $\phi_{y,\star}$ it is checked whether the assumption of an elastic loading step is true, or if a viscoplastic loading step is present:

$$\phi_{y,\star} = \begin{cases} \leq 0, \ \Delta\lambda = 0 \text{ (elastic loading step)} \\ > 0, \ \Delta\lambda > 0 \text{ (elastic-viscoplastic loading step)} \end{cases}. \tag{4.129}$$

For the viscoplastic material, the stresses can exceed the yield stress so that the consistency condition does not apply as it is the case for the rate-independent material. To determine $\Delta\lambda$, the Chaboche model from Eq. (4.117) is used in Eq. (4.82):

$$p_{n+1} = p_n + \Delta t\, \dot{p}_{n+1} = p_n + \underbrace{\Delta t \left\langle \frac{\phi_{y,n+1}}{K} \right\rangle^n}_{\Delta\lambda}. \tag{4.130}$$

Hence, from the last term, the following residual function can be defined for $\Delta\lambda$:

$$f_R = -\Delta\lambda + \Delta t \left\langle \frac{\phi_{y,n+1}}{K} \right\rangle^n = 0. \tag{4.131}$$

Compared to the rate-independent plasticity model, the time increment Δt now enters the integration algorithm explicitly.

f_R is a non-linear equation of $\Delta\lambda$ due to the non-linear overstress law and due to non-linear hardening. It can be solved using Newton's method according to the iteration scheme

$$\Delta\lambda_{j+1} = \Delta\lambda_j - \left(\left.\frac{\partial f_R}{\partial \Delta\lambda}\right|_{\Delta\lambda_j} \right)^{-1} f_R|_{\Delta\lambda_j} \tag{4.132}$$

$$\text{where } \frac{\partial f_R}{\partial \Delta\lambda} = 1 - \Delta t \frac{n}{K} \left(\frac{\phi_{y,n+1}}{K} \right)^{n-1} \frac{\partial \phi_{y,n+1}}{\partial \Delta\lambda}. \tag{4.133}$$

The derivative $\partial\phi_{y,n+1}/\partial\Delta\lambda$ is already given in Eq. (4.96). Convergence in iteration $j+1$ is obtained if the absolute value of $f_R|_{\Delta\lambda_{j+1}}$ gets smaller than a defined tolerance. The updates of the viscoplastic strain, the stress, the accumulated viscoplastic strain and the backstresses are obtained as in case of the rate-independent models by insertion of the determined multiplier $\Delta\lambda$ in Eqs. (4.80) to (4.83).

4.4.1.5 Analytical Solution for Determination of Material Properties

For the determination of the material properties from data of uniaxial material tests, analytical solutions can be derived to compute the stresses from the measured plastic strains. To this end,

$$\phi_y = |\sigma - \alpha| - (R_e + R) = \sigma_v \tag{4.134}$$

is used from Eq. (4.113) that holds for viscoplastic yielding. Rearranging to

$$\underbrace{\frac{\sigma - \alpha}{|\sigma - \alpha|} |\sigma - \alpha|}_{\sigma - \alpha} = \underbrace{\frac{\sigma - \alpha}{|\sigma - \alpha|}}_{N} (R_e + R + \sigma_v), \tag{4.135}$$

where both sides are multiplied by $(\sigma - \alpha) / |\sigma - \alpha|$, and solving for the stress yields:

$$\begin{aligned} \sigma = & \underbrace{N \left(R_e + Hp + Q_\infty \left(1 - e^{-bp}\right)\right)}_{\text{stress in friction element}} \\ & + \underbrace{\sum_{i=1}^{n_\alpha} \left(\alpha_i^0 e^{-\gamma_i (p - p^0)} + \frac{C_i}{\gamma_i} N \left(1 - e^{-\gamma_i (p - p^0)}\right)\right)}_{\text{stress in springs}} \\ & + \underbrace{N K \dot{p}^{\frac{1}{n}}}_{\text{stress in damper}} . \end{aligned} \tag{4.136}$$

Therein, the combined linear and exponential isotropic hardening law and the Armstrong-Frederick kinematic hardening law with multiple backstresses are used. Moreover, the Chaboche model for the viscous overstress is applied. This result shows the additive decomposition of the stress in the parallel connection of the rheological elements in the viscoplastic part of the rheological model in Fig. 4.12.

For the determination of the material properties of the rate-dependent cyclic plasticity model, a cyclic loading history with plastic load reversals is required so that isotropic and kinematic hardening can be identified uniquely. Moreover, the loading history must activate rate-effect by using different strain rates and/or hold times. In Fig. 4.15, a strain-controlled loading history is shown that includes both, different strain rates (0.001, 0.0001 and 0.0001 1/s) as well as hold times with a duration of 1800 s in tension as well as compression. Two different strain amplitudes are applied. The resulting stress response measured in a uniaxial test with the nickel-base superalloy MAR-M247 at 900°C is shown in Fig. 4.15. Higher stresses are observed for the higher strain rates as well as stress relaxation during the hold times in strain. After determination of the material properties of the model (indicated as "model without static recovery" in Fig. 4.15), a good description of the data is obtained. The model describes hardening under cyclic loading as well as rate-dependency very well. Stress relaxation is, however, underestimated. A better description of stress relaxation is obtained by including static recovery in the model (indicated as "model with static recovery" in Fig. 4.15). Static recovery is introduced in Sect. 4.4.3.

4.4.2 *Rate-dependent Yielding: Non-unified Models*

A non-unified model is obtained when in the finite-element programs ABAQUS and ANSYS a rate-independent plasticity model computing the plastic strains and a creep model computing the viscous strains are combined. Hence, in contrast to the unified models, where only one viscoplastic strain exists, the inelastic strains are divided into separate plastic and viscous strains for non-unified models.

4.4.2.1 Rheological Model

Instead of introducing the damper in parallel to the kinematic hardening springs and the isotropic hardening friction element (as it is the case for the unified model shown in Fig. 4.12), the damper is introduced in series to the elastic-plastic model. This is shown in Fig. 4.16. Hence, the mechanical strain in the non-unified model is decomposed into elastic, (rate-independent) plastic and (rate-dependent) viscous strains:

$$\varepsilon = \varepsilon^e + \varepsilon^p + \varepsilon^v. \tag{4.137}$$

The non-unified model is able to describe rate-dependency, i.e. increasing stresses with increasing strain rate, as well as creep and stress relaxation. However, hardening and creeping are decoupled in these models. The evolution of the viscous strains in the damper is depending on the applied stress only. For non-isothermal conditions, where the damper is active at high-temperatures while plasticity is dominant under lower temperatures, isotropic and kinematic hardening might not adequately describe the behavior since it only acts in the plastic part.

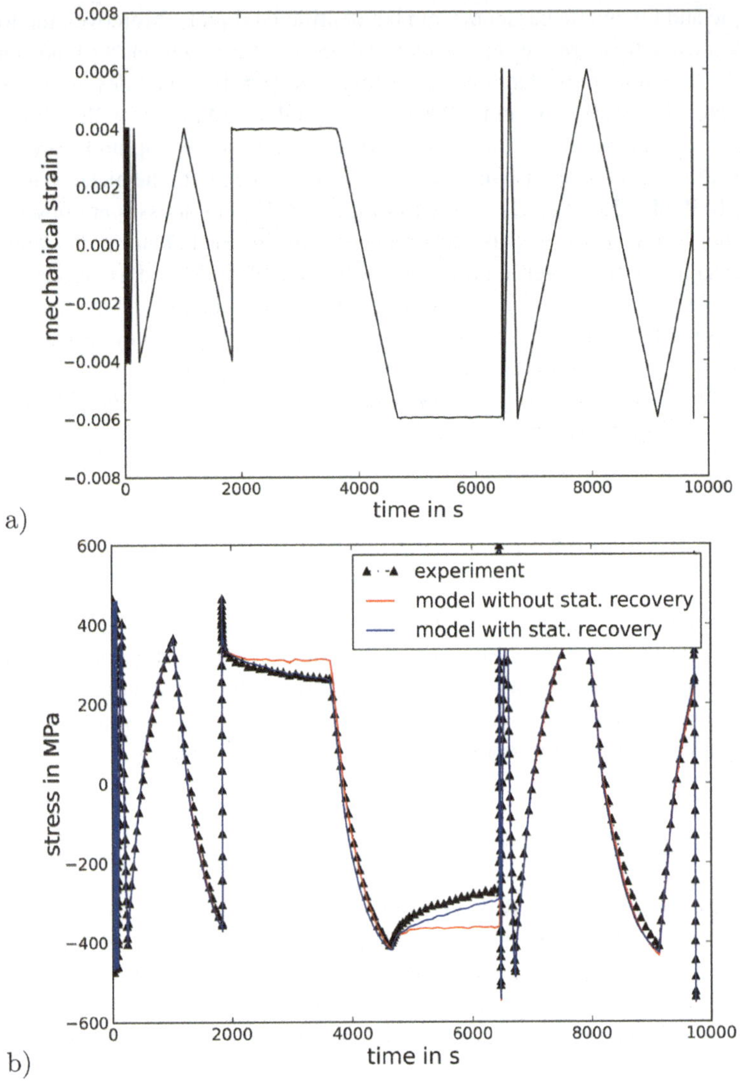

Fig. 4.15 **a** Strain history with cyclic loading, different strain rates and hold times; **b** Measured stress response of MAR-M247 at 900 °C and results obtained with rate-dependent cyclic plasticity models

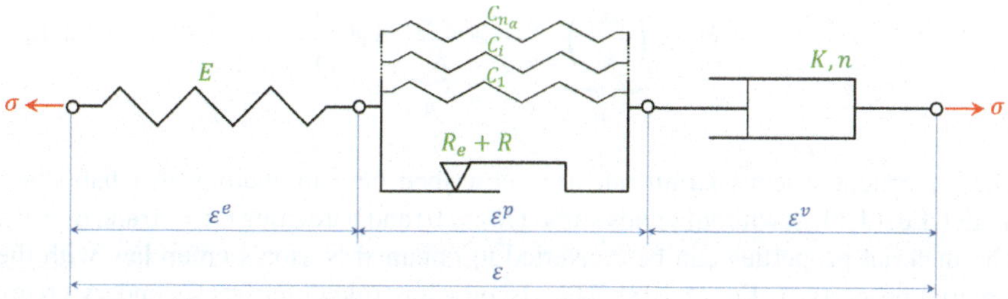

Fig. 4.16 Rheological model of the non-unified rate-dependent isotropic and kinematic hardening elastic-plastic material

4.4.2.2 Mathematical Model

The constitutive equations are summarized in the following without addressing the thermodynamic framework. The law of elasticity is

$$\sigma = E\varepsilon^e = E\left(\varepsilon - \varepsilon^p - \varepsilon^v\right). \tag{4.138}$$

The yield function of the plastic part including isotropic and kinematic hardening reads

$$\phi_y = |\sigma - \alpha| - (R_e + R) \leq 0$$
$$\text{with } \alpha = \sum_{i=1}^{n_\alpha} \alpha_i \text{ and } R = Hp + Q_\infty\left(1 - e^{-bp}\right). \tag{4.139}$$

The flow rule and the evolution equations are given by

$$\dot{\varepsilon}^p = \dot{p}N \text{ where } N = \frac{\sigma - \alpha}{|\sigma - \alpha|} \tag{4.140}$$

$$\dot{\alpha}_i = C_i\dot{\varepsilon}^p - \gamma_i \dot{p}\alpha_i. \tag{4.141}$$

The equivalent plastic strain rate $\dot{p}$ is determined by the consistency condition $\dot{\phi}_y = 0$:

$$\dot{p} = \frac{NE\dot{\varepsilon}}{E + \sum_{i=1}^{n_\alpha}(C_i - N\gamma_i\alpha_i) + \dfrac{\partial R}{\partial p}}. \tag{4.142}$$

Viscous flow in the viscous part of the rheological model is governed by the flow rule

$$\dot{\varepsilon}^{v} = \underbrace{\left(\frac{|\sigma|}{K}\right)^{n}}_{\dot{v}} \frac{\sigma}{|\sigma|} = \underbrace{\frac{1}{K^{n}}}_{A} |\sigma|^{n} \frac{\sigma}{|\sigma|} \tag{4.143}$$

The equivalent viscous strain rate $\dot{v}$ is described here in analogy to Chaboche's model, Eq. (4.117), without yields stress ($R_0 = 0$) and hardening ($R = 0$ and $\alpha = 0$). The material properties can be converted to obtained Norton's creep law with the material property A, Eq. (4.108). The viscous flow rule describes secondary creep, i.e. a constant viscous strain rate for constant stress. Finite-element programs as ABAQUS or ANSYS offer further creep laws that allow for the description of e.g. strain hardening so that primary and secondary creep can be considered in the viscous part of the non-unified rate-dependent model.

4.4.2.3 Numerical Solution with Predictor/Corrector Algorithm

For the numerical solution, the flow rule for the plastic strain, the evolution equations for the non-linear hardening law and the flow rule of the viscous strain have to be solved as coupled system of equations. This is obtained by coupling the integration algorithms presented for the time-independent and the time-dependent model in Sects. 4.3.3 and 4.4.1. If the viscous strain rate is significantly smaller than the plastic strain rate, as typical for creep applications, the equations can be solved decoupled and sequential, assuming that the viscous strain increment does not change the stress significantly.

4.4.2.4 Analytical Solution for Estimation of Material Properties

There is no analytical solution for the coupled system of differential equations that can be applied with uniaxial experimental data to determine the material properties. A separation of the inelastic strain that can be computed from $\varepsilon - \sigma/E$ into plastic and viscous part is not possible from the data without making assumptions. Typically, it is assumed that, e.g. in tensile or cyclic tests with relatively high strain rate, viscous strains can be neglected so that only the solution of the plastic part is relevant and the plastic material properties are determined from that data. Then, for creep tests, it is assumed that only viscous strains are relevant so that the viscous material properties are determined from creep test data alone. It is recommended to check whether the assumption is valid by applying the coupled model to the data.

4.4.3 *Static Recovery of Hardening*

In the previous section, rate-dependent yielding is considered that can result in isotropic and kinematic hardening. At higher temperatures, however, hardening can recover due to thermally activated processes in the material: Assume that a material specimen was hardened due to plastic deformation at room temperature in a tensile test and then heat treated at higher temperature for a certain time. If the specimen is then tested again after the heat treatment, one will find an initial lower hardness than was attained after the first tensile test.

In this section, static recovery of kinematic hardening is considered. Static means that recovery takes place just as a function of time, while dynamic recovery, as described by the Armstrong-Frederick kinematic hardening law (4.42), requires plastic deformations to occur.

4.4.3.1 Rheological Model

In the rheological model, one can assume that the backstresses exerted by the springs in the plastic part of the model can be relaxed due to dampers that are in series with the springs (Fig. 4.17).

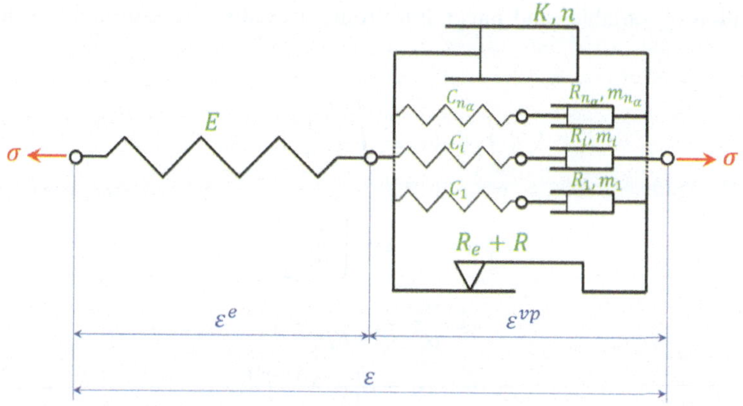

Fig. 4.17 Rheological model of the unified rate-dependent isotropic and kinematic hardening elastic-plastic material with static recovery of kinematic hardening

4.4.3.2 Mathematical Model

Static recovery can be considered in the Armstrong-Frederick kinematic hardening model by adding a static recovery term:

$$\dot{\alpha}_i = \underbrace{C_i \dot{\varepsilon}^{vp}}_{\text{hardening}} - \underbrace{\gamma_i \dot{p} \alpha_i}_{\text{dynamic recovery}} - \underbrace{\left(\frac{|\alpha_i|}{R_i}\right)^{m_i} \frac{\alpha_i}{|\alpha_i|}}_{\text{static recovery}} . \tag{4.144}$$

C_i and γ_i are material properties related to hardening requiring plastic yielding. R_i and m_i are material properties related to static recovery. The static recovery term is not consistent in units as is the case for the viscoplastic Chaboche model in Eq. (4.117). One should theoretically add the unit of reciprocal time. Static recovery of kinematic hardening is available in the finite-element programs ABAQUS and ANSYS. Different representations of the static recovery term are often used, but in most cases the material properties can be converted into one another.

With $m_i = 1$, i.e.

$$\dot{\alpha}_i = C\dot{\varepsilon}^{vp} - \gamma_i \dot{p} \alpha_i - \frac{1}{R_i} \alpha_i , \tag{4.145}$$

an exponential decay of the backstress with time is described: Assuming that there is no plastic yielding, i.e. $\dot{\varepsilon}^{vp} = \dot{p} = 0$, one obtains

$$\dot{\alpha}_i = \frac{\mathrm{d}\alpha_i}{\mathrm{d}t} = -\frac{1}{R_i} \alpha_i . \tag{4.146}$$

Separation of variables and integration from an existing backstress α_i^0 at time t^0 gives:

$$\begin{aligned} \int_{\alpha_i^0}^{\alpha_i} \frac{1}{\bar{\alpha}_i} \mathrm{d}\bar{\alpha}_i &= \int_{t^0}^{t} -\frac{1}{R_i} \mathrm{d}\bar{t} \\ [\ln \bar{\alpha}_i]_{\alpha_i^0}^{\alpha_i} &= -\left[\frac{1}{R_i}\bar{t}\right]_{t^0}^{t} \\ \ln \frac{\alpha_i}{\alpha_i^0} &= -\frac{1}{R_i}\left(t - t^0\right) \\ \alpha_i &= \alpha_i^0 e^{-\frac{1}{R_i}\left(t-t^0\right)} . \end{aligned} \tag{4.147}$$

Hence, the backstress can recover completely as $t \rightarrow \infty$ and R_i controls the recovery rate.

Static recovery can be integrated in the thermodynamic framework by defining the corresponding potential functions: the Helmholtz free energy ψ and the flow potential Ω. The stored Helmholtz free energy does not change if static recovery is included. For combined linear and exponential isotropic hardening and kinematic hardening with $1, .., n_\alpha$ backstresses, the Helmholtz free energy is:

$$\psi = \underbrace{\frac{1}{2} E \varepsilon^{e2}}_{\psi^e} + \underbrace{\frac{1}{2} H p^2 + Q_\infty \left(p + \frac{1}{b} e^{-bp} - \frac{1}{b} \right) + \sum_{i=1}^{n_\alpha} \frac{1}{2} C_i a_i^2}_{\psi^p} . \tag{4.148}$$

This gives the definition of the stress σ and the thermodynamic forces related to isotropic hardening, R, and related to kinematic hardening, the backstresses α_i:

$$\sigma = \frac{\partial \psi}{\partial \varepsilon^e} = E \varepsilon^e = E \left(\varepsilon - \varepsilon^{vp} \right) \tag{4.149}$$

$$R = \frac{\partial \psi}{\partial p} = H p + Q_\infty \left(1 - e^{-bp} \right) \tag{4.150}$$

$$\alpha_i = \frac{\partial \psi}{\partial a_i} = C_i a_i . \tag{4.151}$$

For time-dependent plasticity with static recovery of kinematic hardening, a recovery term is added in the flow potential compared to the potential without static recovery given in Eq. (4.120):

$$\Omega = \frac{K}{n+1} \left\langle \frac{\phi}{K} \right\rangle^{n+1} + \sum_{i=1}^{n_\alpha} \frac{1}{C_i} \frac{R_i}{m_i + 1} \left(\frac{|\alpha_i|}{R_i} \right)^{m_i + 1} . \tag{4.152}$$

The function ϕ is not changed by static recovery compared to Eq. (4.120), namely

$$\phi = \phi_y + \sum_{i=1}^{n_\alpha} \left(\frac{1}{2} \frac{\gamma_i}{C_i} \left(\alpha_i^2 - C_i^2 a_i^2 \right) \right) . \tag{4.153}$$

Hence, the following flow rule and evolution equations are obtained:

$$\dot{\varepsilon}^{vp} = \frac{\partial \Omega}{\partial \sigma} = \underbrace{\frac{\partial \Omega}{\partial \phi} \frac{\partial \phi}{\partial \phi_y}}_{\dot{p} = \left\langle \frac{\phi_y}{K} \right\rangle^n} \underbrace{\frac{\partial \phi_y}{\partial \sigma}}_{N = \frac{\sigma - \alpha}{|\sigma - \alpha|}} = \dot{p} N \tag{4.154}$$

$$\dot{p} = -\frac{\partial \Omega}{\partial R} = -\left\langle \frac{\phi_y}{K} \right\rangle^n \underbrace{\frac{\partial \phi_y}{\partial R}}_{-1} = \left\langle \frac{\phi_y}{K} \right\rangle^n \tag{4.155}$$

$$\dot{a}_i = -\frac{\partial \Omega}{\partial \alpha_i} = -\left\langle \frac{\phi_y}{K} \right\rangle^n \underbrace{\frac{\partial \phi_y}{\partial \alpha_i}}_{-N} + \left\langle \frac{\phi_y}{K} \right\rangle^n \frac{\gamma_i}{C_i} \alpha_i - \frac{1}{C_i} \left(\frac{|\alpha_i|}{R_i} \right)^{m_i} \underbrace{\frac{\partial |\alpha_i|}{\partial \alpha_i}}_{\frac{\alpha_i}{|\alpha_i|}}$$

$$= \dot{\varepsilon}^{vp} - \frac{\gamma_i}{C_i}\dot{p}\alpha_i - \frac{1}{C_i}\left(\frac{|\alpha_i|}{R_i}\right)^{m_i}\frac{\alpha_i}{|\alpha_i|}. \tag{4.156}$$

N is the flow direction that takes the value 1 or -1 and determines the sign of the viscoplastic strain rate. Usually, the kinematic hardening law is expressed in terms of the rate of the backstresses. Using Eqs. (4.151) and (4.156),

$$\dot{\alpha}_i = C_i\dot{a}_i = C_i\dot{\varepsilon}^{vp} - \gamma_i\dot{p}\alpha_i - \left(\frac{|\alpha_i|}{R_i}\right)^{m_i}\frac{\alpha_i}{|\alpha_i|} \tag{4.157}$$

is obtained which is already introduced in Eq. (4.144).

With the law of elasticity (4.149), the isotropic hardening law (4.150), the flow rule (4.154), the equivalent viscoplastic strain rate (4.155) and the evolution equation for the backstresses (4.157), a coupled system of differential equations is obtained that describes the rate-dependent plasticity with isotropic and kinematic hardening.

4.4.3.3 Thermodynamic Consistency

With the internal variables a_i and p and their conjugate forces α_i and R, the local dissipation is given by

$$\begin{aligned}
\mathcal{D}_{loc} &= \sigma\dot{\varepsilon}^{vp} - \sum_{i=1}^{n_\alpha}\alpha_i\dot{a}_i - R\dot{p} \\
&= \sigma\dot{p}N - \sum_{i=1}^{n_\alpha}\alpha_i\left(\dot{p}N - \frac{\gamma_i}{C_i}\dot{p}\alpha_i - \frac{1}{C_i}\left(\frac{|\alpha_i|}{R_i}\right)^{m_i}\frac{\alpha_i}{|\alpha_i|}\right) - R\dot{p} \\
&= \dot{p}\underbrace{(\sigma-\alpha)\,N}_{\frac{(\sigma-\alpha)^2}{|\sigma-\alpha|} = |\sigma-\alpha|} + \dot{p}\sum_{i=1}^{n_\alpha}\left(\frac{\gamma_i}{C_i}\alpha_i^2 + \sum_{i=1}^{n_\alpha}\frac{1}{C_i}\left(\frac{|\alpha_i|}{R_i}\right)^{m_i}\frac{\alpha_i^2}{|\alpha_i|}\right) - R\dot{p} \\
&= \dot{p}\left(|\sigma-\alpha| - R\right) + \dot{p}\sum_{i=1}^{n_\alpha}\frac{\gamma_i}{C_i}\alpha_i^2 + \left(\sum_{i=1}^{n_\alpha}\frac{1}{C_i}\left(\frac{|\alpha_i|}{R_i}\right)^{m_i}\frac{\alpha_i^2}{|\alpha_i|}\right) \geq 0.
\end{aligned} \tag{4.158}$$

The strong restriction is imposed that all three terms should be ≥ 0. Since

$$\dot{p} = \left\langle\frac{\phi_y}{K}\right\rangle^n \geq 0 \tag{4.159}$$

where $K > 0$, the first term with $\phi_y = \sigma_v$ during plastic yielding results in

$$
\begin{aligned}
|\sigma - \alpha| - R &= R_e + \sigma_v \geq 0 \\
|\sigma - \alpha| &= R_e + R + \sigma_v \geq 0.
\end{aligned}
\tag{4.160}
$$

This equation has to be fulfilled for arbitrary strain rates, i.e. also very low strain rates with $\dot{p} \rightarrow 0$ and, thus, $\sigma_v \rightarrow 0$, Hence, the material properties are must be chosen such that $R_e + R \geq 0$. $R_e \geq 0$, since $R = 0$ initially when $p = 0$. Moreover, $R_e \geq -R$ must hold. This is only strictly the case in the combined linear and exponential hardening if $H \geq 0$ and $R_e \geq -Q_\infty$. From the second term, one finds $\gamma_i / C_i \geq 0$ and from the third term $C_i \geq 0$ and $R_i \geq 0$.

4.4.3.4 Numerical Solution with Predictor/Corrector Algorithm

The algorithm as already used for the time-independent and the time-dependent plasticity models in Sects. 4.3.3 and 4.4.1, respectively, can only be applied if an exponential decay in static recovery is considered by choosing $m_i = 1$, see Eq. (4.145). In that case, the evolution of the backstress is linear in the backstress itself and a closed form solution for the backstresses at a new time t_{n+1} can be found by integration:

$$
\begin{aligned}
\alpha_{i,n+1} &= \alpha_{i,n} + \Delta t \dot{\alpha}_{i,n+1} \\
&= \alpha_{i,n} + C_i \Delta\lambda N_{n+1} - \gamma_i \Delta\lambda \alpha_{i,n+1} - \frac{\Delta t}{R_i} \alpha_{i,n+1} \\
&= \frac{1}{1 + \gamma_i \Delta\lambda + \dfrac{\Delta t}{R_i}} \left(\alpha_{i,n} + C_i \Delta\lambda N_{n+1} \right).
\end{aligned}
\tag{4.161}
$$

Compared to the model without static recovery, see Eq. (4.83), only the denominator of the factor in front of the brackets needs to be changed in the complete integration algorithm. Time-dependency is found in the algorithm since the time increment Δt enters the equations explicitly.

If $m_i \neq 1$ is used in the static recovery law, additional equations need to be defined and solved together with the residual Eq. (4.131), see e.g. Seifert et al. (2007), Seifert and Maier (2008).

4.4.3.5 Analytical Solution for Determination of Material Properties

No analytical solutions for the backstresses are now available if static recovery is included. Hence, for the determination of the material properties from uniaxial experimental data, the coupled system of differential equations must be solved. In Fig. 4.15, results obtained with the time-dependent cyclic plasticity model including static recovery are shown for the nickel-base superalloy MAR-M247. The model is used to describe the behavior under cyclic loading with different strain rates and hold times in strain. Due to recovery of backstresses during the hold time, stress relaxation can be described better with static recovery.

4.5 Thermomechanical Loadings

In Sects. 4.3 and 4.4, plasticity models are presented for isothermal conditions that contain a set of material properties $\boldsymbol{p}$ to adapt the material response of the model to experimentally measured data. If non-isothermal, i.e. thermomechanical, loadings are considered, the material properties are depending on temperature since properties related to strength change due to decreasing strength level with increasing temperature and properties related to rate-dependency change since rate-effects increase with temperature. Usually, the material properties are determined from uniaxial material tests that were done at a certain number of $i = 1, ..., n_\theta$ discrete temperature sample points T_i in the relevant temperature range. Then, the typical approach is to linearly interpolate the material properties in temperature:

$$\boldsymbol{p} = \boldsymbol{p}_i + \frac{\boldsymbol{p}_{i+1} - \boldsymbol{p}_i}{T_{i+1} - T_i} (T - T_i) \tag{4.162}$$

$$= \frac{T_{i+1} - T}{T_{i+1} - T_i} \boldsymbol{p}_i + \frac{T - T_i}{T_{i+1} - T_i} \boldsymbol{p}_{i+1} \text{ if } T \in \left[T_i, T_{i+1}\right]. \tag{4.163}$$

If the current temperature is outside the temperature range where parameters are given, the corresponding value of the highest or lowest temperature should be used, respectively.

4.5.1 Temperature-dependent Material Properties

From the experience of the author of this work, physically reasonable temperature-dependencies should be considered to obtain smooth curves when applying the models to thermomechanical loadings. There should be no "zig-zag dependency" on temperature. Achieving physically reasonable temperature-dependencies becomes more and more difficult the more material properties have to be determined and the more temperature sample points are used for the generation of material data. The reason is that during the determination of the material properties for one temperature, the material properties for other temperatures need to be considered as well. Indeed, constraint optimization methods could be used to account for "physically reasonable constraints". However, the results strongly depend on the chosen constraints. Currently, combined experience-based and optimization-based methods are used so that material properties might not be objective, i.e. different persons find different values for the material properties when using the same model and the same temperature-dependent data.

4.5.2 Temperature Rate Terms

4.5.2.1 Kinematic Hardening Laws with Temperature Rate Terms

If material properties are temperature-dependent, the thermodynamic framework results in so-called temperature rate terms in the kinematic hardening law. If the material property C_i of the Armstrong-Frederick kinematic hardening law is depending on temperature, i.e. $C_i = \hat{C}_i(T)$, then one finds from the relation of the backstresses α_i and their corresponding internal variables a_i from Eq. (4.46), namely

$$\alpha_i = C_i a_i, \tag{4.164}$$

the evolution equation for the backstresses

$$\dot{\alpha}_i = C_i \underbrace{\dot{a}_i}_{\dot{\varepsilon}^p - \frac{\gamma_i}{C_i}\dot{p}\alpha_i} + \frac{\partial C_i}{\partial T}\dot{T} \underbrace{a_i}_{\frac{\alpha_i}{C_i}} = C_i \dot{\varepsilon}^p - \gamma_i \dot{p}\alpha_i + \frac{1}{C_i}\frac{\partial C_i}{\partial T}\dot{T}\alpha_i. \tag{4.165}$$

Only for a constant γ_i, the backstress α_i is bounded at each temperature by the saturation value C_i/γ_i of the exponential hardening curve given by the analytical solution in Eq. (4.58). For temperature-dependent γ_i, this is not the case and a "temperature history effect" is introduced in the model. For a strong temperature-dependency of γ_i, the temperature history effect can result in a very unrealistic behavior as shown in Cailletaud et al. (2015). It is recommended to increase the number of backstresses n_α, if with the current number of backstresses a good description of the material is only possible with strongly temperature-dependent γ_i. Alternative formulations of the temperature rate term were proposed so that no temperature history effect occurs for temperature-dependent C_i and γ_i (Ohno and Wang 1991, 1992). However, these approaches were not established.

The temperature-rate term is easily included in the numerical solution of the plasticity model. Numerical integration of Eq. (4.165) using the backward Euler method yields

$$\begin{aligned} \alpha_{i,n+1} &= \alpha_{i,n} + \Delta t \dot{\alpha}_{i,n+1} \\ &= \alpha_{i,n} + C_i \Delta\lambda N_{n+1} - \gamma_i \Delta\lambda \alpha_{i,n+1} + \underbrace{\frac{1}{C_i}\frac{\partial C_i}{\partial T}\Delta T}_{\Delta C_i} \alpha_{i,n+1} \\ &= \frac{1}{1 + \gamma_i \Delta\lambda - \frac{\Delta C_i}{C_i}} \left(\alpha_{i,n} + C_i \Delta\lambda N_{n+1}\right). \end{aligned} \tag{4.166}$$

where all material properties are evaluated at the temperature T_{n+1} at time t_{n+1}. Compared to the isothermal model, see Eq. (4.83), only the denominator of the factor in front of the brackets needs to be changed in the complete integration algorithm.

If static recovery is considered as well in the evolution equation for the backstresses, Eq. (4.157), then the evolution equations becomes with temperature-dependent C_i

$$\dot{\alpha}_i = C_i \dot{\varepsilon}^p - \gamma_i \dot{p} \alpha_i - \left(\frac{|\alpha_i|}{R_i} \right)^{m_i} \frac{\alpha_i}{|\alpha_i|} + \frac{1}{C_i} \frac{\partial C_i}{\partial T} \dot{T} \alpha_i . \tag{4.167}$$

4.5.2.2 Consistency Condition with Temperature Rate Terms

Temperature dependent material properties also need to be considered when evaluating the consistency condition for rate-independent materials, required to determine the multiplier $\dot{\lambda}$ (Burlet and Cailletaud 1986). In case of an isotropic and kinematic hardening material, compare Eq. (4.75), one obtains:

$$\dot{\phi}_y = \underbrace{\frac{\partial \phi_y}{\partial \sigma}}_{N} \dot{\sigma} + \sum_{i=1}^{n_\alpha} \underbrace{\frac{\partial \phi_y}{\partial \alpha_i}}_{-N} \dot{\alpha}_i + \underbrace{\frac{\partial \phi_y}{\partial R_e}}_{-1} \frac{\partial R_e}{\partial T} \dot{T} + \underbrace{\frac{\partial \phi_y}{\partial R}}_{-1} \frac{\partial R}{\partial p} \dot{p} + \underbrace{\frac{\partial \phi_y}{\partial R}}_{-1} \frac{\partial R}{\partial T} \dot{T} = 0. \tag{4.168}$$

The rate form of the law of elasticity with temperature-dependent Young's modulus E is

$$\dot{\sigma} = E \left(\dot{\varepsilon} - \dot{\varepsilon}^p \right) + \frac{\partial E}{\partial T} \dot{T} \underbrace{\left(\varepsilon - \varepsilon^p \right)}_{\frac{\sigma}{E}} . \tag{4.169}$$

Using this result and the evolution equation for the backstress from Eq. (4.165) in Eq. (4.168) above, yields

$$\begin{aligned} \dot{\phi}_y &= N E \left(\dot{\varepsilon} - \dot{\lambda} N \right) + N \frac{1}{E} \frac{\partial E}{\partial T} \dot{T} \sigma \\ &\quad - \sum_{i=1}^{n_\alpha} N \left(C_i \dot{\lambda} N - \gamma_i \dot{\lambda} \alpha_i + \frac{1}{C_i} \frac{\partial C_i}{\partial T} \dot{T} \alpha_i \right) \\ &\quad - \frac{\partial R_e}{\partial T} \dot{T} - \frac{\partial R}{\partial p} \dot{\lambda} - \frac{\partial R}{\partial T} \dot{T} \\ &= N E \dot{\varepsilon} - N^2 E \dot{\lambda} - \dot{\lambda} \sum_{i=1}^{n_\alpha} \left(N^2 C_i - N \gamma_i \alpha_i \right) - \frac{\partial R}{\partial p} \dot{\lambda} \end{aligned}$$

$$+\left(N\frac{1}{E}\frac{\partial E}{\partial T}\sigma - \sum_{i=1}^{n_\alpha} N\frac{1}{C_i}\frac{\partial C_i}{\partial T}\alpha_i - \frac{\partial R_e}{\partial T} - \frac{\partial R}{\partial T}\right)\dot{T} = 0. \tag{4.170}$$

Hence, $\dot{\lambda}$ is given by

$$\dot{\lambda} = \frac{NE\dot{\varepsilon} + \left(N\frac{1}{E}\frac{\partial E}{\partial T}\sigma - \sum_{i=1}^{n_\alpha} N\frac{1}{C_i}\frac{\partial C_i}{\partial T}\alpha_i - \frac{\partial R_e}{\partial T} - \frac{\partial R}{\partial T}\right)\dot{T}}{E + \sum_{i=1}^{n_\alpha}(C_i - N\gamma_i\alpha_i) + \frac{\partial R}{\partial p}}. \tag{4.171}$$

4.6 Conclusion

In this work, time-independent and time-dependent plasticity models are presented that are able to describe hardening under cyclic loadings as well as high-temperature phenomena like rate-dependency, creep, stress relaxation and recovery of hardening. The presented models are available in finite-element programs as ABAQUS and ANSYS so that material properties of the considered materials need to be determined from experimental data to apply the models in finite-element calculations. It is shown in examples in Sect. 4.3.3 that material properties can be non-objective, i.e. different persons determine different material properties of a plasticity model based on the same experimental data. Hence, a good material database that includes material tests with loading histories that activate the relevant phenomena of the plasticity model is important for the determination of the material properties. Such a loading history is e.g. shown in Sect. 4.4.1 with cyclic loading at different strain rates and with hold times in strain.

There are more advanced models, which are capable of describing ratchetting of the material if mean stresses are present and mean stress relaxation in presence of mean strains appropriately. Corresponding kinematic hardening laws were derived e.g. by Chaboche (1991), Ohno and Wang (1993), Jiang and Sehitoglu (1996). Loading conditions that can be used in material tests for the determination of the ratchetting related material properties are investigated in Rekun and Jörg (2016). Other models are able to describe strain range memory effect as e.g. the kinematic hardening model presented in Haupt et al. (1992). All these models are extensions to the Armstrong-Frederick kinematic hardening law. Hence, the models presented in this work form a very good basis to be able to deal with more advanced models.

References

Aktaa, J., & Petersen, C. (2009). Challenges in the constitutive modeling of the thermo-mechanical deformation and damage behavior of EUROFER 97. *Engineering Fracture Mechanics, 76*(10), 1474–1484. MatModels 2007.

Armstrong, P. J., & Frederick, C. O. (1966). *A Mathematical Representation of the Multiaxial Bauschinger Effect* (Vol. 731). Berkley, UK: Central Electricity Generating Board & Berkeley Nuclear Laboratories.

Burlet, H., & Cailletaud, G. (1986). Numerical techniques for cyclic plasticity at variable temperature. *Engineering Computations, 3*(2), 143–153.

Cailletaud, G., Quilici, S., Azzouz, F., & Chaboche, J.-L. (2015). A dangerous use of the fading memory term for non linear kinematic models at variable temperature. *European Journal of Mechanics—A/Solids, 54*, 24–29.

Chaboche, J.-L. (1993). Cyclic viscoplastic constitutive equations, Part I: A thermodynamically consistent formulation. *Journal of Applied Mechanics, 60*(4), 813–821, 12.

Chaboche, J.-L. (1997). Thermodynamic formulation of constitutive equations and application to the viscoplasticity and viscoelasticity of metals and polymers. *International Journal of Solids and Structures, 34*(18), 2239–2254.

Chaboche, J. L. (1986). Time-independent constitutive theories for cyclic plasticity. *International Journal of Plasticity, 2*(2), 149–188.

Chaboche, J. L. (1989). Constitutive equations for cyclic plasticity and cyclic viscoplasticity. *International Journal of Plasticity, 5*(3), 247–302.

Chaboche, J. L. (1991). On some modifications of kinematic hardening to improve the description of ratchetting effects. *International Journal of Plasticity, 7*(7), 661–678.

Chaboche, J. L. (2008). A review of some plasticity and viscoplasticity constitutive theories. *International Journal of Plasticity, 24*(10), 1642–1693. Special Issue in Honor of Jean-Louis Chaboche.

Coleman, B. D., & Gurtin, M. E. (1967). Thermodynamics with internal state variables. *The Journal of Chemical Physics, 47*(2), 597–613.

Fischer, F. D., & Svoboda, J. (2006). A note on the principle of maximum dissipation rate. *Journal of Applied Mechanics, 74*(5), 923–926, 12.

Frederick, C. O., & Armstrong, P. J. (2007). A mathematical representation of the multiaxial bauschinger effect. *Materials at High Temperatures, 24*(1), 1–26.

Hackl, K., & Fischer, F. D. (2008). On the relation between the principle of maximum dissipation and inelastic evolution given by dissipation potentials. *Proceedings of the Royal Society A: Mathematical, Physical and Engineering Sciences, 464*(2089), 117–132.

Haupt, P., Kamlah, M., & Tsakmakis, C. (1992). Continuous representation of hardening properties in cyclic plasticity. *International Journal of Plasticity, 8*(7), 803–817.

Heitmann, H. H., Vehoff, H., & Neumann, P. (1984). Life prediction for random load fatigue based on the growth behavior of microcracks. In S. R. Valluri, D. M. R. Taplin, P. Rama Rao, J. F. Knott, & R. Dubey (Eds.), *Fracture* (Vol. 84, pp. 3599–3606). Pergamon.

Hill, R. (1968). On constitutive inequalities for simple materials-I. *Journal of the Mechanics and Physics of Solids, 16*(4), 229–242.

Hoffmeyer, J., Döring, R., & Vormwald, M. (2001). Kurzrisswachstum bei mehrachsig nichtproportionaler Beanspruchung. *Materialwissenschaft und Werkstofftechnik, 32*(4), 329–336.

Hoffmeyer, J., Döring, R., Seeger, T., & Vormwald, M. (2006). Deformation behaviour, short crack growth and fatigue livesunder multiaxial nonproportional loading. *International Journal of Fatigue, 28*(5), 508–520. Selected papers from the 7th International Conference on Biaxial/Multiaxial Fatigue and Fracture (ICBMFF).

Jiang, Y., & Sehitoglu, H. (1996). Modeling of cyclic ratchetting plasticity, Part I: Development of constitutive relations. *Journal of Applied Mechanics, 63*(3), 720–725.

Jilg, A., & Seifert, T. (2018). A temperature dependent cyclic plasticity model for hot work tool steel including particle coarsening. In *Proceedings of the 21st International ESAFORM Conference on Material Forming: ESAFORM 2018*. American Institute of Physics Conference Series (Vol. 1960, page 170007).

Lemaitre, J., & Chaboche, J.-L. (1990). *Mechanics of Solid Materials*. Cambridge: Cambridge University Press.

Mao, J., Engler-Pinto, C. C., Li, T., Hsieh, J., & Su, X. (2015). Effect of constitutive model on thermomechanical fatigue life prediction. *Procedia Engineering, 133*, 655–668. Fatigue Design 2015, International Conference Proceedings, 6th Edition.

Nissley, D. M. (1995). Thermomechanical fatigue life prediction in gas turbine superalloys—A fracture mechanics approach. *AIAA Journal, 33*(6), 1114–1120.

Oesterlin, H., & Maier, G. (2014). Numerical assessment of thick walled power plant components under creep fatigue load with advanced models for deformation and lifetime. *Materials at High Temperatures, 31*(4), 364–369.

Ohmenhäuser, F., Schwarz, C., Thalmair, S., & Evirgen, H. S. (2014). Constitutive modeling of the thermo-mechanical fatigue and lifetime behavior of the cast steel 1.4849. *Materials & Design, 64*, 631–639.

Ohno, N., & Wang, J. (1992). Nonisothermal constitutive modeling of inelasticity based on bounding surface. *Nuclear Engineering and Design, 133*(3), 369–381.

Ohno, N., & Wang, J.-D. (1993). Kinematic hardening rules with critical state of dynamic recovery, Part I: Formulation and basic features for ratchetting behavior. *International Journal of Plasticity, 9*(3), 375–390.

Ohno, N., & Wang, J. D. (1991). Transformation of a nonlinear kinematic hardening rule to a multi-surface form under isothermal and nonisothermal conditions. *International Journal of Plasticity, 7*(8), 879–891.

Perzyna, P. (1966). Fundamental problems in viscoplasticity. In G. G. Chernyi, H. L. Dryden, P. Germain, L. Howarth, W. Olszak, W. Prager, R. F. Probstein, & H. Ziegler (Eds.), *Advances in Applied Mechanics* (Vol. 9, pp. 243–377). Elsevier.

Rekun, I., & Seifert, T., & Jörg, R. (2016). Determination of stable and robust material properties for the assessment of thermomechanically loaded components of rocket engines with viscoplastic constitutive equations. In *Proceedings of the 14th European Conference on Spacecraft Structures, Materials and Environmental Testing (ECSSMET)*, Toulouse.

Rice, J. R. (1971). Inelastic constitutive relations for solids: An internal-variable theory and its application to metal plasticity. *Journal of the Mechanics and Physics of Solids, 19*(6), 433–455.

Schlesinger, M., Seifert, T., & Preussner, J. (2017). Experimental investigation of the time and temperature dependent growth of fatigue cracks in inconel 718 and mechanism based lifetime prediction. *International Journal of Fatigue, 99*, 242–249. Recent Developments in Thermo-mechanical Fatigue.

Seifert, T. (2006). Ein komplexes LCF-Versuchsprogramm zur schnellen und günstigen Werkstoff-parameteridentifizierung. In M. Borsutzki & S. Geisler (Eds.), *Proceedings Werkstoffprüfung 2006* (pp. 409–414). Bad Neuenahr.

Seifert, T., & Maier, G. (2008). Consistent linearization and finite element implementation of an incrementally objective canonical form return mapping algorithm for large deformation inelasticity. *International Journal for Numerical Methods in Engineering, 75*(6), 690–708.

Seifert, T., & Riedel, H. (2010). Mechanism-based thermomechanical fatigue life prediction of cast iron. Part I: Models. *International Journal of Fatigue, 32*(8), 1358–1367.

Seifert, T., Schenk, T., & Schmidt, I. (2007). Efficient and modular algorithms in modeling finite inelastic deformations: Objective integration, parameter identification and sub-stepping techniques. *Computer Methods in Applied Mechanics and Engineering, 196*(17–20), 2269–2283.

Seifert, T., Schweizer, C., Schlesinger, Möser, M., & Eibl, M. (2010). Thermomechanical fatigue of 1.4849 cast steel—Experiments and life prediction using a fracture mechanics approach. *International Journal of Materials Research, 101*(8), 942–950. Dedicated to Professor Dr. Hermann Riedel on the occasion of his 65th birthday.

Seifert, T., Hazime, R., Chang, C.-C., & Hu, C. (2019). Constitutive modeling and thermomechanical fatigue life predictions of A356–T6 aluminum cylinder heads considering ageing effects. In *WCX SAE World Congress Experience*. SAE International.

Simo, J. C., & Hughes, T. J. R. (1998). *Computational Inelasticity*. New York: Springer.

Wilhelm, F., Spachtholz, J., Wagner, M., Kliemt, C., & Hammer, J. (2014). Simulation of the viscoplastic material behaviour of cast aluminium alloys due to thermal-mechanical loading. *Journal of Materials Science and Engineering A, 4*(1), 56–64.

Yaguchi, M., Yamamoto, M., & Ogata, T. (2002). A viscoplastic constitutive model for nickel-base superalloy, Part 2: Modeling under anisothermal conditions. *International Journal of Plasticity, 18*(8), 1111–1131.

Chapter 5
Damage and Failure of Composite Materials

Ramesh Talreja

Abstract This article is a summary of six lectures given by the author at the webinar, "Advanced Theories for Deformation, Damage and Failure in Materials" organized by the International Centre for Mechanical Sciences, Udine, Italy, May 3–7, 2021. The topics of the lectures dealt with mechanisms of damage and failure, continuum and synergistic damage mechanics, failure theories for anisotropic homogenized solids, and physical modeling of failure in composites containing manufacturing defects. Finally, future research directions were discussed. An attempt is made here to bring together those topics in a coherent manner to provide a comprehensive exposition of the field of damage and failure in composite materials.

5.1 Introduction

The field of damage and failure modeling for fiber-reinforced composite materials has evolved from phenomenological descriptions of strength to mechanisms-based multi-scale computational simulations for assessing failure described as a critical response state. The current trend is to explore correlations between descriptors of microstructural details and critical loading states associated with failure using artificial intelligence methodologies. This exposition will review the observations of failure events in composite materials under different elementary loading modes and their combinations. This will be done first for a unidirectionally fiber-reinforced layer, abbreviated as a UD composite, followed by a discussion of additional failure events in laminates where the UD composite forms a basic unit stacked in different orientations.

R. Talreja (✉)
Department of Aerospace Engineering and Department of Materials Science and Engineering, Texas A&M University, College Station, Texas 77843, USA
e-mail: talreja@tamu.edu

H. Altenbach and A. Ganczarski (eds.), *Advanced Theories for Deformation, Damage and Failure in Materials*, CISM International Centre for Mechanical Sciences 605, https://doi.org/10.1007/978-3-031-04354-3_5

5.2 Failure Modes in UD Composites

For the sake of systematic description of experimental observations, the failure modes will be categorized as fiber failure modes in axial tension and axial compression, matrix and fiber/matrix interface failure modes in transverse tension, transverse compression, and in-plane shear, followed by a discussion of failure modes in combined loading. In this section, only observations will be described, leaving modeling to a later section. Also, the failure mode observations will be described briefly highlighting essential features. For more information the reader should look up reviews of the topic in the literature.

5.2.1 Fiber Failure Mode in Axial Tension

A study that reports very clear observations of failure events in a UD composite in axial tension is by Aroush et al. (2006). The authors used high-resolution synchrotron X-ray tomography to produce images showing single and multiple fiber breaks in a quartz/epoxy model composite. A series of reported images is shown in Fig. 5.1. As seen, the failure process consists of single fiber breaks, called singlets, followed by doublets, N-plets, etc., until a critical fiber breakage cluster is formed causing catastrophic failure of the composite. The roles of matrix and fiber-matrix interface in the tensile failure process are not clear in observations of this nature. For this, inferences are based on modeling and simulation studies that will be discussed later.

5.2.2 Fiber Failure Mode in Axial Compression

Figure 5.2 from Jumahat et al. (2010) is illustrative of the main mechanism in compression failure of UD composites. The schematic in Fig. 5.2 describes the formation of a kink band from initially misaligned fibers, Fig. 5.2a, of misalignment angle ϕ_0. The fibers deform by micro-buckling, Fig. 5.2b, aided by matrix plastic deformation, leading to formation of a kink band of width w spanning fiber breakage at two points along fibers, Fig. 5.2c. Closer details of fiber breaks in kink bands can be seen in Fig. 5.3 which shows a scanning electron microscopy (SEM) image (Narayanan and Schadler, 1999).

5.2.3 Matrix and Fiber/Matrix Interface Failure Mode in Transverse Tension

Experimental observations of this failure mode are often made by testing cross ply laminates in axial tension, thereby inducing the 90-degree layers in transverse tension.

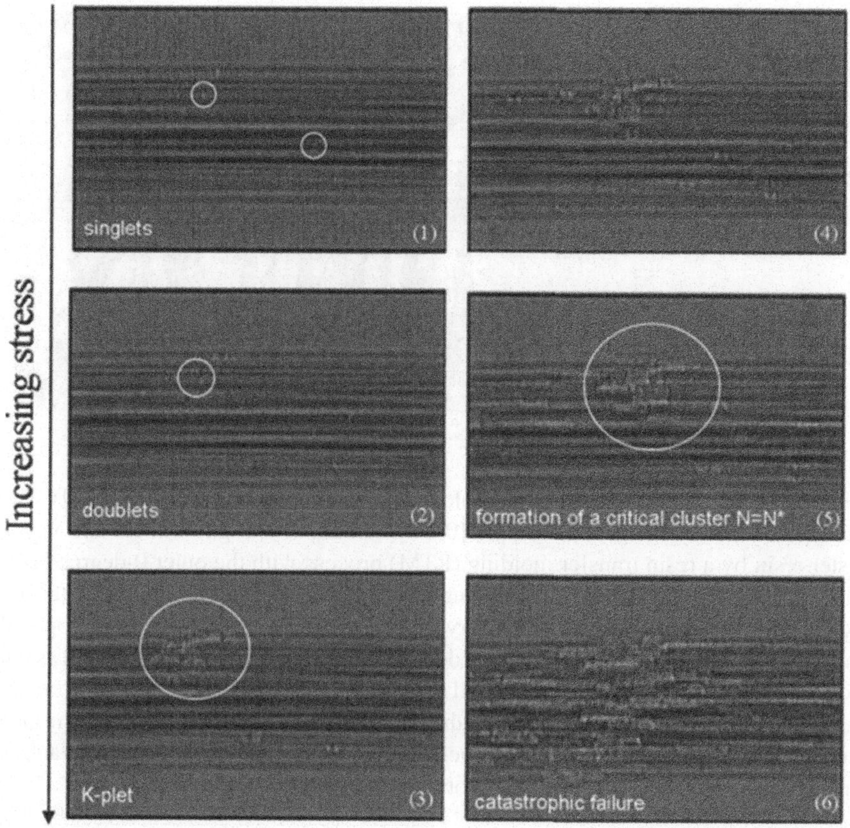

Fig. 5.1 X-ray radiograph sequence showing fiber breaks in a quartz-epoxy UD composite in axial tension (Aroush et al., 2006)

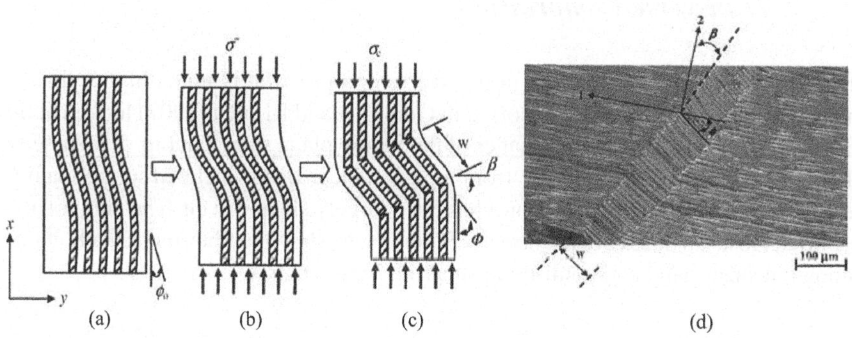

Fig. 5.2 Schematic description of kink band formation **a–c**, and **d** an optical micrograph showing details of a kink band (Jumahat et al., 2010)

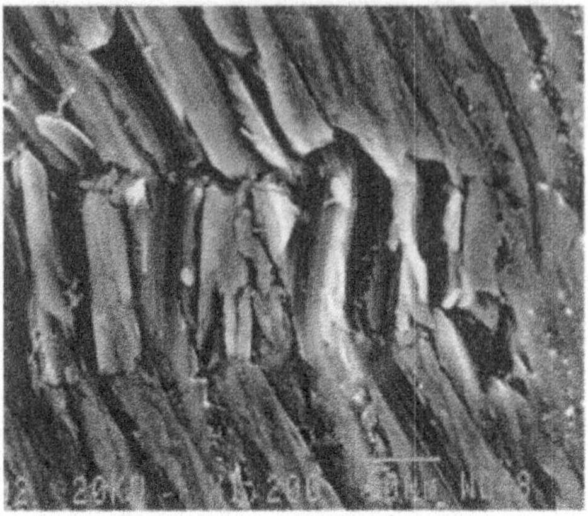

Fig. 5.3 An SEM image of a kink band formed in a single tow carbon/epoxy composite (Narayanan and Schadler, 1999)

Figure 5.4 shows an illustrative example of this (Gamstedt and Sjögren, 1999). The authors of this study manufactured a $[0/90]_s$ laminate with glass fibers and vinyl-ester resin by a resin transfer molding (RTM) process with the outer 0-degree layer thickness of 0.5 mm and the inner 90-degree layers of 1.0 mm. They applied a tensile load in the 0-degree direction and observed the edge of the specimen under an optical microscope. They found that the fibers debonded from the matrix initially, and as the load was increased, some of the debond cracks coalesced to form a transverse crack, as depicted in Fig. 5.4a. On further loading, a coalesced crack showed crack surface separation, indicating a continuous crack, as seen in Fig. 5.4b. Similar observations have been reported later by other authors for carbon-epoxy composites.

5.2.4 Matrix and Fiber/Matrix Interface Failure Mode in Transverse Compression

The strength of UD composites in transverse compression is much higher than in transverse tension. One study reported in González and LLorca (2007) looked at the failure process by taking SEM images of the failure plane shown in Fig. 5.5. As seen, the failure appears to have come from a plane (shear band) inclined to the loading axis at 56 degrees, suggesting that it has been largely shear driven. The authors also reported close-up images taken near the failure plane showing fiber/matrix debonding along the shear band and confirming shearing of the matrix.

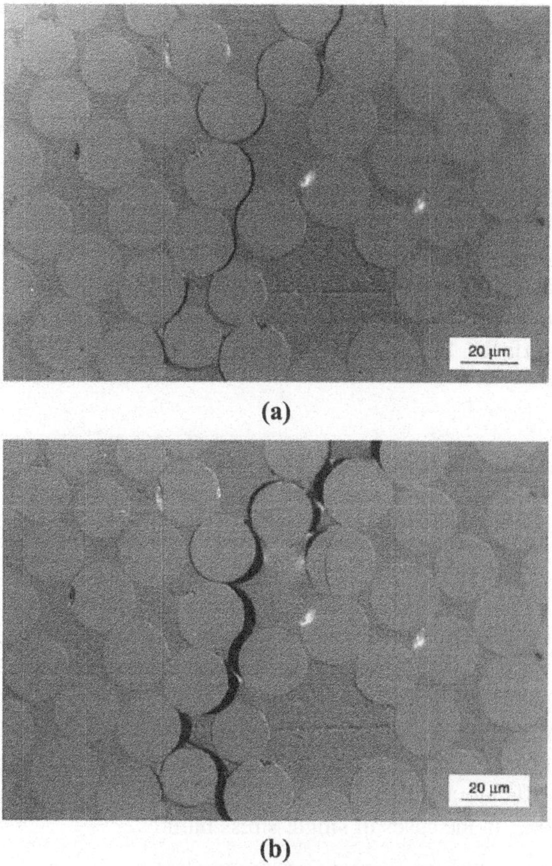

Fig. 5.4 **a** Fiber/matrix debonding, and **b** coalescence of debonds into a transverse crack (Gamstedt and Sjögren, 1999)

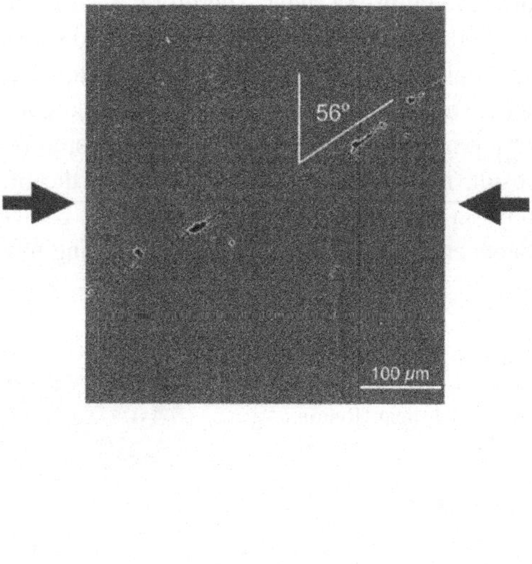

Fig. 5.5 SEM image of the lateral surface of a carbon-epoxy UD composite taken before the failure load showing formation of a failure plane inclined to the loading direction indicated by arrows (González and LLorca, 2007)

5.2.5 *Matrix and Fiber/Matrix Interface Failure Mode in In-plane Shear*

There are practical difficulties in introducing pure shear stress in UD composites due to stress gradients induced by gripping of specimens. A study attempted to induce shear stress directly at the fiber level by cutting fibers under load that caused the cut fibers to retract, thereby inducing equal and opposite forces locally, Fig. 5.6 (Redon, 2000). As seen in the figure, the cracks in the matrix are inclined to the fiber direction and turn towards the fibers at the fiber/matrix interfaces. Other studies have confirmed the shear-induced cracking of the matrix by observing shear hackles in the fracture surfaces.

5.2.6 *Failure in Combined Loading*

The purpose of failure theories is to describe, if not predict, the occurrence of failure in UD composites in simultaneous application of different stresses. For homogenized thin layers of UD composites that make up the basic unit in multidirectional laminates, the stress state consists of axial (fiber-direction) normal stresses, transverse normal stresses, and in-plane shear stress. Testing limitations usually allow a pairwise combination of these stresses. Thus, eight combined loading cases can be examined for failure initiation and progression. These will be described next. It is noted, however, that testing in combined loading makes observation of failure events difficult, and therefore, much less clarity of failure mechanisms has been achieved than in the cases of single-stress failures.

Axial Tension and In-plane Shear. Fiber/matrix interfaces are usually weak and prone to failure in shear. Therefore, axial splitting tends to occur at low values of the applied shear stress making it difficult to study the effect of shear on failure induced by axial tension. At the same time, axial tension can cause fiber breaks, which induces local stress concentration and stress gradients at the fiber/matrix interfaces making observations of the effects of axial tension on shear-induced failure difficult. As a result, not much has been reported in the literature on combined effects of axial tension and in-plane shear on failure of UD composites. To discern any interaction between failure caused by the two loading modes individually from the final failure

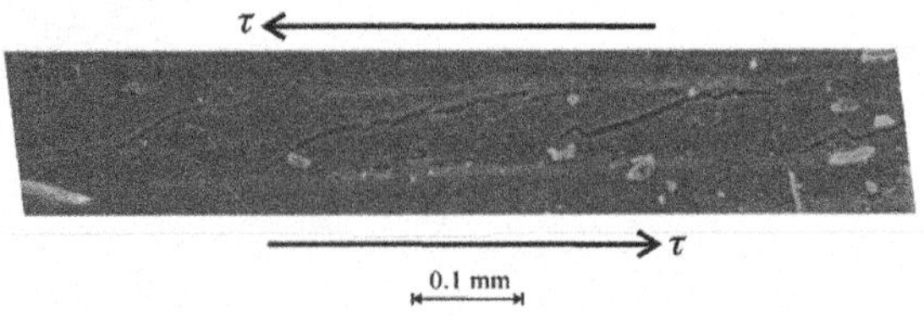

Fig. 5.6 Cracking in the inter-fiber region resulting from axial shear (Redon, 2000)

results is also uncertain because of the significant scatter in the strength data that is inherent in the tensile failure.

Axial Tension and Transverse Tension. The testing and observation difficulties in this case are like the case of combining axial tension and in-plane shear because of the fiber/matrix interface failure occurring at low values of transverse tension in polymer matrix composites. This limits the range of combined loading in which interaction between failures in the two loading modes can be observed. The final failure (strength) data reported by Hinton et al. (2004) suggest failure either by fiber failure in axial tension or by unstable crack growth along fibers in transverse tension.

Axial Tension and Transverse Compression. The specimen dimensions and fixtures for transverse load introduction in this case prevent making clear observations of failure events. The failure mechanisms in the two individual loading modes are significantly different which suggests little interaction between them. However, it is conceivable that fiber failures which are highly statistically governed when axial tension is applied can become less so under the influence of shear bands induced by transverse compression.

Axial Compression and In-plane Shear. A study by Vogler and Kyriakides (2001) found that the unstable growth of kink bands occurring in axial compression can be stabilized if in-plane shear is applied. Figure 5.7 depicts the stable kink band growth process under combined loading. The degree of fiber breakage was found to be less than when only axial compression is applied. While this study was focused on compression failure and has clarified the moderating effect of in-plane shear, it is not clear how the failure induced by in-plane shear itself is affected by the presence of axial compression.

Axial Compression and Transverse Tension. The effect of transverse tension on failure by axial compression has not been studied experimentally because of fiber/matrix interface failing at low transverse tension. At the same time, if and how the failure due to transverse tension is affected by kink bands has not been studied. It can be expected that any fiber/matrix debonding under transverse tension is likely to reduce the lateral support to the fibers and thereby cause earlier fiber microbuckling. The splitting failures occurring near wavy layers under axial compression reported in Davidson and Waas (2017) are likely to grow when transverse tension is applied.

Axial Compression and Transverse Compression. Oguni et al. (2000) reported the effect of biaxial compression by studying failure in a E-glass/-vinylester UD composite. The observations showed that a failure mode transition occurred from axial splitting to kink band formation as uniaxial compression changed to biaxial compression. Under axial compression kink bands were found to be affected by axial splitting while under proportional biaxial compression only kink bands were observed.

Transverse Tension and In-plane Shear. Quaresimin and Carraro (2014) studied failure in thin-walled cylinders of glass/epoxy containing circumferential fibers subjected to axial tension and torsion applied cyclically. The first failure event was

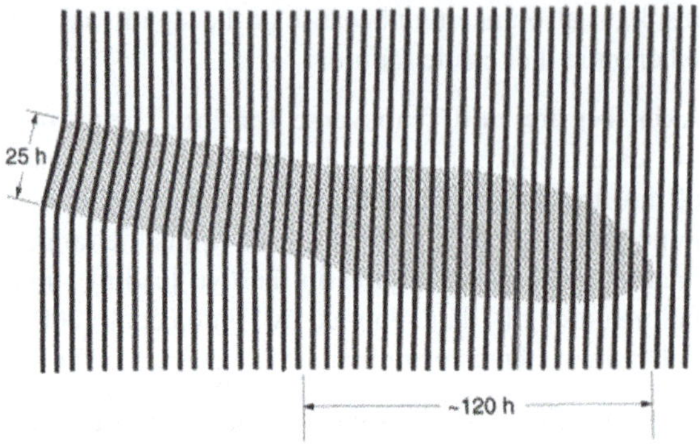

Fig. 5.7 A schematic depiction of a kink band growing in combined axial compression and in-plane shear (Vogler and Kyriakides, 2001). The dimensions shown are in terms of the fiber diameter h

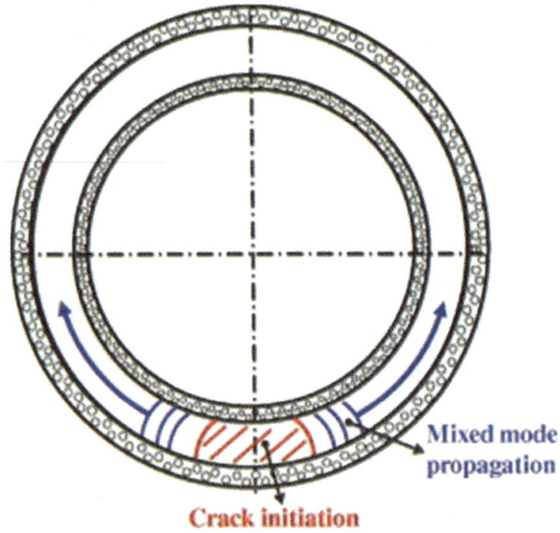

Fig. 5.8 Crack initiation and its mixed mode propagation along circumferential fibers in a tubular specimen subjected to cyclic axial tension and torsion (Quaresimin and Carraro, 2014)

assessed to be crack initiation along fibers from preexisting manufacturing defects. To stabilize the crack growth small amount of glass fabrics was added on both sides of the UD composite. The crack growth was found in mixed mode as depicted in Fig. 5.8.

Transverse Compression and In-plane Shear. Both loading modes when applied individually initiate matrix inelasticity that leads to cracking in matrix and at the fiber/matrix interfaces. Combining the two loading modes will enhance the deformation and failure processes. A systematic interaction of the effects has not been studied experimentally.

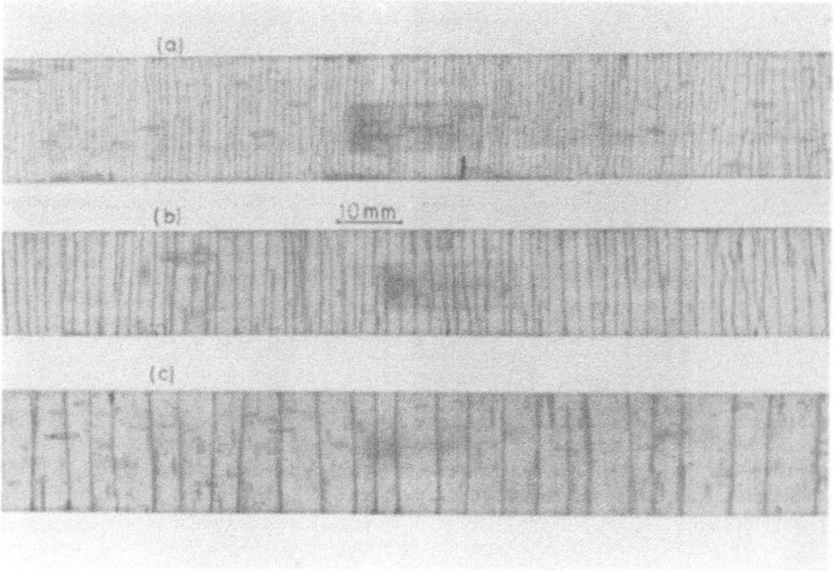

Fig. 5.9 Transverse cracking in specimens with transverse-ply thickness of **a** 0.75 mm, **b** 1.5 mm and **c** 2.6 mm, strained at 1.6% (Garrett and Bailey, 1977)

5.3 Failure Modes in Laminates

The combinations in multi-directional laminates are many when lamina orientation and sequence (layup) and lamina thickness are taken as parameters. However, practical laminates in most cases are limited to balanced laminates and quasi-isotropic laminates. Cross ply laminates are used in experimental studies as a simple case of the effects of constraint by a neighboring lamina on a lamina undergoing cracking. In the following, the failure modes in cross ply laminates will be reviewed followed by description of additional features observed in other (general) laminates.

5.3.1 Cross Ply Laminates

Figure 5.9 (Garrett and Bailey, 1977) depicts the multiple transverse cracking process which results from the ply constraint (deformation of transverse plies restricted by the presence of axial plies) and is a characteristic feature of cracking in plies of laminates. The images shown illustrate how the transverse crack spacing increases as the thickness of the outer axial plies is decreased. In the limit, as the thickness of the outer plies tend of zero, the crack spacing tends to infinity, i.e., the laminate approaches a UD composite that develops a single transverse crack.

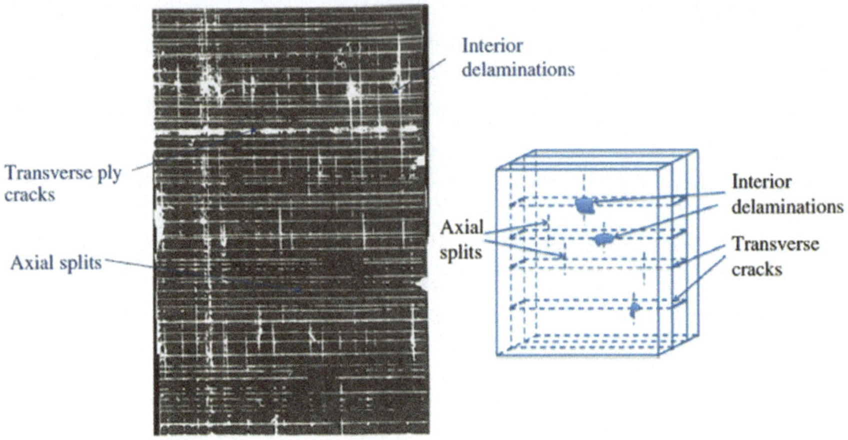

Fig. 5.10 X-ray radiograph of a carbon/epoxy cross ply laminate subjected to tension-tension cyclic loading (Reifsnider and Jamison, 1982). The sketch to the right shows a three-dimensional depiction for clarity

The multiple cracking shown in Fig. 5.9 is seen in quasi-static loading, i.e., in monotonically increasing load. However, if a load below the failure load is applied repeatedly, other local failure mechanisms become visible. Figure 5.10 (Reifsnider and Jamison, 1982) depicts the additional mechanisms of axial splits and interior delamination that are observed under cyclic loading.

5.3.2 *General Laminates*

A systematic study (Varna et al., 1999) reported the constraint effect on ply cracking by varying the off-axis angle while keeping the axial plies as constraining plies. In that study, laminates of $[0/+\theta/-\theta/0_{1/2}]_s$ layup tested in axial tension showed that as the angle θ was decreased from 90°, the initiation of ply cracks was delayed, and the crack spacing reached at the same applied load was also higher. In fact, for θ approaching 0°, ply cracks were not seen before failure of the laminate. Axial splits and interior delamination cracks also appear to diminish as the off-axis angle reduces from 90°.

5.4 Damage and Failure

Composite materials show significantly different deformation and failure behavior compared to metals. Therefore, to properly model the mechanical response of composite materials, one must use a terminology that is specific to this material system and relates to the characteristic features of the underlying mechanisms that govern the observed response. This is generally not the case in the literature and as a result confusion in interpretation of composite materials performance exists. Talreja and Waas (2022) have discussed at length the issues related to the use of terminology such as "strength" in describing the load carrying capacity of composite materials and have pointed to the misinterpretations that follow from such usage. Here, we shall lay down the definitions of terms such as damage and failure in the context of composite materials to avoid any confusion in the treatment of the topics to follow.

Damage refers to a material state where a multitude of entities formed by irreversible (energy dissipative) mechanisms exists. Thus, it does not refer to a single crack but refers collectively to the presence of multiple cracks, e.g., those shown in Figs. 5.9 and 5.10.

Failure refers to occurrence of certain criticality in the response of the material to an imposed condition such as a mechanical load or a temperature change. To specify failure, it is generally necessary to add a qualifier, e.g., in "a failure event", "a failure mechanism", "a failure mode", "initiation of failure", etc. A criticality (or a critical state) must be defined to specify failure. One example to illustrate a criticality is the occurrence (or initiation) of inelasticity in an initially elastic composite. Another example of criticality is the loss of integrity, e.g., by separation of material in a region. In this case, a load cannot be transferred across the region. This loss of load carrying capacity is often described as "total failure".

Fracture and failure are sometimes incorrectly used synonymously. **Fracture** is a special case of failure that describes separation of material resulting from an unstable growth of a crack. In the failure modes described above, fracture occurs only in the case of a UD composite loaded in tension at an angle to the fiber direction.

5.4.1 Damage Mechanics

The field of damage mechanics deals with the response of a material in a state of damage. The response includes deformational response, e.g., changes in stiffness due to damage, as well as evolution of the state of damage. Damage mechanics can be a part of the larger field of failure analysis that deals with analysis of criticalities in the materials response that fall within the definition of failure as described above.

Continuum damage mechanics (CDM) and its derivative, **synergistic damage mechanics** (SDM), is a field where continuum concepts such as homogenization are used to treat the presence of a multitude of entities that collectively define damage. Figure 5.11 is a schematic description of the CDM concept. As depicted there, a composite material is viewed as containing two types of entities of microstructural scale embedded in a matrix, one described as stationary microstructure and the other as evolving microstructure. The stationary microstructure consists of fibers and particles that do not change permanently under externally applied loads, while the evolving microstructure consists of entities such as cracks that form by energy dissipative mechanisms and can evolve (change irreversibly) by appropriate driving forces. The continuum method of homogenization is applied in two steps: Step 1 where the stationary microstructure is smeared into the matrix to form a continuum, and step 2 where the evolving entities are replaced by a field describing the state of damage of the continuum formed at the end of state 1. The continuum of step 1 homogenization will be called a composite solid, which is generally anisotropic, and the continuum of step 2, i.e., a fully homogenized continuum, is characterized by the state of damage. The characterization of the damage state is conducted by a specified averaging procedure applied on the evolving entities lying within a representative volume element (RVE) associated with a generic point P of the fully homogenized continuum. An individual evolving entity is called a damage entity in the damage characterization procedure proposed here where two vectors are used, the unit vector n describing the damage entity orientation and the vector a that stands for a chosen influence the presence of the damage entity has on its surroundings. The proposed damage characterization is described next.

5.4.2 Damage Characterization

Beginning with his first contribution in the damage mechanics area (Talreja and Kelly, 1985) this author has pursued a physically based damage characterization that

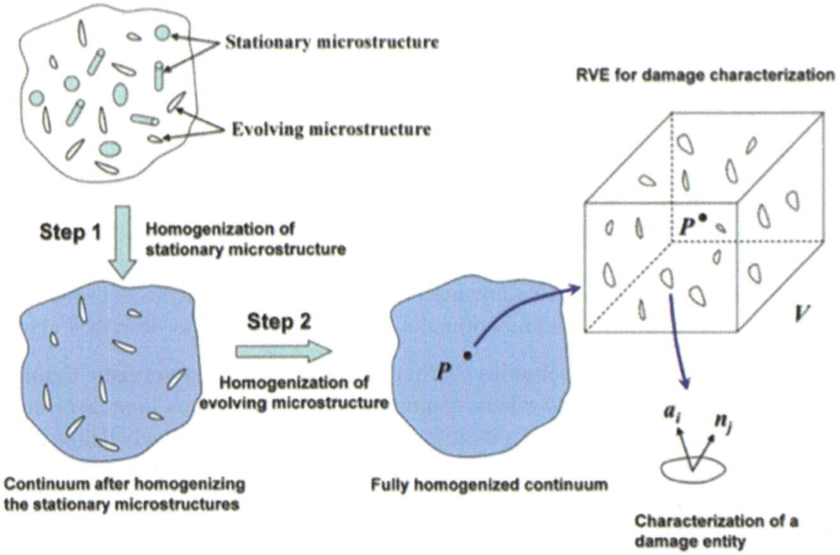

Fig. 5.11 A schematic depiction of the continuum damage mechanics concept

does not leave behind essential features of damage in the context of the material response of interest while keeping in sight practical implementation of the resulting methodology. As argued in Talreja and Kelly (1985) and elaborated further in Talreja (1990, 1996), a scalar-valued characterization would be inadequate in view of the oriented nature of the internal surfaces formed. Although a vector-valued characterization was employed at first, the issue of ambiguity of the sense of a vector could be addressed more elegantly by using a second order tensor instead. Referring to the previous works cited above for more details a brief description of the damage as second order tensors is given below.

Following Fig. 5.11, where the components a_i and n_i of the influence vector $\boldsymbol{a}$ and the unit normal vector $\boldsymbol{n}$, respectively, are associated with a damage entity, a dyadic product of these vectors, integrated over the surface S of the damage entity is denoted damage entity tensor, and given by

$$\boldsymbol{d} = \int_S \boldsymbol{a} \otimes \boldsymbol{n} \mathrm{d}S \tag{5.1}$$

with the dyadic (tensor) product $\otimes$. The physical significance of this characterization is that it represents the oriented nature of the presence of internal surfaces within the volume of a composite with damage. As illustrated by the examples of damage discussed above, common internal surfaces are cracks (flat or curved) generated by interface and matrix failures. The unit normal vector at a point on the damage entity carries the information on orientation of the surface (with respect to the frame of reference), while the other vector represents an appropriate influence induced by activation of the considered point on the surface. This influence is generally also directed in nature. For the case of mechanical response, the appropriate influence would be the displacement of the activated point on the damage entity surface. For a non-mechanical response, such as thermal or electrical conductivity, the perturbation induced by an internal surface can also be cast as a vector-valued quantity.

Integrating the dyadic product in Eq. (5.1) over the damage entity surface provides the total net effect of the entity. For example, if the entity is a flat crack, then taking a as the displacement vector in the integral gives the crack surface separation times the crack surface area. This product may be viewed as an affected volume associated with the crack. For a penny-shaped crack with the two surfaces that separate symmetrically about the initial crack plane, the sole surviving term of the damage entity tensor, Eq. (5.1), represents an ellipsoidal shaped volume.

Referring once again to Fig. 5.11, the RVE associated with a generic point P carries a sufficiently large number of the discrete damage entities to represent the collective effect on the homogenized constitutive response at the point. The number of damage entities needed for this representation, and the consequent RVE size, depend on the distribution of the entities. For instance, if the entities are sparsely distributed, then the RVE size would be large, while for a densely distributed case a small RVE would suffice. Furthermore, for uniformly distributed entities of the same geometry, a repeating unit cell containing a single entity can replace the RVE, while for the cases of nonuniform distribution of unequal entities, the RVE size will increase until a statistically homogeneous representation is attained. This implies that further increasing the RVE size will have no impact on the averages of the selected characteristics. As an example, if the selected characteristic is the affected matrix volume by a damage entity, as mentioned above, then the average value of this quantity will vary as the RVE size increases and will approach a constant value at a certain RVE size. The minimum RVE size beyond which no appreciable change in the considered average is found may be taken as the needed RVE. It is apparent that the RVE is not unique but is subject to the choice made for the formulation of the constitutive response of a continuum with damage. Consequently, there is no unique constitutive theory of a continuum with damage; however, the use of the concept of an internal state in each theory requires specifying RVE in a consistent manner and assuring that the conditions for its existence are present.

From the cases of damage mechanisms reviewed above it can be noted that in composite laminates the damage tends to occur as sets of parallel cracks within the plies, each oriented along fibers in the given ply. It is therefore convenient to separate each set of ply cracks according to its orientation, referred to a fixed frame of reference, and assign it a damage mode number. Denoting damage mode by $\alpha = 1, 2, \ldots, n$, a damage mode tensor can be defined as

$$\boldsymbol{D}^{(\alpha)} = \frac{1}{V}\sum_{k_\alpha}\left(\boldsymbol{D}^{(\alpha)}\right)_{k_\alpha} \tag{5.2}$$

where k_α is the number of damage entities in the αth mode, and V is the RVE volume. As noted above, if the ply cracks of a given orientation are uniformly spaced, then the RVE will reduce to the unit cell containing one crack. For nonuniform distribution of ply cracks, V must be large enough to provide a steady average of the damage mode tensor components.

As defined by Eq. (5.2) the damage mode tensor will in general be asymmetrical. Decomposing the influence vector, a, along directions normal and tangential to the damage entity surface S gives,

$$\boldsymbol{a} = a\boldsymbol{n} + b\boldsymbol{m} \tag{5.3}$$

where $\boldsymbol{n}$ and $\boldsymbol{m}$ are unit normal and tangential vectors on S.

Using Eq. (5.3) in (5.2) the damage entity tensor can be written in two parts as

$$\boldsymbol{d} = \boldsymbol{d}^1 + \boldsymbol{d}^2 = \int_S a\boldsymbol{n} \otimes \boldsymbol{n}\mathrm{d}S + \int_S b\boldsymbol{m} \otimes \boldsymbol{n}\mathrm{d}S \tag{5.4}$$

The damage mode tensor for a given mode can now be written as (dropping mode number α for convenience),

$$\boldsymbol{D} = \boldsymbol{D}^1 + \boldsymbol{D}^2 = \frac{1}{V}\sum_k \left(\boldsymbol{d}^1\right) + \frac{1}{V}\sum_k \left(\boldsymbol{d}^2\right) \tag{5.5}$$

This separation of the damage mode tensor in two parts allows simplifying the analysis so that dealing with asymmetric tensors is avoided. For instance, for damage entities consisting of flat cracks, the two parts of the damage mode tensor represent the two crack surface separation modes. If the assumption can be made that only the symmetric crack surface separation (known as mode I or crack opening mode in fracture mechanics) is significant, then the second term in Eq. (5.5) can be neglected. This will render the damage mode tensor symmetrical, and it can then be written as

$$\boldsymbol{D} = \boldsymbol{D}^1 = \frac{1}{V}\sum_k \left[\int_S a\boldsymbol{n} \otimes \boldsymbol{n} \mathrm{d}S\right]_k \tag{5.6}$$

The consequence of this assumption was examined by Varna (2013) for one class of laminates, and it was found that not including the crack sliding displacement (CSD) for ply cracks inclined to the laminate symmetry directions results in errors in estimating degradation of average elastic properties of laminates. However, these errors were found to be small in absolute values while being significant in percentages. In fact, for those ply crack orientations where CSD dominates, the cracks are difficult to initiate until high loads close to failure load are applied.

For cases where the damage entity surfaces conduct tangential displacements only (e.g., CSD by flat cracks), it is possible to formulate the damage mode tensor as a symmetric tensor. One example of this is sliding of fiber/matrix interface in ceramic matrix composites (Talreja, 1991).

With stress, strain, and damage, all expressed as symmetric second order tensors, a constitutive theory can now be formulated to have a convenient, usable form. Such a formulation is described next.

5.4.3 CDM Framework for Materials Response

Referring once again to Fig. 5.11, a formulation of the constitutive response of a homogenized continuum with damage will now be discussed. In view of the observed behavior of common composite materials such as glass/epoxy and carbon/epoxy, only elastic response will be considered. Theoretical treatment of elastic response of anisotropic solids is classical and can be found in textbooks. Incorporating damage is, however, not a simple extension of the classical theory of elasticity. So far, the efforts made in this respect can be categorized as micro-damage mechanics (MIDM) and macro-damage mechanics (MADM), the latter often described as continuum damage mechanics (CDM). A framework for CDM to be described here is based on thermodynamics and is naturally suited for thermo-mechanical response. It can be extended to incorporate non-mechanical effects, such as electrical and magnetic, as well as chemical. Every extension, however, comes with the price of having to determine associated response coefficients (material constants) by certain identification procedures. In the treatment presented here, the task of determining material constants is reduced by using judiciously selected micromechanics. This way of combining micromechanics with CDM generates useful synergism, justifying the characterization of the combined approach as synergistic damage mechanics (SDM). We begin with the conventional CDM framework first.

At the foundation of CDM are the first and second laws of thermodynamics. Additionally, use is made of the concept of an internal state, which is identified here as the evolving microstructure depicted in Fig. 5.11. As discussed before, this microstructure is homogenized into a damage field characterized by the set of damage mode tensors $\boldsymbol{D}^{(\alpha)}$. The collection of all variables resulting from thermodynamics with internal state can now be placed in two categories: state variables and response functions. The former for the case of small deformation is given by the strain tensor $\boldsymbol{\varepsilon} = (\nabla \boldsymbol{u})^{\mathrm{sym}}$, where $\boldsymbol{u}$ is the displacement vector, ∇ is the nabla operator and $(\ldots)^{\mathrm{sym}}$, is the symmetric part of the tensor, the absolute temperature T, the temperature gradient $\boldsymbol{g} = \nabla T$, and damage mode tensors $\boldsymbol{D}^{(\alpha)}$. The response functions are the Cauchy stress tensor $\boldsymbol{\sigma}$, the specific Helmholtz free energy ψ, the specific entropy η, and the heat flux vector $\boldsymbol{q}$.

Following Truesdell's principle of equipresence, which states that all state variables should be present in all response functions unless thermodynamics or other relevant considerations preclude their dependency, we write,

$$\boldsymbol{\sigma} = \boldsymbol{\sigma}\left(\boldsymbol{\varepsilon}, T, \boldsymbol{g}, \boldsymbol{D}^{(\alpha)}\right) \tag{5.7}$$

$$\psi = \psi\left(\boldsymbol{\varepsilon}, T, \boldsymbol{g}, \boldsymbol{D}^{(\alpha)}\right) \tag{5.8}$$

$$\eta = \eta\left(\boldsymbol{\varepsilon}, T, \boldsymbol{g}, \boldsymbol{D}^{(\alpha)}\right) \tag{5.9}$$

$$\boldsymbol{q} = \boldsymbol{q}\left(\boldsymbol{\varepsilon}, T, \boldsymbol{g}, \boldsymbol{D}^{(\alpha)}\right) \tag{5.10}$$

By applying the second law of thermodynamics, expressed in the form of the Clausius-Duhem inequality, it can be shown that

$$\boldsymbol{\sigma} = \rho \frac{\partial \psi}{\partial \boldsymbol{\varepsilon}} \tag{5.11}$$

$$\eta = -\frac{\partial \psi}{\partial T} \tag{5.12}$$

$$\frac{\partial \psi}{\partial \boldsymbol{g}} = \boldsymbol{0} \tag{5.13}$$

where ρ is the mass density.

The collection of response functions, Eqs. (5.7)-(5.10) and the associated interrelations, Eqs. (5.11)-(5.13) form the framework for a rational description of constitutive behavior of heterogeneous solids containing evolving internal surfaces (damage). This framework is based on the homogenization depicted in Fig. 5.11, the concept of RVE for characterization of the evolving internal state, and the two laws of thermodynamics. It is possible to extend the framework by including other energy dissipative

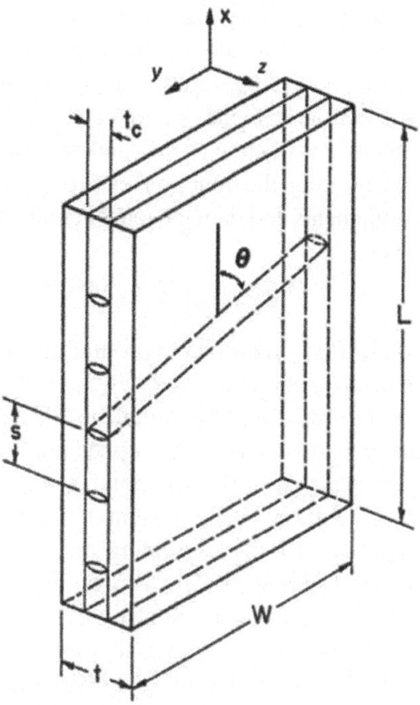

Fig. 5.12 An orthotropic composite laminate containing one array of ply cracks

entities than the internal surfaces formed by atomic bond breakages, e.g., metal crystalline slip in plasticity, and polymer morphological changes in viscoelasticity and aging.

In view of the experimental data, which are mostly available for polymer matrix composites at room temperature, the thermo-mechanical framework will be developed further for the mechanical response. Thus, for isothermal conditions ($T = 0, \boldsymbol{g} = \boldsymbol{0}$) the set of response functions is reduced to the following function.

$$\psi = \psi(\boldsymbol{\varepsilon}, \boldsymbol{D}^{(\alpha)}) \tag{5.14}$$

Note that stress is derivable from the Helmholtz free energy function according to Eq. (5.11). Thus, dealing with this scalar-valued function as the sole response function for a given internal state of damage provides a favorable situation for further development of the theory.

Consider an initially orthotropic composite laminate with one array of intralaminar cracks as illustrated by Fig. 5.12. Denoting this damage mode by $\alpha = 1$, the damage mode tensor components can be written as

$$\boldsymbol{D}^{(1)} = \frac{\kappa t_c^2}{st\cos\theta} \otimes \boldsymbol{n} \tag{5.15}$$

where Eq. (5.6) is used in which $V = L \cdot W \cdot t$, where L, W and t are the length, width, and thickness of the representative volume V of the laminate, respectively, $S = t_c(W/\cos\theta)$, and $a = \kappa \cdot t_c$, where κ is an assumed constant of proportionality between the influence vector magnitude (here, crack opening displacement) and crack dimension t_c (same as the thickness of cracked ply). Finally, $n_i = (\cos\theta, \sin\theta, 0)$.

Choosing the function in Eq. (5.14) as a polynomial function of strain tensor components and damage tensor components, both symmetrical second order tensors, gives

$$\psi = P(\varepsilon_1, \varepsilon_2, \ldots, \varepsilon_6, D_1, D_2, \ldots, D_6) \tag{5.16}$$

where P stands for a polynomial function and the tensor components are expressed in the Voigt notation, $\varepsilon_1 = \varepsilon_{11}, \varepsilon_2 = \varepsilon_{22}, \varepsilon_3 = \varepsilon_{33}, \varepsilon_4 = \varepsilon_{23}, \varepsilon_5 = \varepsilon_{13}, \varepsilon_6 = \varepsilon_{12}$, and similarly for the $\boldsymbol{D}$ components. The superscript on $\boldsymbol{D}$ components indicating the damage mode has been dropped here for convenience.

Expansion of the polynomial function (5.16) can in general have infinite terms, which will obviously present an impractical situation. One way to restrict the functional form is by expanding the polynomial in terms that account for the initial material symmetry. This is done in the polynomial invariant theory by using the so-called integrity bases. Such bases have been developed for scalar functions of various vector and tensor variables. For the case of two symmetric second order tensors, such as in Eq. (5.16), the integrity bases for orthotropic symmetry are given by Adkins (1959),

$$\begin{aligned}
&\varepsilon_1, \varepsilon_2, \varepsilon_3, \varepsilon_4^2, \varepsilon_5^2, \varepsilon_6^2, \varepsilon_4, \varepsilon_5, \varepsilon_6,\\
&D_1, D_2, D_3, D_4^2, D_5^2, D_6^2, D_4, D_5, D_6,\\
&\varepsilon_4 D_4, \varepsilon_5 D_5, \varepsilon_6 D_6,\\
&D_4\varepsilon_5\varepsilon_6, D_5\varepsilon_6\varepsilon_4, D_6\varepsilon_4\varepsilon_5,\\
&\varepsilon_4 D_5 D_6, \varepsilon_5 D_6, \varepsilon_6 D_4 D_5
\end{aligned} \tag{5.17}$$

For the sake of applying the constitutive theory to thin laminates where only in-plane strains are of interest, and for small strains, the expansion of the function (5.16) can be restricted to no more than quadratic terms in strain components ε_1, ε_2 and ε_6. To what extent the damage tensor components are to be taken in the expansion depends on the nature and amount of information that can be acquired for evaluation of the material constants that will appear in the polynomial function. This issue will be discussed later. To begin with the simplest possible case, we will include only linear terms in D_1, D_2 and D_6, which are the non-zero components for intralaminar cracks. Thus,

$$\begin{aligned}
\rho\psi = {} & c_1\varepsilon_1^2 + c_2\varepsilon_1\varepsilon_2 + c_3\varepsilon_1^2 D_1\varepsilon + c_4\varepsilon_1^2 D_2 + c_5\varepsilon_1\varepsilon_6 D_6 + c_6\varepsilon_2^2 + c_7\varepsilon_2^2 D_1 + c_8\varepsilon_2^2 D_2\\
& + c_9\varepsilon_2\varepsilon_6 D_6 + c_{10}\varepsilon_6^2 + c_{11}\varepsilon_6^2 D_1 + c_{12}\varepsilon_6^2 D_2 + c_{13}\varepsilon_1\varepsilon_2 D_1 + c_{14}\varepsilon_1\varepsilon_2^2 D_2\\
& + P_0 + P_1(\varepsilon_p, D_q) + P_2(D_q)
\end{aligned} \tag{5.18}$$

where $c_i, i = 1, 2, \ldots, 14$, are material constants, P_0 is a constant, P_1 is a linear function of strain and damage tensor components, and P_2 is a linear function of the damage tensor components. $P_0 = 0$ if the free energy value in the undeformed and undamaged state is set to zero. Letting the unstrained material of any damage state to be stress-free, one gets $P_1 = 0$, on using Eq. (5.11).

Equation (5.11) expressed in the Voigt notation is given by

$$\sigma_p = \rho \frac{\partial \psi(\varepsilon_q, D_r)}{\partial \varepsilon_p} \tag{5.19}$$

A differential in stress can now be written as

$$\mathrm{d}\sigma_p = \rho \frac{\partial \psi}{\partial \varepsilon_p \partial \varepsilon_q} \mathrm{d}\varepsilon_q + \rho \frac{\partial \psi}{\partial \varepsilon_p \partial D_r} \mathrm{d}D_r = C_{pq} \mathrm{d}\varepsilon_q + K_{pr} \mathrm{d}D_r \tag{5.20}$$

where

$$C_{pq} = \rho \frac{\partial \psi}{\partial \varepsilon_p \partial \varepsilon_q} \tag{5.21}$$

is the stiffness matrix when $\mathrm{d}D_r = 0$, i.e., constant damage. Thus, the elastic modulus at any point on the stress-strain curve is the secant modulus, not the tangent modulus.

Using Eqs. (5.18) and (5.21), one obtains

$$C_{pq} = C_{pq}^0 + C_{pq}^1 \tag{5.22}$$

where

$$C_{pq}^0 = \begin{bmatrix} 2c_1 & c_2 & 0 \\ c_2 & 2c_6 & 0 \\ 0 & 0 & 2c_{10} \end{bmatrix} \tag{5.23}$$

and

$$C_{pq}^1 = \begin{bmatrix} 2c_3 D_1 + 2c_4 D_2 & c_{13} D_1 + c_{14} D_2 & c_5 D_6 \\ c_{13} D_1 + c_{14} D_2 & 2c_7 D_1 + 2c_8 D_2 & c_9 D_6 \\ c_5 D_6 & c_9 D_6 & 2c_{11} D_1 + 2c_{12} D_2 \end{bmatrix} \tag{5.24}$$

It can be noted here that Eqs. (5.22)-(5.24) show linear dependence of the stiffness properties on damage tensor components. This is the consequence of including only linear terms in these components in the polynomial expansion of the free energy function, Eq. (5.18). Including higher order terms will add additional constants c_i, which will need to be evaluated. The evaluation procedure is described below, but it is remarked here that the formulation of constitutive response is in no way restricted only to linear dependence on the chosen damage measure.

The evaluation of the material constants $c_i (i = 1, \ldots, 14$ for one intralaminar cracking mode) will first be illustrated by the case of transverse cracks, i.e., $\theta = 0$ in Fig. 5.12. For this case, $D_2 = D_6 = 0$, and D_1 is given by

$$D_1 = \frac{\kappa t_c^2}{st} \tag{5.25}$$

and the C^1_{pq} matrix takes the form

$$C^1_{pq} = \frac{\kappa t_c^2}{st}\begin{bmatrix} 2c_3 & c_{13} & 0 \\ c_{13} & 2c_7 & 0 \\ 0 & 0 & 2c_{11} \end{bmatrix} \tag{5.26}$$

Thus, with the transverse cracking mode the elastic response of the initially orthotropic laminate retains this symmetry. The number of constants that need evaluation is now eight, four of which (c_1, c_2, c_6 and c_{10}) correspond to the initial elastic response for which the constants can either be measured experimentally or calculated by the laminate plate theory using ply properties. The remaining four constants can be deduced from measurements of the four independent elastic constants, the Young's moduli E_1 and E_2, in the axial and transverse directions, respectively, the major Poisson's ratio ν_{12} and the shear modulus G_{12}, all at a given value of D_1. Expressions for these constants linearized in D_1 are as follows.

$$\begin{aligned} E_1 &= E_1^0 + 2\frac{\kappa t_c^2}{st}\left[c_3 + c_7(\nu_{12}^0)^2 - c_{13}\nu_{12}^0\right] \\ E_2 &= E_2^0 + 2\frac{\kappa t_c^2}{st}\left[c_7 + c_3(\nu_{12}^2)^2 - c_{13}\nu_{12}^0\right] \\ \nu_{12} &= \nu_{12}^0 + \frac{\kappa t_c^2}{st}\left[\frac{1-\nu_{12}^0\nu_{21}^0}{E_2^0}\right](c_{13} - 2c_7\nu_{12}^0) \\ G_{12} &= G_{12}^0 + 2\frac{\kappa t_c^2}{st}c_{11} \end{aligned} \tag{5.27}$$

where quantities with superscript 0 correspond to initial (undamaged) state, and

$$\nu_{21}^0 = \nu_{12}^0\frac{E_2^0}{E_1^0}$$

The unknown constants in Eq. (5.27) can now be expressed as

$$\begin{aligned} \kappa c_3 &= A_1(1 - 2\nu_{12}^0\nu_{21}^0) + A_2(\nu_{12}^0)^2 + 2A_3E_2^0\nu_{12}^0 \\ \kappa c_7 &= -A_1(\nu_{12}^0)^2 + A_2 + 2A_3E_2^0\nu_{21}^0 \\ \kappa c_{11} &= A_4 \\ \kappa c_{13} &= -2A_1\nu_{12}^0(\nu_{21}^0)^2 + 2A_2\nu_{12}^0 + 2A_3E_2^0(1 + \nu_{12}^0\nu_{21}^0) \end{aligned} \tag{5.28}$$

where

$$\begin{aligned} A_1 &= Q(E_1 - E_1^0) \\ A_2 &= Q(E_2 - E_2^0) \\ A_3 &= Q(\nu_{12} - \nu_{12}^0) \\ A_4 &= \frac{Q}{(1 - \nu_{12} - \nu_{12}^0)^2}(G_{12} - G_{12}^0) \end{aligned} \tag{5.29}$$

with

$$Q = \frac{ts}{2t_c^2(1 - \nu_{12}^0 \nu_{21}^0)^2} \tag{5.30}$$

From Eqs. (5.28)-(5.30) it is seen that in the case of elastic response linearized in damage components the material constants c_i are not evaluated but instead their products with the crack opening parameter κ are determined. While the values of c_i are fixed for a given composite laminate (that has been homogenized), the parameter κ depends on the ability of the cracks to perform surface displacements under applied mechanical impulse. Thus, this parameter may be viewed as a measure of the constraint to the crack surface separation imposed by the material surrounding the crack. One way to treat this is by considering a crack of a given size embedded in an infinite isotropic material, in which case the crack surface separation is unconstrained and can be calculated by fracture mechanics methods. When the laminate geometry is finite and its symmetry is different from isotropic, the κ parameter will take a value less than that for the infinite isotropic medium. This consideration allows us to assign κ an undermined value, say κ_0, for a reference laminate under reference loading conditions, and evaluate a change from this value for another crack orientation. This approach is discussed further below.

5.4.4 Synergistic Damage Mechanics (SDM)

The observation that the κ parameter (hitherto referred to as constraint parameter) may be viewed as a carrier of the local effects on damage entities within a RVE, while the ci-constants are material constants, led to several studies for exploring prediction of elastic property changes due to damage in different modes. To be sure, the elastic properties are the averages over appropriate RVEs. At first it was found that from changes in E_1 and ν_{12} due to transverse cracking in $[0/90_3]_s$ glass/epoxy laminates reported in Highsmith and Reifsnider (1982) and assuming no changes in E_2, the constants calculated by the procedure described above, Eqs. (5.28)-(5.30), gave the values $\kappa c_3 = -6,712$ GPa, $\kappa c_7 = -0,770$ GPa, and $\kappa c_{13} = -4,455$ GPa. Using these values changes in E_1 for the same glass/epoxy of $[0/90]_s$ configuration could be predicted with good accuracy. Also, in $[0/\pm 45]_s$ laminate of the same glass/epoxy, the change in E_1 could be predicted by setting $D_1 = D_2$ (a good approximation, supported by crack density data). These results have been reported in Talreja (1990).

Later, a systematic study of the effect of constraint on the κ parameter was done by experimentally measuring the crack opening displacement (COD) in $[\pm\theta/90_2]_s$ laminates (Varna et al., 1999) for different θ values. By relating these values to the COD at $\theta = 0$ normalized by a unit applied strain, the predictions of E_1 and ν_{12} for different θ could be made. Another study of the constraint effects was made by examining $[0/\pm\theta_4/0_{1/2}]_s$ laminates, where the ply orientation θ was varied. Once again, using experimentally measured COD for $\theta = 0$ as the reference, the κ parameter for other ply orientations was evaluated from the COD values and E_1 and ν_{12} for different θ were predicted (Varna et al., 1999).

While the experimental studies supported the idea of using the κ parameter as a carrier of local constraints, the scatter in test data and the cost of testing do not make the experimental approach attractive. Therefore, another systematic study of $[0_m/\pm\theta_n/\theta_{m/2}]_s$ laminates was undertaken (Singh and Talreja, 2008) where computational micromechanics was employed instead of physical testing. An elaborate parametric study of the κ parameter allowed developing a master curve for elastic property predictions.

Another study (Singh and Talreja, 2013a) examines damage modes consisting of transverse ply cracks as well as inclined cracks of different orientations in $[0_m/\pm\theta_n/90_r]_s$ and $[0_m/90_r/\pm\theta_n]_s$ laminates. The SDM approach is developed and its predictions are compared with available experimental data for $[0/90/-45/+45]_s$ laminate.

The treatment above used only elastic response to illustrate CDM/SDM formulation. Polymer matrix composites are found to have time dependent (viscoelastic) response at elevated temperatures. For the linear viscoelastic case, Kumar and Talreja (2001) developed a methodology for predicting the material constants at fixed damage using the correspondence principle. If the viscoelastic deformation becomes nonlinear, the correspondence principle does not apply in which case the internal state can be characterized by variables that represent the molecular morphology changes in addition to the damage variables. An approach along these lines was developed by Ahci and Talreja (2006) where a mixed experimental and computational procedure for identification of the material constants was proposed and implemented.

5.4.5 Remarks on Characterization of Damage

At this point it would be appropriate to remind the reader that damage as defined here is a material state consisting of a multitude of internal surfaces in a solid formed by energy dissipative (irreversible) processes. Thus, the proposed damage characterization must be consistent with this definition. In composite materials the internal surfaces are oriented due to the oriented nature of the fibers as well as of the interfaces between fibers and matrix and between layers of UD composites. Furthermore, the response of the material to an imposed loading (mechanical, thermal, etc.) depends on the material state, which by evolving to another state will change the response. The formulation of the material response presented here is based on internal state variables and is a rational framework consistent with the laws of thermodynamics. However, it is in no way a unique framework as it depends on how the damage is characterized. There are other approaches in the literature where the internal state is characterized differently, e.g., as a discrete set of scalar variables representing internal structural rearrangements (Rice, 1971) in the context of metal plasticity. The important point is that the chosen characterization of the internal state is made a part of the material response framework subjected to the restrictions of the laws of thermodynamics and any material symmetry requirements. Unfortunately, there is a host of approaches in the literature that quantify damage arbitrarily as a number

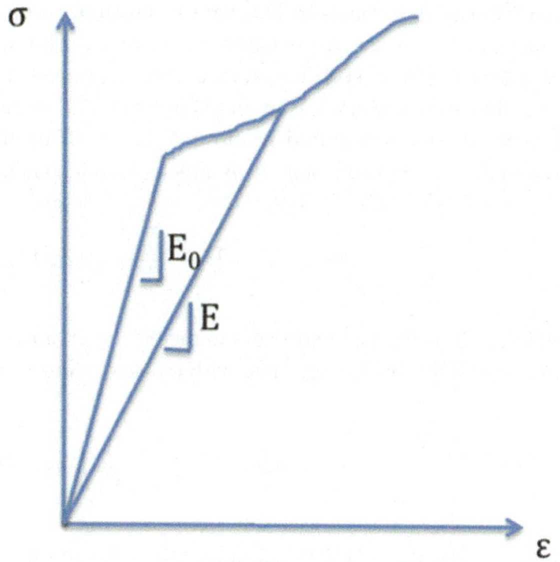

Fig. 5.13 Schematic illustration of the elastic response in the current state of damage, given by the elasticity constant E, while the initial (undamaged) state has constant E_0

between 0 and 1 (or another chosen value) or express damage in terms of a normalized response change, e.g., in an elasticity constant (often the Young's modulus). Such approaches should be viewed with caution and not taken as legitimate damage mechanics treatments.

5.4.6 Evolution of Damage

So far, we have described the elastic response of a composite with a fixed damage. In internal variable thermodynamics this is expressed as response of a constrained equilibrium state (Rice, 1971), i.e., in measuring the material's response its evolving internal state is constrained from evolving. This is easily done in the case of a composite with damage by measuring its stress-strain response under unloading when no further damage is induced and the response represents the damage state at the stress from where unloading was done, as illustrated in Fig. 5.13. The evolution from a previous material state is treated by the so-called kinetic theory. The kinetic theory can be made a part of the constitutive framework described above by adding a damage rate function to the set of response functions, Eqs. (5.7)–(5.10). However, an evaluation of the damage rate function along the lines of, say, the Helmholtz free energy function, i.e., by expanding it as a polynomial will lead to more phenomenological constants. Instead, another approach where the damage evolution is treated at a local scale, i.e., at the scale of the damage entities, without smearing out the damage entities is preferred. Such an approach is described next.

An Energy Approach to Damage Evolution. In this approach, a virtual crack is introduced between two existing cracks and work required to close this new crack is obtained. If the work for crack closure is greater than the critical energy release rate, then this new crack is assumed to form. Following the approach by Varna et al. (1999), the work required to close N cracks from the state of $2N$ cracks (crack spacing, $s = 2l$) to the state of N cracks (crack spacing, $s = l$), can be derived as

$$W_{2N \to N} = 2N \frac{\left(\sigma_{xx0}^{90}\right)^2}{E_2} t_{90}^2 [2\tilde{u}(l/2) - \tilde{u}(l)] \tag{5.31}$$

where σ_{xx0}^{90} is the far-field stress in the 90° layer, and $\tilde{u}(l)$ represents the normalized average COD for damage state with crack spacing s, and is defined as

$$\tilde{u}(s) = \frac{E_2}{\sigma_{xx0}^{90}(t_{90})^2} \int_0^{t_{90}} u(z, s)\mathrm{d}z \tag{5.32}$$

where u represents COD variation along the ply thickness (z-direction) evaluated at a given crack spacing s. The formation of new cracks is assumed to occur when $W_{2N \to N}$ is greater than or equal to the total surface energy of N newly created cracks, i.e.,

$$W_{2N \to N} \geq 2 \cdot N \cdot 2t_{90} \cdot G_c \tag{5.33}$$

where G_c is the critical energy release rate. Substituting $W_{2N \to N}$ from Eq. (5.31) in Eq. (5.33), the criterion for crack multiplication is derived as

$$\frac{(\sigma_{xx0}^{90})^2}{2E_2} t_{90}[\tilde{u}(s/2) - \tilde{u}(s)] \geq G_c \tag{5.34}$$

If the new crack does not form midway between two existing cracks, but at distance $l_1/2$ from one crack and $l_2/2$ from the other crack, then the cracking criterion is given by

$$\frac{(\sigma_{xx0}^{90})^2}{2E_2} t_{90}[2\tilde{u}(l_1/2) - \tilde{u}(l_1) + 2\tilde{u}(l_2/2) - \tilde{u}(l_2)] \geq G_c \tag{5.35}$$

To predict evolution of ply cracking in multiple off-axis plies, Singh and Talreja (2010) modified the fracture mechanics-based analysis described above. In this analysis, the work required to close N cracks in State 1 with crack spacing s, and $2N$ cracks in State 2 with crack spacing $s/2$ (Fig. 5.14), can be written as

$$W_{N \to 0} = N \frac{1}{\sin\theta}(t_\theta)^2 \frac{1}{E_2}[(\sigma_{20}\theta)^2 \tilde{u}_n^\theta(s) + 2\tilde{u}(l_1/2) - \tilde{u}(l_1)(\sigma_{120}\theta)^2 \tilde{u}_t^\theta(s)] \tag{5.36}$$

$$W_{2N\to 0} = N\frac{1}{\sin\theta}(t_\theta)^2\frac{1}{E_2}[(\sigma_{20}\theta)^2\tilde{u}_n^\theta(s/2) + 2\tilde{u}(l_1/2) - \tilde{u}(l_1)(\sigma_{120}\theta)^2\tilde{u}_t^\theta(s/2)] \tag{5.37}$$

where θ and t_θ denote orientation and total thickness, respectively, of the cracked off-axis plies, and $\tilde{u}_n^\theta$, $\tilde{u}_t^\theta$ are the normalized average crack opening displacement (COD) and crack sliding displacement (CSD), respectively. These are evaluated as

$$\tilde{u}_n^\theta(s) = \frac{1}{(t_\theta)^2}\frac{E_2}{\sigma_{20}^\theta}\int_{-t_\theta/2}^{t_\theta/2} u_n^\theta(z,s)\mathrm{d}z, \qquad \tilde{u}_t^\theta(s) = \frac{1}{(t_\theta)^2}\frac{E_2}{\sigma_{120}^\theta}\int_{-t_\theta/2}^{t_\theta/2} u_t^\theta(z,s)\mathrm{d}z \tag{5.38}$$

In the above equations, u_n and u_t represent the relative opening and sliding displacement of the cracked surfaces, respectively, and overbars indicate their averages. The new cracks in off-axis plies will form when the work required in going from State 1 to State 2 exceeds a critical value of the work, i.e., if

$$W_{2N\to N} = W_{2N\to 0} - W_{N\to N} \geq NG_c\frac{t_\theta}{\sin\theta} \tag{5.39}$$

where G_c here is the critical value of energy required for multiple ply crack formation within the given laminate. It should be noted here that Gc should not be taken strictly as the critical energy rate defined in the usual sense (i.e., for a single crack growth) in linear elastic fracture mechanics, but rather as a fitting parameter to represent multiple crack formation. Therefore, it should be evaluated by fitting model predictions to experimental data for multiple cracking in a representative laminate, such as a selected cross-ply laminate; see Singh and Talreja (2013a). Furthermore, for cracking in a general off-axis ply a multi-mode criterion should be utilized, given as

$$\left(\frac{w_\mathrm{I}}{G_\mathrm{Ic}}\right)^M + \left(\frac{w_\mathrm{II}}{G_\mathrm{IIc}}\right)^N \geq 1 \tag{5.40}$$

where

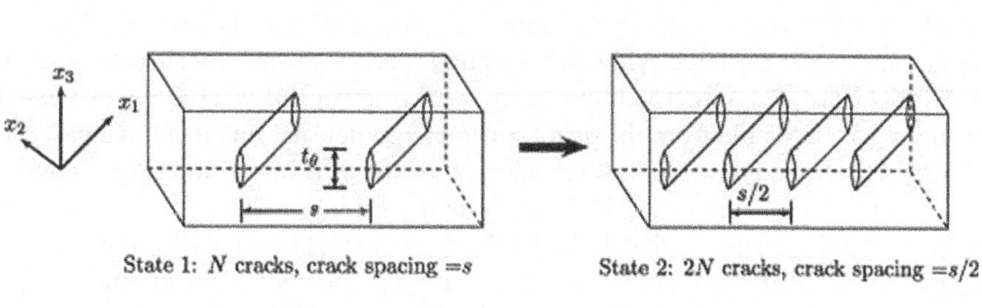

Fig. 5.14 Multiplication of cracks in multidirectional laminates. The figure illustrated cracks in plies of one orientation (Singh and Talreja, 2010)

$$w_{\mathrm{I}} = \frac{(\sigma_{20}^{\theta})^2 t_\theta}{E_2}\left[2\tilde{u}_n^\theta\left(\frac{s}{2}\right) - \tilde{u}_n^\theta(s)\right], \qquad w_{\mathrm{II}} = \frac{(\sigma_{120}^{\theta})^2 t_\theta}{E_2}\left[2\tilde{u}_t^\theta\left(\frac{s}{2}\right) - \tilde{u}_t^\theta(s)\right] \tag{5.41}$$

and G_{Ic} and G_{IIc} are the critical energy release rates in Mode I (opening mode) and Mode II (crack sliding mode), respectively. The multi-mode exponents M and N in Eq. (5.40) depend on the material system, e.g., for glass/epoxy laminates, one uses $M = 1$, $N = 2$. For further details, see Singh and Talreja (2010, 2013a) where predictions of cracking evolution are compared with experimental data.

5.5 Modeling of Failure

Attempts at modeling of failure in composite materials began much before the mechanisms underlying failure could be observed. The review of the failure modes in Sect. 5.2 above is the status of observations that continues to evolve as refined techniques such as micro-focus X-ray tomography are used. At the same time, computational simulations are revealing detailed characteristics of failure that were not possible to find with analytical formulations in the past. The early treatments of failure in composites were an outgrowth of failure in metals using the concept of strength that has shown to be inadequate for composites (Talreja and Waas, 2022). In the following review of failure modeling a historical development of essential theories will be sketched followed by recommendations for future development. Previous articles on the subject (Talreja, 2014, 2016) should be consulted for more details.

5.5.1 *Phenomenological Failure Theories for UD Composites*

The first notable effort to describe failure of a homogenized unidirectional UD composite was by Azzi and Tsai (1965), which has come to be known as the Tsai-Hill theory. Hill (1948) proposed a generalization of the yield criterion for isotropic metals to orthotropic metals by assuming six independent yield stresses, three for normal stresses in the three symmetry directions and other three for shear stresses in the three planes of symmetry. Importantly, while the von Mises criterion expressed in stresses is derivable from the distortional energy density, this is not the case for the Hill criterion. Thus, Hill's generalization of the von Mises criterion is simply a mathematical generalization. Hill in fact motivated his criterion from the need to describe yielding in metals that develop texture from sheet forming by rolling, resulting in the orthotropic symmetry. The Azzi-Tsai adaptation of the Hill criterion to UD composites was based on assumed similarity between the rolling direction in metal sheets and the reinforcing direction in unidirectional composites. The final equation, referred to as the Tsai-Hill criterion, was derived by equating the six yield stresses in the Hill criterion to the corresponding "strengths" of a UD composite, and further assuming transverse isotropy in the composite cross-sectional plane. Since the UD composites used as layers in laminates are usually thin, the criterion is commonly used in its two-dimensional version, given by

$$\frac{\sigma_{xx}^2}{X^2} + \frac{\sigma_{yy}^2}{Y^2} - \frac{\sigma_{xx}\sigma_{yy}}{X^2} + \frac{\tau_{xy}^2}{T^2} = 1 \tag{5.42}$$

where σ_{xx} and σ_{yy} are the normal stresses in the fiber and transverse directions, respectively, τ_{xy} is the in-plane shear stress, and X, Y and T are the strength (maximum attained) values of σ_{xx} and σ_{yy} and τ_{xy}, respectively. For unequal strengths in tension and compression a variation of Eq. (5.42) is easily derived.

Although initial verification of Eq. (5.42) was indicated for the case of UD composites loaded in tension at various angles to the fiber direction, many cases later did not support the criterion.

The next noteworthy development of failure criteria for UD composites appeared in a paper by Tsai and Wu (1971) based on a formulation proposed by Gol'denblat and Kopnov (1965, 1971a, b). The proposed formulation is a scalar function expressed as a polynomial of all stress tensor components. The coefficients of the polynomial terms represent the strength constants. The Tsai-Wu modification of this function was a simplification of it to a quadratic expression that allowed an interpretation as a quadric surface using analytic geometry. Thus, the ellipsoidal surface of "failure" of a UD composite under in-plane stresses took the form

$$F_p\sigma_p + F_{pq}\sigma_p\sigma_q = 1 \tag{5.43}$$

where indices p and q take values 1, 2 and 6, representing the indices xx, yy and xy, respectively, in the usual notation, and the summation convention for repeated indices is implied. The coefficient terms in the equation represent strength values of the corresponding stresses. Coefficients of the linear terms are non-zero when the strengths for positive and negative stress components differ. Thus, F_6 is always zero, and for the same reason, F_{16} and F_{26} vanish. Of the remaining coefficients, F_1, F_2, F_{11}, F_{22} and F_{66} can be expressed in terms of the normal strength values in the fiber and transverse directions and the shear strength in the plane of the composite. Finally, the coefficient F_{12} can in principle be obtained by a biaxial test, but the non-uniqueness of this procedure is a source of ambiguity in the criterion. It can be argued that since the Tsai-Wu criterion, Eq. (5.43), is not connected to the specifics of the failure in composites, its applicability will always be limited by the uncertainty of determining the F_{12} value.

At this point it can be said that the two criteria discussed above are severely handicapped by their inherent limitations: The Tsai-Hill criterion is motivated by an incorrect failure mechanism (yielding, which is not the failure mechanism in UD composites with a non-metallic matrix), and the Tsai-Wu criterion has no specifics of a failure mechanism, rendering it simply a quadratic curve-fitting framework.

Hashin (1980) pointed out that the single ellipsoid represented by Eq. (5.43) leads to physically unacceptable interactions between stress components in some cases. He suggested to introduce piecewise smooth surfaces instead to describe critical states and additionally to recognize that composite failure involving fiber breakage was governed by stresses differently than when the failure occurs in the matrix only. Assuming the two failure modes to be independent, Hashin proceeded to formulate criteria for them separately. Thus, for example, the fiber failure mode when the fiber axial stress is tensile, was assumed to be given by

$$\left(\frac{\sigma_1}{X}\right)^2 + \left(\frac{\sigma_6}{T}\right)^2 = 1 \tag{5.44}$$

Thus, the transverse normal stress was assumed not to affect the fiber failure mode and the quadratic form of the failure criterion was retained.

For the matrix failure mode, Hashin (1980) proposed the notion of a failure plane not intersecting fibers in a UD composite. Assuming the fiber direction stress not to have influence on this plane, Hashin proposed that the other stress components would determine the inclination of the plane. He outlined the difficulties involved in determining the angle of this inclination and proposed that some (unknown) extremum principle would govern it. Notwithstanding Hashin's word of caution, Puck and Schürmann (1998) launched an elaborate procedure to develop failure criteria incorporating the matrix failure plane idea. The resulting so-called Puck failure criteria for a UD composite require determination of seven material constants.

The use of a failure plane as a basis for developing phenomenological failure theories has continued since the Puck criterion was found to fare well in the world-wide failure exercise (Hinton et al., 2004) which compared different failure theories against a set of test data. This is likely because the Puck criterion requires seven calibration constants, which gives it a good ability to describe variation of strength under different conditions of combined loading. Among the latest efforts to develop failure theories based on failure planes is that by Gu et al. (2020) where a modified nonlinear Mohr-Coulomb criterion is used for UD composites assuming these to be homogeneous transversely isotropic solids. The controversies regarding correctness of phenomenological failure criteria have also continued, see Li (2020) and Gu et al. (2021), where most cited failure criteria are examined. Gu et al. (2021) conclude that none of the examined criteria were able to describe test data under all combinations of stresses.

Without conducting a detailed and exhaustive scrutiny of all proposed failure theories that operate on the homogenized description of composites, one can reasonably put forth the argument that the inherent limitation in such theories would still be the inability to incorporate the conditions for initiation of the first failure event and for the sequential development of subsequent failure events. These failure events necessarily occur at scales that have been obliterated in the homogenization. The final failure event associated with criticality (e.g., unstable progression of failure events or transition to another failure mode) cannot generally be analyzed without delineation of the subcritical path of the prior failure events. These arguments point to an alternative approach to failure assessment of composites where resort is made to the size-scales of the composite constituents and the configurations in which the

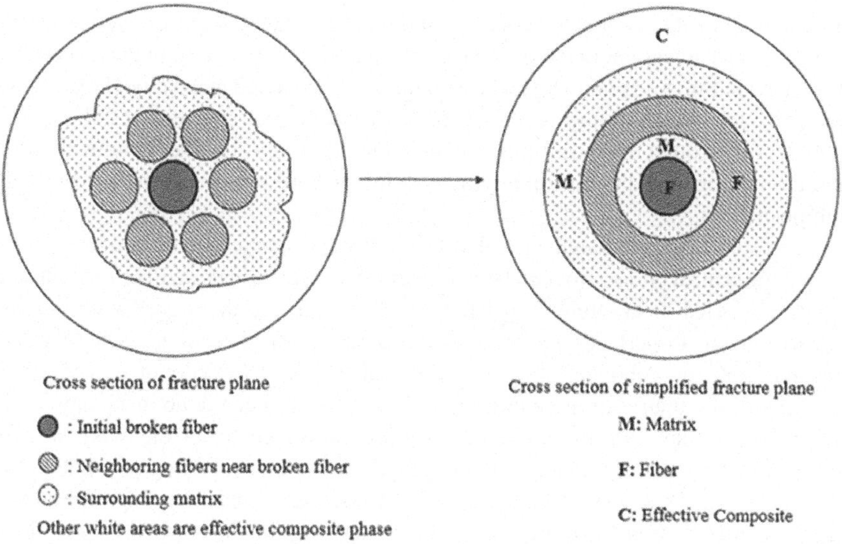

Fig. 5.15 A fiber cluster of focus for the composite failure analysis (left) is modeled as a five-cylinder axisymmetric assembly (right) (Nedele and Wisnom, 1994)

constituents have been put together. The approach of multiscale nature to structural integrity of composites will be discussed next.

5.5.2 *Physical Modeling of Failure*

Physical modeling in contrast to phenomenological strength formulations described above captures understanding of the observed failure mechanisms into models. Such models operate at two scales: a scale where the failure events occur and a scale where description of the structural failure is given. In this context, the following subsections describe multi-scale approaches with respect to individual failure modes discussed in Sect. 5.2 above. See also Talreja (2014, 2016).

Fiber Failure Mode in Axial Tension. The observed failure process in this case, as described in Sect. 5.2.1 above, suggests that the first scale at which a failure event occurs is the fiber diameter. Either a broken fiber exists from a manufacturing process, or it results from loading as breakage at the weakest point. The next event is failure of fibers about a broken fiber caused by the stress enhancement induced by the first broken fiber. Observations indicate localized failure in clusters of fibers, one or more of which grow unstably, triggering ultimate failure of the UD composite.

Most literature has focused on analyzing conditions for formation of a cluster of broken fibers. The scale at which stress and failure analysis is conducted consists of a representative region containing sufficient fibers to simulate a fiber-cluster failure.

An example of this is in Nedele and Wisnom (1994) where a representative region is modeled as an assembly of five concentric cylinders. This axisymmetric model is illustrated in Fig. 5.15. The central broken fiber is surrounded in the model by a cylinder of matrix, which in turn is surrounded by a cylinder of smeared-out fibers. The cylinder of the outer matrix is placed to maintain the composite fiber volume fraction and the homogenized composite cylinder lying beyond represents the surrounding composite.

The Nedele-Wisnom analysis calculated the stress enhancement in fibers neighboring the central broken fiber and the probability of failure in these fibers accounting for the statistical fiber strength distribution. The matrix in the model was assumed as either elastic or perfectly plastic and consideration was given to a frictional stress transfer from the debonded length of the broken fiber. No account was taken of cracking of the matrix in the interfiber region. This has been done in Zhuang et al. (2018a, b). In that work the broken fiber placed at the center of the axisymmetric model initiates matrix cracking or fiber-matrix debonding depending on whether the broken fiber exists before loading is applied or if it breaks under loading. The subsequent failure events consist of failure of fibers neighboring the existing broken fiber. The stress enhancement in the neighboring fibers is affected by the matrix crack emanating from the broken fiber end or because of kinking out of the debond crack. The stress enhancement on the neighboring fibers extends over an axial distance and its intensity depends on the extent to which the matrix crack has grown and if the broken central fiber has debonded from the matrix, in which case, the extent of debonding matters.

Before the findings of Swolfs et al. (2015); Zhuang et al. (2016) had concluded that the effect of the presence of a matrix crack had negligible effect on stress concentration, ineffective length, cluster development and failure strain of a UD composite in axial tension. The Swolfs et al. (2015) analysis assumed a matrix crack to lie around a single fiber break within a UD composite and compared this scenario with having no matrix crack. The heavy constraint imposed by the stiff fibers on the separation of the crack surfaces essentially resulted in having no crack. It should be obvious that without a matrix crack growing a single fiber break cannot result in a broken fiber cluster formation. Furthermore, failure of the composite can only result from either a fiber cluster growing unstably or multiple fiber clusters joining through fiber/matrix interface cracking to form a failure plane that grows unstably. Experimental data suggest this behavior and indicate a significant role in it of matrix cracking.

Fiber Failure Mode in Axial Compression. The earliest model for compressive failure of UD composites was proposed by Rosen (1964), which assumed two micro-buckling modes in initially straight fibers, as depicted in Fig. 5.16. The out-of-phase mode, called extensional mode, gave higher compressive strength than the in-phase mode, called shear mode. The prediction of compressive strength by the shear mode of fiber micro-buckling, $\sigma_c = G$, the shear modulus of the UD composite, was found to over-estimate the experimentally found values.

Focus on compressive failure modeling shifted to kink bands when Budiansky (1983) proposed a model for kink band formation. He derived an expression for the compressive strength, assuming inextensible fibers, as

$$\sigma_c = G + E_T \tan \beta \tag{5.45}$$

where, E_T is the transverse Young's modulus of the UD composite and β is as defined in Fig. 5.2. It is noted that when $\beta = 0$, the Budiansky prediction equals the Rosen prediction for shear mode of fiber micro-buckling.

Based on a post-buckling analysis of elastic composite laminates, Budiansky (1983) argued that the compressive strength was not sensitive to defects such as fiber misalignment. However, he showed that if the composite had significant plasticity in shear, then the initial fiber misalignment had a significant effect. Based on perfect plasticity in shear, he derived the ratio of the new (plasticity-based) failure stress, σ_s, to the elasticity-based failure stress σ_c, as

$$\frac{\sigma_s}{\sigma_c} = \frac{1}{(\phi/\phi_s) + 1} \tag{5.46}$$

Thus, if the misalignment angle ϕ is, say 2°, and the shear induced rotation, i.e., shear strain, is $\phi_s = 0.002$, then the compressive failure stress is predicted to be 1/18 of the elastic value σ_c.

Substituting σ_c from Eq. (5.46) into Eq. (5.45) and using $G = \tau_y/\phi_s$, where τ_y is the shear yield stress, one obtains,

$$\sigma_s = \frac{\tau_y + (E_T/\phi_s)\tan \beta}{\phi + \phi_s} \tag{5.47}$$

It is seen that for $\beta = 0$, the compression failure stress becomes

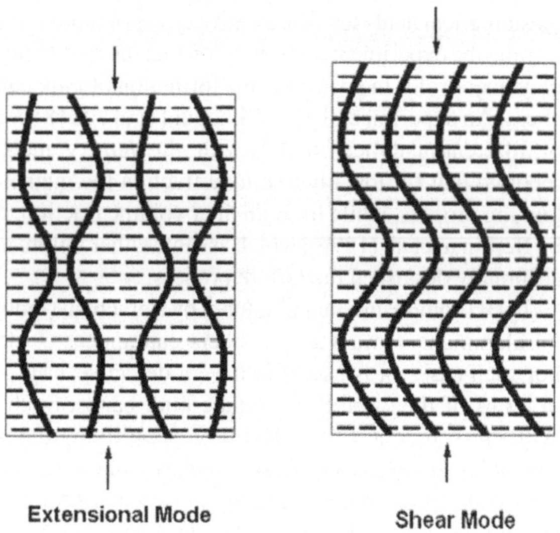

Fig. 5.16 Extensional (out-of-phase) and shear (in-phase) micro-buckling modes postulated by Rosen (1964)

$$\sigma_s = \frac{\tau_y}{\phi + \phi_s} \tag{5.48}$$

At the point when the shear induced rotation is incipient, i.e., when $\phi_s = 0$, the elastic kink band formation stress becomes

$$\sigma_c = \frac{\tau_y}{\phi} \tag{5.49}$$

This formula was proposed by Argon (1972) based on instability in the localized rotation of the initially misaligned fibers. In his model, the stress τ_y is the shear strength of the UD composite rather than the shear yield stress. The instability in the fiber rotation angle may be viewed as caused by the perfect (constant-stress) plasticity in the shear response of the UD composite in the Budiansky model. In other words, the Argon formula is for the beginning of the kink band formation process. In this sense, it gives a lower bound to the compression failure stress.

Note that the kink band rotation angle β appearing in Eq. (5.47) has been found by Budiansky and Fleck (1993) to not have a large effect on the compression failure stress for most angles found in experimental observations. Also, the fiber misalignment angle is comparatively significantly more important.

Niu and Talreja (2000) conducted an elastic microbuckling analysis using a shear deformation beam model and derived the Rosen formula. In the elastic-plastic microbuckling analysis with fiber misalignment they derived the Budiansky-Argon kinking formula. Thus, a unified analysis for microbuckling and kink band formation was developed.

To obtain further insight into the compression failure process numerous computational simulations have been done, e.g., in Bishara et al. (2017) and Zhou et al. (2018), incorporating fiber fracture, matrix yielding and fiber/matrix interface cracking in the numerical analyses. The relative effect of these parameters has been clarified confirming the experimental observation that the matrix plasticity and fiber misalignment play major roles in governing the formation of kink bands and the consequent failure stress. An experimental study (Leopold et al., 2018) compared the failure predictions of different analytical models with glass/epoxy and carbon/epoxy composites and reported that for low fiber volume fraction the matrix properties governed the fiber kinking process, while for high fiber volume fraction, the fiber microbuckling model by Berbinau et al. (1999), which assumes fiber failure as initiator of kink bands, more accurately described the failure process.

In UD composites manufactured by a practical (industrial) manufacturing process, misalignment or waviness of fibers can appear in clusters or regions ("patches"). For such cases, it is useful to have a defect severity measure that rank the regions in terms of their ability to initiate kink bands. Such a measure was developed in Wilhelmsson et al. (2019) based on simplifying the misaligned regions as ellipses and taking analogy with stress singularity measures for elliptical cracks. Figure 5.17 depicts the geometrical similarity between arbitrarily shaped oriented crack planes in a brittle solid and regions of wavy fibers in a UD composite that have different sizes and orientations. Simplifying the cracks as well as the wavy regions by ellipses and using the stress intensity factors for elliptical cracks as a guide, the following expression was proposed for the critical compressive stress.

$$\sigma_{\mathrm{cr}}^{\mathrm{c}} = \tau_{\mathrm{c}} D_1 D_2 D_{\mathrm{s}} \tag{5.50}$$

where τ_c is the critical shear stress (shear yield stress, eventually modified by the strain hardening behavior) and D_1, D_2 and D_s are the fiber waviness related factors. The factor D_s is the defect size, which here would be inverse of the maximum misalignment angle, while D_1 and D_2 can be taken to represent characteristics of the areas affected by the misaligned fibers. Following expressions are proposed for the defect factors

$$D_1 = f_1(A), \qquad D_2 = f_2(a, \alpha), \qquad D_{\mathrm{s}} = \frac{1}{\theta} \tag{5.51}$$

where A is the area of the ellipse, a and α are the half-minor axis and the angle of inclination made by the major axis with the direction of the compressive stress, respectively.

Functions f_1 and f_2 may be determined either by the local stress analysis of the elliptical region or by an experimental correlation with the compression strength data. For further details, see Wilhelmsson et al. (2019).

Matrix and Fiber/Matrix Interface Failure Mode in Transverse Tension. The physical observations of this failure mode described in Sect. 5.2.3 suggest that a transverse crack forms by coalescence of smaller cracks. These smaller cracks are in the matrix and at the fiber/matrix interfaces. For clarity, an SEM image in Fig. 5.18 is shown (Wood and Bradley, 1997). This image shows the early stage of the failure process before the formation of a single transverse crack. Coalescence of three debond cracks can be seen bridged by cracking the matrix in the inter-fiber region. This seems to suggest that fiber/matrix debonding is the first failure event followed by matrix cracking resulting from kinking out of the debond cracks.

The fiber/matrix debonding has been a subject of modeling in the literature by various approaches. The approaches that consider a single fiber embedded in the matrix give questionable results because the stress state at the interface in that case differs significantly from that in a UD composite where multiple fibers exist. Also, approaches where the interfaces are given assumed properties that cannot be determined by independent tests introduce uncertainties that cast doubts on the results. For instance, assuming interface strength or toughness when the interface is a plane of no material has obvious discrepancy. Similarly, assuming cohesive zones at interfaces that are assigned properties that can only be extracted (back calculated) from consequential results questions the uniqueness of those results.

An entirely different approach to fiber/matrix debonding in UD composites with epoxy as a matrix emerged from the works of Asp et al. (1995, 1996a, b). At first, Asp et al. (1995) showed that the matrix within a UD composite under transverse tension has local stress states that approach equi-triaxial (hydrostatic) tension. Under such stress states an epoxy matrix fails by a dilatation process that leads to transverse failure of the composite at very low strains that agree with experimental data. A

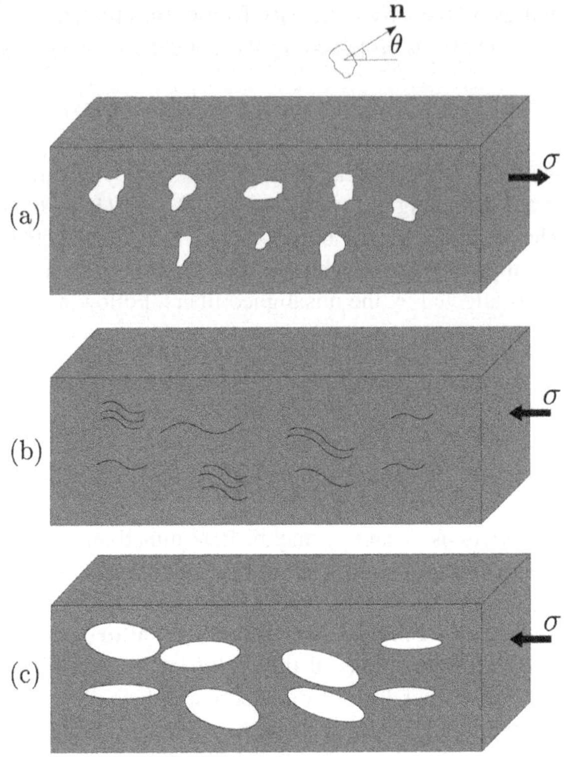

Fig. 5.17 Schematic depiction of the defect severity concept for a UD composite with regions of wavy fibers. **a** A brittle solid in tension with oriented cracks, and regions of wavy fibers (**b**) replaced by oriented ellipses (**c**), from Wilhelmsson et al. (2019)

poker-chip test previously used for rubbers was utilized in conjunction with stress analysis to verify the observed low failure strains. Asp et al. (1996a) then developed a criterion for brittle cavitation as a critical value of the dilatation energy density. This criterion was applied to transverse failure of UD composites and the predictions were supported by test data for different epoxy matrix materials (Asp et al., 1996b). The importance of this approach lies in the fact that the uncertainty in modeling of the fiber/matrix interface properties is eliminated by instead using the critical value of the dilatation energy density for the matrix epoxy by an independent test. The finite element computations show that the maximum hydrostatic tension occurs close to the fiber/matrix interface where the brittle cavitation on unstable growth (expansion) will lead to fiber/matrix debonding.

Several studies by others in the literature have validated the Asp et al. approach, e.g., by Sato et al. (2014) which concluded that the initiation of dilatation induced failure occurs first in off-axis plies of laminates with off-axis inclination (with respect to the axial loading direction) of greater than 60 degrees. Studies by Elnekhaily and Talreja (2018) and by Sudhir and Talreja (2019) have applied the dilatation energy density criterion to investigate the formation of transverse cracks in UD composites with nonuniform fiber distributions. A molecular dynamic study of epoxies subjected

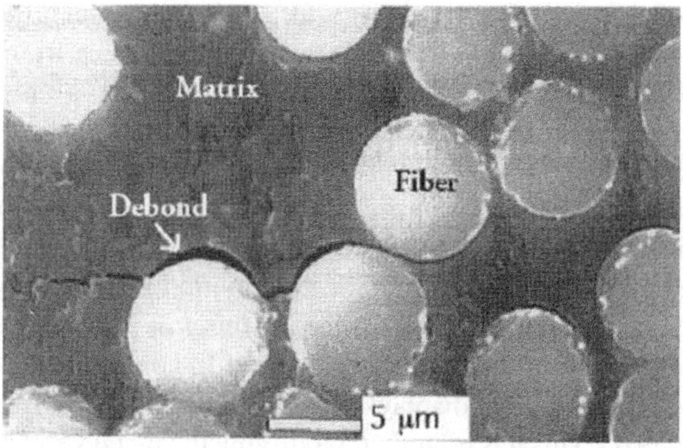

Fig. 5.18 A micrograph showing fiber/matrix interface debond cracks and their coalescence in the early stage of the formation of a transverse crack (Wood and Bradley, 1997)

to multiaxial stress states has further increased the validity of the dilatation induced cavitation as a critical mechanism for crack formation (Neogi et al., 2018).

As noted above, the formation of a transverse crack involves coalescence of fiber/matrix debonds. This stage of the cracking process has been treated by Zhuang et al. (2018a, b) by assuming that a small debond crack has formed by brittle cavitation in the matrix and by then studying the growth of the debond crack along the circular interface followed by its kink-out into the matrix. The influence of various parameters such as the inter-fiber spacing and existence of multiple debonds on the debond linkup has been clarified.

The fiber-matrix debonding mechanism is also possible without the matrix cavitation failure as a precursor. This will be the case if favorable conditions for cavitation do not exist or if the fiber-matrix interface is sufficiently weak or is sufficiently weakened by defects. In that case, the radial tensile stress, and possibly combined with the shear stress on the fiber surface, will break the fiber-matrix interface bonds, initiating debonding. For modeling purposes, it is difficult to know what interface failure properties to use, as these depend on the actual quality of the bond formed during the composite manufacturing process, which is not the same for model composites with one or few filaments that are used for experimental determination of the bond characteristics. The interface bond strength cannot be determined accurately by theoretical means. Several experimental methods have therefore been devised (for a review, see Zhandarov and Mäder 2005). These are either stress-based (strength) or energy-based (toughness) methods. As noted above, the characteristics of the fiber-matrix interface obtained by the experimental methods are commonly under simple conditions (e.g., single-fiber tests). Using the material constants thus obtained for interfaces that are under constrained conditions within composites would require caution. Also, applying the interface toughness criterion for evaluating initiation of fiber-matrix debonding faces difficulties due to the uncertainty of knowing the flaw size and its variability.

For ductile failure to occur under transverse tension of a UD composite, conditions must exist for inelastic deformation of the polymer matrix within the composite. Since the inelastic deformation requires shear stress to drive it, the distortional part of the strain energy density at a point must be sufficient to initiate what may be called "yielding" (although this term originates from crystalline metals behavior). For isotropic metals, the initiation of yielding is satisfactorily given by the critical value of the distortional energy density obtained experimentally for the given metal. Equivalently, the yield criterion for metals can be expressed in terms of the second invariant of the deviatoric stress tensor, as is the case for the von Mises criterion. The initiation of yielding in polymers is governed by molecular phenomena that differ significantly from the dislocation motion underlying yielding of crystalline metals. Still, it is common to describe the onset of inelastic deformation in polymers by the approaches used for metal yielding. In contrast to metals, the inelastic response of glassy polymers displays pressure sensitivity, as discussed by Rottler and Robbins (2001). This has prompted modifying the metal yield criteria by including the hydrostatic stress, e.g., by adding to the threshold of the octahedral shear stress τ_0 and a constant α times the mean pressure p, as

$$\tau_{oct} = \tau_0 - \alpha p \tag{5.52}$$

where p is the average of the three principal stresses, $p = -(\sigma_1 + \sigma_2 + \sigma_3)/3$.

Equation (5.52) is the modified von Mises yield criterion, which in energy terms states that the dilatational energy density contributes also to the shear-driven onset of inelastic deformation in polymers. Additionally, temperature and strain rate are also found to affect the inelastic deformation (Arruda et al., 1995).

In a polymer matrix within a UD composite, the stress triaxiality is generally high except in resin-rich regions. The inelastic deformation will thus tend to occur away from the fiber-matrix interfaces. Once initiated, the inelastic deformation can lead to shear banding before crack formation. Estevez et al. (2000) have studied the crack formation process in glassy polymers by considering the competition between shear banding and crazing. Based on their study it can be stated that the role of the distortional part of the strain energy density at a point is to localize inelastic deformation in shear bands, while the dilatational component is responsible for cavitation leading to craze formation, craze widening and breakdown of craze fibrils. The mix of the two energy components determines the ease or difficulty of ductile crack formation in the matrix within the composite.

Matrix and Fiber/Matrix Interface Failure Mode in In-plane Shear. The stress-strain behavior of UD composites under in-plane shear is found to be nonlinear. Part of this nonlinearity is attributed to the inelastic deformation of the matrix and the other part to the formation of the type of cracks shown in Fig. 5.6. Unless defects in the matrix lie with potential weak planes inclined to the fiber direction, assuming these cracks to form due to the tensile stress acting normal to those planes would not be reasonable. Puck and Schürmann (2002) offered, however, the explanation that these cracks form at 45o to the fiber direction under the action of the tensile principal stress, assuming a pure shear stress in the matrix between the fibers. Actual experimental evidence, such as in Fig. 5.6, shows these cracks to be not at 45° and not to be straight. These cracks are curved ("sigmoidal", as described in Redon

(2000), multiple, and tend to link up near the fiber surfaces, eventually forming a wavy crack (described in observations as a transverse crack) with its plane running overall parallel to the fibers. Along the lines of Puck and Schürmann (2002); Carraro and Quaresimin (2014) assumed the shear-induced crack to be formed by the largest local (tensile) principal stress, which they calculated numerically. The direction of the local maximum principal stress differed from 45° to the fiber axis, assumed in Puck and Schürmann (2002), because of the triaxiality of the local stress field.

Matrix and Fiber/Matrix Failure Mode in Transverse Compression. As noted in Sect. 5.2.4 above, observed failure of a UD composite under transverse compression indicates that failure occurs on a plane inclined to the loading direction. This prompts the suggestion that the failure mode is a manifestation of the matrix yielding in compression governed by cohesion and friction acting on the plane. The criterion for this type of yielding is the classical Mohr-Coulomb criterion, expressed by

$$\tau = c - \sigma \tan \phi \tag{5.53}$$

where τ is the shear stress at yielding acting on the critical plane, σ is the normal stress on that plane, and c and ϕ are material constants representing cohesion and friction angle, respectively. This criterion is like Eq. (5.52) if c is considered yield stress in pure shear and ϕ is viewed as the effect of hydrostatic stress. The angle made by the inclined failure plane with respect to the normal to the compression loading direction is given by

$$\beta = \frac{\pi}{4} + \frac{\phi}{2} \tag{5.54}$$

In a UD composite, when compression is applied normal to the fibers, the inclination of the failure plane will be altered by the triaxility of the local stress state (González and LLorca, 2007; Pinho et al., 2006).

The formation of the failure plane in a UD composite involves progression from yielding at the most favorable point to connectivity of the yield (failure) planes at subsequent points of yielding. This connectivity is supposedly by bands in which ductile deformation and failure mechanisms take place. These aspects cannot be treated in a homogenized composite but require representation of the microstructure for detailed local stress field calculations. González and LLorca (2007) have presented an approach of this nature. They include in their analysis also the effect of fiber/matrix interface failure via an assumed cohesive zone model.

Failure Modes in Combined Loading. Having considered the individual failure modes in single loads, i.e., tension or compression in fiber or transverse directions, or shear in the plane of the UD composite, the next area of inquiry is concerning failure when these loads are combined. The combined effect is perceived to be an "interaction", meaning mutually enhanced degradation (or, exceptionally, strengthening).

The notion of interaction in combined loading has its history in yielding of metals. It is conceivable in metals that each loading mode applied separately, or two or more

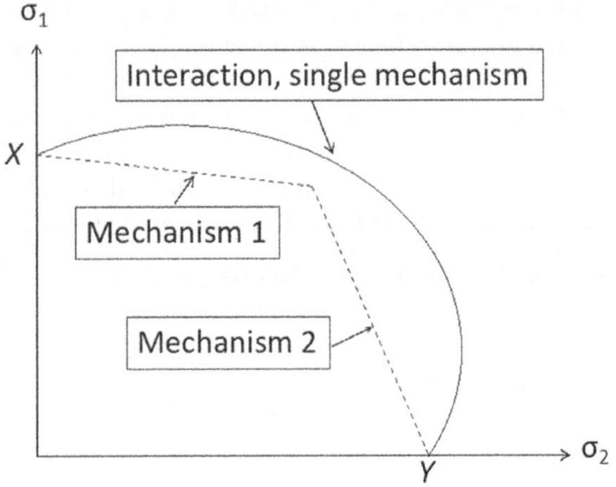

Fig. 5.19 Failure interaction under combined loading. The continuous curve is typical of interaction when a single mechanism operates, while two different mechanisms result in discontinuity in the interaction

loading modes applied together, may contribute to energy needed to drive the mechanisms underlying yielding. These mechanisms are basically related to dislocation mobility and therefore the relevant strain energy density involved is distortional. This is the key reason for the von Mises criterion's ability to successfully describe the combined effect of the normal and shear stresses. The ensuing quadratic expression of the criterion indicates the nature of the combined action of the individual stress components in initiating yielding.

For combined action of the stress components in contributing to the failure of a composite there is no fundamental justification like that for metal yielding. First, there is no single failure mode in composites such as metal yielding to warrant a quadratic combination of the partial effect of each stress component. The conditions governing each failure mode are different, as discussed above. Therefore, the only reasonable course of action in describing the combined effects is to consider each failure mode separately and analyze how the individual effects combine for that failure mode. Figure 5.19 illustrates the nature of combined loading effects in metals versus in UD composites. Contrary to metal yielding, the failure modes in two single loading cases in UD composites are distinctly different. A single continuous curve of interaction in UD composites is therefore not expected.

A general remark before considering the combined effects of loading modes in UD composites is in order. Since each loading mode generally induces a separate failure mechanism, it would be convenient and appropriate to consider a given failure mode as being driven primarily by one loading mode, denoted as the "dominant" loading mode, and any other loading mode that is acting simultaneously with this loading mode as the "modifying" loading mode. Thus, for example, the remotely applied tension is the dominant loading mode for tensile failure of fibers and an in-plane shear would be the modifying loading mode for this failure mode.

Let us begin with the fiber failure mode in tension, discussed in Sects. 5.2.1 and 5.5.2 (**Fiber Failure Mode in Axial Tension**) above. The essential feature of this failure mode is the local load sharing from one failed fiber to its neighboring intact fibers. The properties of the matrix and the fiber-matrix interface as well as the microstructural parameters such as the inter-fiber spacing play roles in this load transfer process. If the remotely applied transverse tension (or compression) and in-plane shear stress are combined with the axial tension, then these additional loads will affect the local load transfer by changing the local stress field. Without fully analyzing these modifying effects it is commonly assumed that the transverse load (tension or compression) has no effect and that the in-plane shear contributes to the fiber failure mode in tension in the form of a quadratic interaction, as in Eq. (44). In fact, this equation also suggests that if the in-plane shear is the dominant loading mode, then the mechanisms of shear cracks leading to transverse crack formation, discussed in Sects. 5.2.5 and 5.5.2 (**Matrix and Fiber/Matrix Interface Failure Mode in In-plane Shear**) above, will be modified by the imposed axial tension in the same way. Obviously, this assertion is not supported by any analysis at the microstructural level.

Consider next the fiber failure mode in compression, discussed in Sects. 5.2.2 and 5.5.2 (**Fiber Failure Mode in Axial Compression**) above. Although different mechanisms underlying this failure mode are possible, in the common case of glass/epoxy and carbon/epoxy, the mechanism involved is plastic microbuckling leading to kink-band formation. The roles of the local shear stress and fiber misalignment have been identified as the important factors governing this mechanism. If transverse normal stress (tensile or compressive) and the in-plane shear stress are applied in addition to the axial compression, then their effects must be considered in terms of any changes induced in the local shear stress and in the fiber rotation. Experimental studies have focused on combining the axial compression with the in-plane shear (Jelf and Fleck, 1994; Vogler et al., 2000). In their experiments, both studies found that the fiber failure mode in compression remained unchanged by the additional application of the in-plane shear. If this is the case, then the contribution of the in-plane shear can be analyzed in terms of the modification induced in the local shear associated with the kink-band formation. It must be realized, however, that if the applied axial compression does not trigger the plastic microbuckling, then increasing the in-plane shear would likely result in shear cracks followed by transverse cracks, the mechanisms that come into play when the in-plane shear alone is applied. The role of the axial compression in these mechanisms is yet not clear.

As far as failure in the matrix polymer is concerned, all combinations of loads applied—tension and compression normal to fibers and in-plane shear—induce mechanisms of yielding and crack formation. In going from the unreinforced polymer to the same polymer within a UD composite, care must be exercised in accounting for the stress triaxiality induced by the presence of fibers and the potential failure surfaces at the fiber-matrix interfaces. In theories developed for homogenized composites, such as by Puck and Schürmann (1998), these aspects cannot be accounted for directly. Various assumptions and curve-fitting parameters must be introduced related to the fracture planes within the matrix between the fibers. Although such theories facilitate simple procedures for design, fundamental objections to their validity remain. For instance, what triggers the formation of the fracture planes within composites where the local stress fields are essentially triaxial? In other words, if no specific weak planes exist beforehand in the matrix, then what is the mechanism involved in getting to the flat planes of fracture from discrete failures at points? It is true that fracture planes are observed in experiments, but closer examination of the planes often suggest some precursor mechanism(s). For instance, in the case of compression normal to fibers, the observed hackles near the fibers on the inclined fracture planes indicate the role of the local stresses in initiating discrete failures.

Beyond the UD Composite Failure. UD composites form the building blocks in many composite structures, e.g., in stacked configurations in multidirectional laminates. Failure ensued in a single layer is often the beginning and not the end of the structural failure. In going beyond the UD composite failure, two aspects are noteworthy, one, that the UD composite failure within a layer is necessarily under combined loading, and two, that this failure is constrained by the presence of the other differently oriented layers, as described in Sect. 5.3 above. When failure in a UD composite lying within a laminate occurs, it is not an ultimate failure in the sense of breaking apart as separate pieces, but rather as initiation of a crack that has its plane aligned with the fiber direction in the UD composite layer. This failure, described in the early literature as "first ply failure", has a progression consisting of multiple parallel cracks, which increase in number, reducing their mutual spacing, until a saturation state is reached. Failure in other plies progresses similarly, in a sequence determined by the criticality of their stress states.

The field of progressive failure in composite laminates has still not matured despite decades of modeling efforts. The challenges are enormous because a multitude of possible paths exist from the first failure within a ply until the total loss of integrity by final separation of plies. In fact, seeking to predict a continuous degradation of integrity, like the "residual strength" related to the effect of a single crack in metals, may not be an effective way to address composite laminate performance. The field of damage mechanics is comparatively more mature and provides a way to assess "residual stiffness" at least in the case of multiple ply cracking without extensive delamination. If the growth of delamination is of concern, then resort can be made to fracture mechanics methods that are also sufficiently mature to provide bases for integrity degradation.

5.6 Future Directions for Research

Further progress in modeling of damage and failure in composite materials for the purpose of assessing their performance in structures made with practical manufacturing methods will come from a new paradigm in research. Such a paradigm is presented in Fig. 5.20. The starting point in this paradigm, described as a manufacturing sensitive design paradigm, is the manufacturing method employed, e.g., an autoclave-based curing of a part prepared with prepreg tape-laying process, a part made with resin transfer molding (RTM) or a modified version of it such as vacuum-

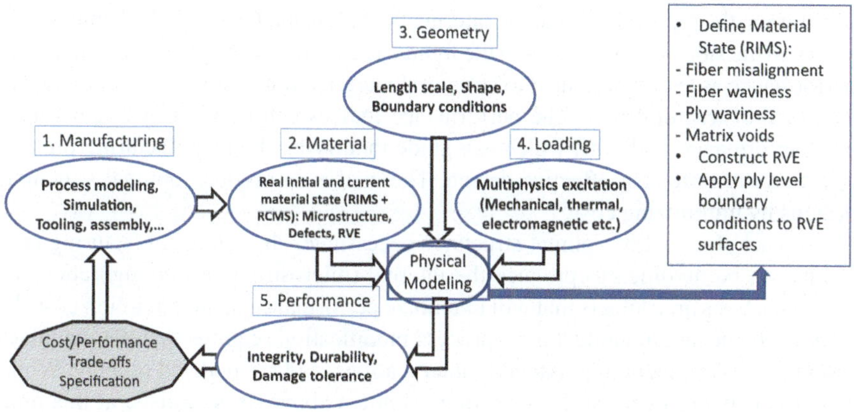

Fig. 5.20 A flow chart for manufacturing sensitive design of composite structures

assisted RTM (VARTM), or any of the many resin infusion processes. The complete manufacturing of a part before entering it in service can also involve tooling (machining of holes, cutouts, etc.), assembly by co-curing and adhesive bonding, etc. Thus, the internal structure of a composite part may have spatial distribution of geometrical features that are a result of all such operations. Theoretically, this distribution and the geometrical variation of the features should be possible to predict by modeling and simulation given the process history. However, while progress is being made in this direction, the goal is not yet within sight.

Until the internal structure of a composite can be predicted, the alternative available, at least partly, is to observe it with techniques of different resolution such as optical microscopy and X-ray computed tomography. The next step will then be to relate the observed features to the material response characteristics, e.g., for deformational (stress-strain) response, commonly called "stiffness", or for failure response, commonly called "strength", or for resistance to growth of a crack, commonly called "fracture toughness". In any case, a more general concept than that of traditional material properties defined on homogenized internal structure is needed. We have proposed here the concept of a "material state" that should be described by statistical descriptors that carry information on the internal structure elements of relevance to the desired material response. These elements can be, for instance, the (homogenized) properties of the constituents, their spatial distributions, and any manufacturing induced entities such as voids. Since some or all these elements can change under service, the material state descriptors must reflect those changes. Therefore, we introduce the terminology of real initial material state (RIMS) and real current material state (RCMS), which is the evolved material state induced by irreversible changes.

Continuing with the flow chart in Fig. 5.20, the material state descriptors along with the information on geometry at the length scale of interest for failure analysis, and the imposed loading, enter in the physical modeling of the local failure. For illustration, the type of information entering the failure analysis of a UD composite is shown in the side box to the right in the figure. The output of the physical modeling is performance assessment related to structural integrity, durability, damage tolerance, etc., of the structural part. The performance indices can be used in conducting a cost/performance trade-off, which can guide modifying the manufacturing process to achieve a more cost-effective design. The goal is to produce optimal structural designs by minimizing cost.

To reach the goal of optimal structures necessitates that the manufacturing processes can be modified to produce the internal (microstructural) configurations of heterogeneities and defects that will maximize performance at the minimum cost. To achieve the needed manufacturing process modification requires that the manufacturing processes, including assembly interventions, can be modeled reliably. While progress is being made in this direction, as noted above, it is realistic at this time to observe the internal composite structure and simulate it by stochastic methods to create representation of its features relevant to initiation and progression of failure.

5.7 Concluding Remarks

This article has summarized the six lectures cited in the abstract. The scope of the subject of damage and failure of composite materials is broad and different approaches have been initiated and developed to more or less extent over the approximately five decades of activities in the field. The exposition here has not attempted to review in detail all approaches but has tried to convey the author's own philosophy pursued since his first publication on continuum damage mechanics in 1985. The future directions of research outlined above are also a product of that philosophy. Those who attended the lectures and others who have interest in the field should take this article in the spirit of one thought process among many and use it accordingly.

References

Adkins, J. (1959). Symmetry relations for orthotropic and transversely isotropic materials. *Archives of Rational Mechanics Analysis, 4*(1), 193–213.

Ahci, E., & Talreja, R. (2006). Characterization of viscoelasticity and damage in high temperature polymer matrix composites. *Composites Science and Technology, 66*(14), 2506–2519.

Argon, A. S. (1972). Fracture of composites. In H. Herman (Ed.), *Treatise on Materials Science & Technology* (Vol. 1, pp. 79–114). Elsevier.

Aroush, D. R.-B., Maire, E., Gauthier, C., Youssef, S., Cloetens, P., & Wagner, H. D. (2006). A study of fracture of unidirectional composites using in situ high-resolution synchrotron X-ray microtomography. *Composites Science and Technology, 66*(10), 1348–1353.

Arruda, E. M., Boyce, M. C., & Jayachandran, R. (1995). Effects of strain rate, temperature and thermomechanical coupling on the finite strain deformation of glassy polymers. *Mechanics of Materials, 19*(2), 193–212.

Asp, L. E., Berglund, L. A., & Gudmundson. (1995). Effects of a composite-like stress state on the fracture of epoxies. *Composites Science and Technology, 53*(1), 27–37.

Asp, L. E., Berglund, L. A., & Talreja, R. (1996). A criterion for crack initiation in glassy polymers subjected to a composite-like stress state. *Composites Science and Technology, 56*(11), 1291–1301.

Asp, L. E., Berglund, L. A., & Talreja, R. (1996). Prediction of matrix-initiated transverse failure in polymer composites. *Composites Science and Technology, 56*(9), 1089–1097.

Azzi, V. D., & Tsai, S. W. (1965). Anisotropic strength of composites. *Experimental Mechanics, 4*(9), 283–288.

Berbinau, P., Soutis, C., & Guz, I. A. (1999). Compressive failure of 0° unidirectional carbon-fibre-reinforced plastic (CFRP) laminates by fibre microbuckling. *Composites Science and Technology, 59*(9), 1451–1455.

Bishara, M., Rolfes, R., & Allix, O. (2017). Revealing complex aspects of compressive failure of polymer composites – Part I: Fiber kinking at microscale. *Composite Structures, 169*, 105–115. In Honor of Prof. Leissa.

Budiansky, B. (1983). Micromechanics. *Computers & Structures, 16*(1), 3–12.

Budiansky, B., & Fleck, N. A. (1993). Compressive failure of fibre composites. *Journal of the Mechanics and Physics of Solids, 41*(1), 183–211.

Carraro, P. A., & Quaresimin, M. (2014). A damage based model for crack initiation in unidirectional composites under multiaxial cyclic loading. *Composites Science and Technology, 99*, 154–163.

Davidson, P., & Waas, A. M. (2017). The effects of defects on the compressive response of thick carbon composites: An experimental and computational study. *Composite Structures, 176*, 582–596.

Elnekhaily, S. A., & Talreja, R. (2018). Damage initiation in unidirectional fiber composites with different degrees of nonuniform fiber distribution. *Composites Science and Technology, 155*, 22–32.

Estevez, R., Tijssens, M. G. A., & Van der Giessen, E. (2000). Modeling of the competition between shear yielding and crazing in glassy polymers. *Journal of the Mechanics and Physics of Solids, 48*(12), 2585–2617.

Gamstedt, E. K., & Sjögren, B. A. (1999). Micromechanisms in tension-compression fatigue of composite laminates containing transverse plies. *Composites Science and Technology, 59*(2), 167–178.

Garrett, K. W., & Bailey, J. E. (1977). Multiple transverse fracture in 90° cross-ply laminates of a glass fibre-reinforced polyester. *Journal of Materials Science, 12*(1), 157–168.

Gol'denblat, I. I., & Kopnov, V. A. (1965). Strength of glass-reinforced plastics in the complex stress state. *Polymer Mechanics, 1*(1), 54–59.

Gol'denblat, I. I., & Kopnov, V. A. (1971). General theory of criteria of strength for isotropic and anisotropic materials. *Strength of Materials, 3*(2), 184–188.

Gol'denblat, I. I., & Kopnov, V. A. (1971). General criteria of the strength of isotropic media. *Polymer Mechanics, 7*(2), 212–220.

González, C., & Lorca, J. (2007). Mechanical behavior of unidirectional fiber-reinforced polymers under transverse compression: Microscopic mechanisms and modeling. *Composites Science and Technology, 67*(13), 2795–2806.

Gu, J., Li, K., & Su, L. (2020). Modified nonlinear Mohr-Coulomb fracture criteria for isotropic materials and transversely isotropic UD composites. *Mechanics of Materials, 151*, 103649.

Gu, J., Chen, P., Su, L., & Li, K. (2021). A theoretical and experimental assessment of 3D macroscopic failure criteria for predicting pure inter-fiber fracture of transversely isotropic UD composites. *Composite Structures, 259*, 113466.

Hashin, Z. (1980). Failure criteria for unidirectional fiber composites. *Transaction ASME Journal of Applied Mechanics, 47*(2), 329–334.

Highsmith, A. L., & Reifsnider, K. L. (1982). Stiffness-reduction mechanisms in composite laminates. In K. L. Reifsnider (Ed.), *Damage in Composite Materials* (pp. 103–117). Philadelphia: ASTM International.

Hill, R. (1948). A theory of the yielding and plastic flow of anisotropic metals. *Proceedings of the Royal Society of London. Series A. Mathematical and Physical Sciences, 193*(1033), 281–297.
Hinton, M. J., Kaddour, A. S., & Soden, P. D. (2004). A further assessment of the predictive capabilities of current failure theories for composite laminates: comparison with experimental evidence. *Composites Science and Technology, 64*(3), 549–588. Failure criteria in fibre reinforced polymer composites Part C: Additional theories conclusions and recommendations.
Jelf, P. M., & Fleck, N. A. (1994). The failure of composite tubes due to combined compression and torsion. *Journal of Materials Science, 29*(11), 3080–3084.
Jumahat, A., Soutis, C., Jones, F. R., & Hodzic, A. (2010). Fracture mechanisms and failure analysis of carbon fibre/toughened epoxy composites subjected to compressive loading. *Composite Structures, 92*(2), 295–305.
Kumar, R. S., & Talreja, R. (2001). Linear viscoelastic behavior of matrix cracked cross-ply laminates. *Mechanics of Materials, 33*(3), 139–154.
Leopold, C., Harder, S., Philipkowski, T., Liebig, W. V., & Fiedler, B. (2018). Comparison of analytical approaches predicting the compressive strength of fibre reinforced polymers. *Materials, 11*(12), 2517.
Li, S. (2020). A reflection on the Mohr failure criterion. *Mechanics of Materials, 148*, 103442.
Narayanan, S., & Schadler, L. S. (1999). Mechanisms of kink-band formation in graphite/epoxy composites: A micromechanical experimental study. *Composites Science and Technology, 59*(15), 2201–2213.
Nedele, & Wisnom, M.R. (1994). Stress concentration factors around a broken fibre in a unidirectional carbon fibre-reinforced epoxy. *Composites, 25*(7), 549–557. Third International Conference on Interfacial Phenomena in Composite Materials.
Neogi, A., Mitra, N., & Talreja, R. (2018). Cavitation in epoxies under composite-like stress states. *Composites Part A: Applied Science and Manufacturing, 106*, 52–58.
Niu, K., & Talreja, R. (2000). Modeling of compressive failure in fiber reinforced composites. *International Journal of Solids and Structures, 37*(17), 2405–2428.
Oguni, K., Tan, C. Y., & Ravichandran, G. (2000). Failure mode transition in unidirectional e-glass/vinylester composites under multiaxial compression. *Journal of Composite Materials, 34*(24), 2081–2097.
Pinho, S. T., Robinson, P., & Iannucci, L. (2006). Fracture toughness of the tensile and compressive fibre failure modes in laminated composites. *Composites Science and Technology, 66*(13), 2069–2079.
Puck, A., Schürmann, H. (1998). Failure analysis of FRP laminates by means of physically based phenomenological models. *Composites Science and Technology, 58*(7), 1045–1067. This article represents the authors' contributions to a world-wide exercise to confirm the state-of-the-art for predicting failure in composites, organised by Hinton and Soden.
Puck, A., & Schürmann, H. (2002). Failure analysis of FRP laminates by means of physically based phenomenological models. *Composites Science and Technology, 62*(12), 1633–1662.
Quaresimin, M., & Carraro, P. A. (2014). Damage initiation and evolution in glass/epoxy tubes subjected to combined tension-torsion fatigue loading. *International Journal of Fatigue, 63*, 25–35.
Redon, O. (2000). Fatigue damage development and failure in unidirectional and angle-ply glass fibre/carbon fibre hybrid laminates. Technical Report Risø-R-1168, Risø National Laboratory, Roskilde, Denmark.
Reifsnider, K. L., & Jamison, R. (1982). Fracture of fatigue-loaded composite laminates. *International Journal of Fatigue, 4*(4), 187–197.
Rice, J. R. (1971). Inelastic constitutive relations for solids: An internal-variable theory and its application to metal plasticity. *Journal of the Mechanics and Physics of Solids, 19*(6), 433–455.
Rosen, B. W. (1964). Mechanics of composite strengthening. In *Fiber Composite Materials*. American Society for Metals. Proceedings of the Seminar on Fiber Composite Materials, Philadelphia, PA, USA, 17–18 October 1964.

Rottler, J., Robbins, & Mark, O. (2001). Yield conditions for deformation of amorphous polymer glasses. *Physical Review E, 64*, 051801.

Sato, Y., Okabe, T., Higuchi, R., & Yoshioka, K. (2014). Multiscale approach to predict crack initiation in unidirectional off-axis laminates. *Advanced Composite Materials, 23*(5–6), 461–475.

Singh, C. V., & Talreja, R. (2008). Analysis of multiple off-axis ply cracks in composite laminates. *International Journal of Solids and Structures, 45*(16), 4574–4589.

Singh, C. V., & Talreja, R. (2010). Evolution of ply cracks in multidirectional composite laminates. *International Journal of Solids and Structures, 47*(10), 1338–1349.

Singh, C. V., & Talreja, R. (2013). A synergistic damage mechanics approach to mechanical response of composite laminates with ply cracks. *Journal of Composite Materials, 47*(20–21), 2475–2501.

Sudhir, A., & Talreja, R. (2019). Simulation of manufacturing induced fiber clustering and matrix voids and their effect on transverse crack formation in unidirectional composites. *Composites Part A: Applied Science and Manufacturing, 127*, 105620.

Swolfs, Y., McMeeking, R. M., Verpoest, I., & Gorbatikh, L. (2015). Matrix cracks around fibre breaks and their effect on stress redistribution and failure development in unidirectional composites. *Composites Science and Technology, 108*, 16–22.

Talreja, R. (1990). Internal variable damage mechanics of composite materials. Damage and Failure of Anisotropic Solids. In J. P. Boehler (Ed.), *Yielding* (pp. 509–533). London: Mechanical Engineering Publications.

Talreja, R. (1991). Continuum modelling of damage in ceramic matrix composites. *Mechanics of Materials, 12*(2), 165–180.

Talreja, R. (1996). A synergistic damage mechanics approach to durability of composite systems. In A. H. Cardon, H. Fukuda, & K. Reifsnider (Eds.), *Progress in Durability Analysis of Composite Systems* (pp. 117–129). Rotterdam: A.A. Balkema.

Talreja, R. (2014). Assessment of the fundamentals of failure theories for composite materials. *Composites Science and Technology, 105*, 190–201.

Talreja, R. (2016). Physical modelling of failure in composites. *Philosophical Transactions of the Royal Society A: Mathematical, Physical and Engineering Sciences, 374*(2071), 20150280.

Talreja, R., & Kelly, A. (1985) . A continuum mechanics characterization of damage in composite materials. *Proceedings of the Royal Society of London. A. Mathematical and Physical Sciences, 399*, 195–216 (1817).

Talreja, R., & Waas, A. M. (2022). Concepts and definitions related to mechanical behavior of fiber reinforced composite materials. *Composites Science and Technology, 217*, 109081.

Tsai, S. W., & Wu, E. M. (1971). A general theory of strength for anisotropic materials. *Journal of Composite Materials, 5*(1), 58–80.

Varna, J. (2013). Modelling mechanical performance of damaged laminates. *Journal of Composite Materials, 47*(20–21), 2443–2474.

Varna, J., Joffe, R., Akshantala, N. V., & Talreja, R. (1999). Damage in composite laminates with off-axis plies. *Composites Science and Technology, 59*(14), 2139–2147.

Vogler, T. J., & Kyriakides, S. (2001). On the initiation and growth of kink bands in fiber composites: Part I. Experiments. *International Journal of Solids and Structures, 38*(15), 2639–2651.

Vogler, T. J., Hsu, S.-Y., & Kyriakides, S. (2000). Composite failure under combined compression and shear. *International Journal of Solids and Structures, 37*(12), 1765–1791.

Wilhelmsson, D., Talreja, R., Gutkin, R., & Asp, L. E. (2019). Compressive strength assessment of fibre composites based on a defect severity model. *Composites Science and Technology, 181*, 107685.

Wood, C. A., & Bradley, W. L. (1997). Determination of the effect of seawater on the interfacial strength of an interlayer E-glass/graphite/epoxy composite by in situ observation of transverse cracking in an environmental sem. *Composites Science and Technology, 57*(8), 1033–1043. 6th International Conference On Composite Interfaces.

Zhandarov, S., & Mäder, E. (2005). Characterization of fiber/matrix interface strength: Applicability of different tests, approaches and parameters. *Composites Science and Technology, 65*(1), 149–160.

Zhou, L., Zhao, L., Liu, F., & Zhang, J. (2018). A micromechanical model for longitudinal compressive failure in unidirectional fiber reinforced composite. *Results in Physics, 10*, 841–848.

Zhuang, L., Talreja, R., & Varna, J. (2016). Tensile failure of unidirectional composites from a local fracture plane. *Composites Science and Technology, 133*, 119–127.

Zhuang, L., Pupurs, A., Varna, J., Talreja, R., & Ayadi, Z. (2018). Effects of inter-fiber spacing on fiber-matrix debond crack growth in unidirectional composites under transverse loading. *Composites Part A: Applied Science and Manufacturing, 109*, 463–471.

Zhuang, L., Talreja, R., & Varna, J. (2018). Transverse crack formation in unidirectional composites by linking of fibre/matrix debond cracks. *Composites Part A: Applied Science and Manufacturing, 107*, 294–303.

The manufacturer's authorised representative in the EU is Springer Nature Customer Service Centre GmbH, Europaplatz 3, 69115 Heidelberg, Germany. If you have any concerns regarding our products, please contact ProductSafety@springernature.com

Printed and bound by CPI Group (UK) Ltd, Croydon, CR0 4YY
30/09/2026
02225239-0009